Fourth Revolution and the Bottom Four Billion

Fourth Revolution and the Bottom Four Billion

Making Technologies Work for the Poor

Nir Kshetri

University of Michigan Press
Ann Arbor

Published in the United States of America by the
University of Michigan Press
Printed and bound by CPI Group (UK) Ltd, Croydon, CR0 4YY

First published April 2023

A CIP catalog record for this book is available from the British Library.

Library of Congress Cataloging-in-Publication data has been applied for.

ISBN 978-0-472-07589-8 (hardcover : alk. paper)
ISBN 978-0-472-05589-0 (paper : alk. paper)
ISBN 978-0-472-90321-4 (open access ebook)

DOI: https://doi.org/10.3998/12205632

The University of Michigan Press's open access publishing program is made possible thanks to additional funding from the University of Michigan Office of the Provost and the generous support of contributing libraries.

Contents

List of Illustrations vii
List of Tables ix
List of In Focus Boxes xi
Preface xiii
Abbreviations xv

Part 1: Overview of the Fourth Revolution and the Bottom 4 Billion

Chapter 1. The Fourth Revolution and the Bottom 4 Billion: Key Underlying Concepts and Developments 3

Part 2: Major 4R Technologies

Chapter 2. Artificial Intelligence 37
Chapter 3. Blockchain 72
Chapter 4. Remote Sensing and Satellite Imagery 103
Chapter 5. Internet of Things 127

Part 3: The 4R in Economic and Social Developments

Chapter 6. Healthcare and Pandemic Preparedness 157
Chapter 7. Agriculture 179
Chapter 8. Finance, Banking, and Insurance 203

Part 4: The Way Forward

Chapter 9. Enablers, Opportunities, Barriers, and Challenges 231

Chapter 10. Economic Development Implications 256

Chapter 11. Political, Social, and Ethical Implications 284

Chapter 12. Conclusion and Recommendations 312

Glossary 337

About the Author 347

Index 349

Digital materials related to this title can be found on the Fulcrum platform via the following citable URL: https://doi.org/10.3998/mpub.12205632

Illustrations

Figure 1.1. Average cost of an IoT sensor and cost per 10 gigabytes of a hard drive (US$) 10
Figure 2.1. How machine learning works 44
Figure 3.1. An illustration of the use of blockchain to trace a diamond in a supply chain 76
Figure 3.2. Time and cost required to enforce contracts in different geographic regions and countries 79
Figure 4.1. Key steps in remote sensing 109
Figure 5.1. Percentage of the world's population subscribing to or covered by cell phones and cellular mobile networks 128
Figure 5.2. How IoT works 131
Figure 9.1. Per capita computing power (in GFLOPS) (2020) 248
Figure 10.1. A comparison of the economies of Angola and Kenya (GDP, PPP, billion, constant 2017 international $) 258
Figure 10.2. Percentage of population covered by a cellular network: Angola and Kenya (2020) 259
Figure 10.3. Percentage of population with access to various technologies and services: Angola and Kenya (2020) 260
Figure 10.4. Internet use in key economic activities: Angola and Kenya (2020) (% of population) 261
Figure 12.1. Estimated mobile subscriptions by region and technology (%, 2026) 326

Tables

Table 1.1. Classification of world population by income and ability to engage in economic exchange 6
Table 2.1. AI use in diverse industries in developing economies 52
Table 2.2. Examples of data-labeling companies operating in LMICs 58
Table 2.3. Classification of countries with respect to data availability and strictness of privacy law 65
Table 3.1. Blockchain in LMICs: Some applications currently in use or in development 81
Table 4.1. Usefulness satellite imagery to predict, evaluate, respond to, and monitor the effects and recovery from pandemics in remote communities 113
Table 5.1. The IoT industry in regions with high concentrations of poor people 129
Table 5.2. The IoT's roles in including features in a contract to create and capture value 143
Table 7.1. 4R technologies' effect on contact, contract, and control for smallholder farmers 193
Table 8.1. Examples of blockchain-based crop insurance in LMICs 221
Table 9.1. Opportunities for the B4B created by major 4R technologies 238
Table 12.1. Examples of the 4R's effects on diverse industries and markets 316

In Focus Boxes

In Focus 1.1. Sight Diagnostics' spin-off ZzappMalaria uses AI to eliminate malaria 11
In Focus 1.2. BanQu's "economic passport" for the poverty market 15
In Focus 1.3. Afreximbank's initiative to facilitate SME's market access 17
In Focus 1.4. Alibaba's 4R solutions for small business 23
In Focus 2.1. Aravind Eye Hospital uses AI-based solutions to help visually impaired individuals 38
In Focus 2.2. TechnoServe uses machine learning to identify cashew trees in Benin 41
In Focus 2.3. China's techno-utilitarian approach targeting the Uyghur population 65
In Focus: 2.4. Cambridge Analytica's beta testing in Africa before U.S. and U.K. elections 68
In Focus 4.1. Bext360's tracking of coffee supply chains through satellite images, blockchain, and AI 111
In Focus 4.2. Satellite technology and forced labor in the marine fishing industry 124
In Focus 5.1. Togo's government teams up with SunCulture and Bboxx to offer IoT-based irrigation and solar pumping to farmers 131
In Focus 5.2. Usangu Logistics uses IoT-enabled solution to fight oil pilfering 135
In Focus 5.3. PayGo Energy's IoT solution for LPG cylinders 137
In Focus 5.4. IoT-enabled Koko points to deliver ethanol in Kenya's capital 146
In Focus 6.1. Sanku's Project Healthy Children uses IoT solutions for small-scale fortification 159

In Focus 6.2. StoveTrace's IoT-based clean-burning stoves 161
In Focus 6.3. Babylon Health offers video appointments and diagnosis in LMICs 166
In Focus 6.4. RxAll's AI-based handheld device to fight fake drugs 170
In Focus 6.5. Indian government uses blockchain to fight counterfeit drugs 171
In Focus 6.6. Uthabiti's blockchain solutions to enhance traceability and trackability of pharmaceutical supply chain in Kenya 172
In Focus 7.1. Intelligent sensors, data analysis, and automated digital greenhouse increase productivity: An experiment 180
In Focus 7.2. Agrilife connects smallholder farmers with financial institutions and suppliers 181
In Focus 7.3. Agryo connects lenders to farmers 183
In Focus 7.4. 4R technologies in the Colombian cacao industry 185
In Focus 7.5. BanQu and Anheuser-Busch promote supply chain transparency 188
In Focus 7.6. Demetria's AI-based solution to determine coffee-bean quality 195
In Focus 7.7. E-Livestock Global brings visibility to the Zimbabwean cattle industry 197
In Focus 7.8. Japan Tobacco International integrates blockchain into its ARISE initiative 198
In Focus 8.1. Interswitch's blockchain system to provide end-to-end visibility 204
In Focus 8.2. Mexico's Konfio's online financial services for SMEs 209
In Focus 8.3. MyBucks's AI-based system provides financial services to refugees in Malawi 210
In Focus 8.4. Migo uses machine learning to make lending decisions 211
In Focus 8.5. OKO's agricultural insurance 214
In Focus 8.6. Sarafu in Kenya 223
In Focus 9.1. Flutterwave's digital payment-processing solution 241
In Focus 10.1. Blockchain and smart contracts distance Cambodian farmers from poverty 265
In Focus 11.1. Blockchain in Colombia's school meal procurement 289
In Focus 11.2. Huawei's AI surveillance 302
In Focus 12.1. Circulor's blockchain-based cobalt supply chain traceability solution 330

Preface

Radical technologies such as artificial intelligence (AI), biotechnology, blockchain, big data, fifth-generation (5G) telecommunications, genome editing, the internet of things, and remote sensing are rapidly emerging to bring significant social and economic transformations in low- and middle-income countries (LMICs). These technologies are facilitating low-income consumers' access to diverse products and services, such as light-emitting diode (LED) solar home systems, clean-burning liquefied petroleum gas (LPG), bank loans to buy agricultural inputs, and low-cost crop insurance. The rapid pace of development of these technologies means that what is coming is surely even bigger.

While these technologies and the transformations they have introduced—referred to in this book as the Fourth Revolution (4R)—have been extensively written about, the focus has been primarily on high-income countries. This topic in the context of low- and middle-income countries, where most of world's bottom 4 billion (B4B) people live, has been addressed only superficially. This research gap is unfortunate, because low-income people and small businesses in LMICs may benefit more significantly from 4R technologies. This book tries contributes to filling this gap by explaining what the Fourth Revolution means for the B4B, providing a detailed analysis and description of major 4R technologies, delving into economic and social developments associated with the 4R, and highlighting the opportunities, challenges, and implications.

Regarding the ideas, concepts, and content presented in this book, I am grateful to several people for their comments, suggestions, support, and encouragement. I would like to express deep appreciation to Elizabeth Sherburn Demers, editorial director at University of Michigan Press, who inspired me to undertake this project. Elizabeth shepherded the project with the greatest of care and professionalism through its various phases. I thank Haley Winkle, editorial associate to the political and social sciences editors in the University of Michigan Press's Acquisitions Department, for providing administrative support and coordination for this project. Special thanks are also due to anonymous University of Michigan Press reviewers for their useful comments and excellent suggestions. Finally, I could not have done it without the love and support of my wife and best friend, Maya.

Abbreviations

4G	fourth generation
4IR	Fourth Industrial Revolution
4R	Fourth Revolution
5G	fifth generation
ACRE	Agriculture and Climate Risk Enterprise
AfCFTA	African Continental Free Trade Agreement
AgTech	Agricultural Technology
AI	artificial intelligence
API	application programming interface
AR	augmented reality
AU	African Union
B2B	business-to-business
B4B	bottom 4 billion
BaaS	blockchain as a service
BIS	Bureau of Industry and Security
CAD	computer-aided design
CATL	Contemporary Amperex Technology Co. Limited
CCP	Chinese Communist Party
CLMS	Child Labor Monitoring System
COLCO	Colombian Cocoa Control System
CSM	cylinder smart meter
DE	Digital Earth
DRC	Democratic Republic of the Congo
EcoProMIS	Ecological Productivity Management Information System
ESA	European Space Agency
ESG	environmental, social, and governance
ETC	Easy Trading Connect

EU	European Union
FAO	Food and Agriculture Organization
FCC	Federal Communications Commission
FGP	flying geese paradigm
FinTech	financial technology
G4AW	geodata for agriculture and water
GB	gigabyte
GCC	Gulf Cooperation Council
GDPR	General Data Protection Regulation
GEE	Google Earth engine
GFW	Global Fishing Watch
GIF	generic insurance framework
GPS	Global Positioning System
GPT	general purpose technology
G20	Group of Twenty
GSM	Global System for Mobile
GSMA	GSM Association
GVA	gross value added
HOT	Humanitarian OpenStreetMap Team
HP	Hewlett-Packard
HSS	Hoefsloot Spatial Solutions
ICO	initial coin offering
ICT	Information and Communications Technology
IDC	International Data Corporation
IFC	International Finance Corporation
IGF	Internet Governance Forum
IoT	internet of things
IPFS	Interplanetary File System
IT	information technology
ITU	International Telecommunication Union
JAXA	Japan Aerospace Exploration Agency
JTI	Japan Tobacco International
KES	Kenyan Shilling
KYC	Know Your Customer
LC	letter of credit
LCM	liquid crystal monitor
LDC	least developed country
LED	light-emitting diode
LMICs	low- and middle-income countries
LoRa	long range

LPG	liquefied petroleum gas
LTE	Long-Term Evolution
LTE-M	Long Term Evolution for Machines
MFI	microfinance institution
ML	machine learning
MNC	multinational corporation
MPH	miles per hour
MSME	micro, small, and medium enterprises
MT	middle-income trap
NASA	National Aeronautics and Space Administration
NB	Narrow Band
NBD	National Bank of Dubai
NeurIPS	neural information processing systems
NGO	nongovernmental organization
NLP	natural language processing
OBP	Oracle's blockchain platform
OECD	Organisation for Economic Co-operation and Development
P2P	peer-to-peer
PAPSS	Pan African Payment and Settlement System
PoS	proof of stake
PoH	proof of history
PoW	proof of work
PPP	purchasing power parity
QoS	quality of service
R&D	research and development
RBI	Reserve Bank of India
RFID	radio-frequency identification
SAB	South African Breweries
SEC	Securities and Exchange Commission
SIM	Subscriber Identity Module
SMEs	small and medium-sized enterprises
SMIC	Semiconductor Manufacturing International Corporation
SMS	short message service
SSA	Sub-Saharan Africa
SSC	South-South Cooperation
SVS	stock visibility solution
TFP	total factor productivity
UBA	United Bank for Africa
UNICEF	United Nations Children's Fund
USB	Universal Serial Bus

USSD	Unstructured Supplementary Service Data
VR	virtual reality
WAN	Wide-Area Network
WFP	World Food Program
WHO	World Health Organization
WII	Weather Index Insurance
WIPO	World Intellectual Property Organization

PART 1

Overview of the Fourth Revolution and the Bottom 4 Billion

CHAPTER 1

The Fourth Revolution and the Bottom 4 Billion

Key Underlying Concepts and Developments

Radical innovations and technologies have brought about a fundamental shift in the global economy. This phenomenon, which is referred to in this book as the Fourth Revolution (4R), follows the earlier three revolutions, which were led by steam engine, electric power, and information technology (IT). Other roughly interchangeable terms such as *Fourth Industrial Revolution* (4IR) and *Industry 4.0* also describe this phenomenon. Some even view this shift as a continuation of the Third Revolution, or the Third Industrial Revolution.

The 4R can be attributed to disruptive technologies such as three-dimensional (3D) printing, fifth-generation (5G) telecommunications, augmented reality (AR), artificial intelligence (AI), autonomous vehicles, biotechnology, blockchain, big data, genome editing, the internet of things, nanotechnology, quantum computing, remote sensing (satellite imagery and drones), robotics (including collaborative robots or "co-bots"), virtual reality (VR), and wearables. The confluence of technologies related to the 4R have made it possible to take actions and make decisions that can have a profound impact on our economy, health and well-being, environment, and social relations.

Data is a key driver of the 4R. Indeed, in 2017, data became the world's most valuable commodity, overtaking oil.[1] According to a March 2021 report by multinational professional services firm PwC, Apple was the world's big-

1. "The world's most valuable resource is no longer oil, but data," *The Economist*, May 6, 2017, https://www.economist.com/leaders/2017/05/06/the-worlds-most-valuable-resource-is-no-longer-oil-but-data.

gest business by market capitalization. Moreover, among the ten most valuable public companies, five were in the technology sector (Apple, Microsoft, Alphabet, Facebook, and Tencent) and three in the consumer discretionary sector (Amazon, Tesla, and Alibaba), which also relies heavily on data.[2] Of note is also that many of this sector's services are free, but users are required to provide their data to access these services. Business models of companies in this sector rely on collecting and monetizing users' data.[3]

Many books and articles have been written on the 4R. However, the focus has been on the middle class and rich people mainly in the developed world. This topic in the context of low- and middle-income countries (LMICs), where most of the bottom 4 billion people live, has been addressed only superficially. This is understandable, as most benefits of the 4R in the near terms will accrue in rich countries. However, a lesser known, rarely described phenomenon concerns the 4R technologies' increasingly widespread use in addressing health, social, humanitarian, economic, and environmental problems facing low-income people and other marginalized groups. The use by and impact on these groups, specifically on the B4B, are the focus of this book. The book seeks to describe and analyze the roles of 4R technologies in the socioeconomic transformations of the B4B.

As a result of their widespread availability, falling costs, and increasingly transformative potential, 4R technologies are having significant social and economic impact in low- and middle-income countries. Just like mobile telephony or mobile banking, 4R technologies may present an opportunity for these countries to leapfrog legacy IT systems. Much as data is a driver of the 4R, it will be key to the socioeconomic development of the B4B. In this way, 4R technologies have the potential to eliminate the causes of inequality, poverty, and injustice for the B4B. This book uncovers various mechanisms by which these technologies can increase the ability of the B4B to fully engage in productive economic activities and market exchange.

Before moving further, some clarifications are in order. This book does not take the B4B is a literal sense, instead focusing mainly on the B4B populations living in LMICs. The stories related to the use of 4R technologies in promoting socioeconomic development are central in representing those nations. The emphasis of the book is on applications in the three areas—healthcare and pandemic preparedness; agriculture; and finance, banking,

2. PWC, *Global Top 100 companies by market capitalisation*, May 2021, https://www.pwc.com/gx/en/audit-services/publications/assets/pwc-global-top-100-companies-2021.pdf.

3. "The world's most."

and insurance. Nonetheless, examples draw from several other industries and economic sectors to illustrate how 4R technologies can help the B4B. The book does not provide an in-depth analysis of the effect of culture on the adoption of 4R technologies.

This chapter introduces key underlying concepts and developments related to the 4R, the nature and characteristics of the B4B, and key 4R technologies. It also highlights some notable uses of 4R technologies for the B4B and explores different value propositions of these technologies for low- and high-income people.

Understanding the Nature and Characteristics of the B4B

Unfortunately, no clear universal understanding has been reached for defining the bottom 4 billion, so different criteria may lead to substantively different answers. In this book, I use B4B as a broad concept rather than a precise term. Various indicators related to income, wealth, and ability to participate in the market economy are best suited for identifying the B4B.

Per capita income is a simple, straightforward way to identify the B4B. In this regard, in 2018, 50% of the world's population, or about 3.8 billion people, lived in households with discretionary expenditures that are enough to be considered "middle class" or "rich." Such households can *afford some luxury* products such as vacations or restaurant meals.

About the same number lived in households that were poor or vulnerable to poverty.[4] The B4B, and the subjects of this book, falls into the latter group and can be roughly mapped to the population that lacks internet access. In 2019, 65% of the world's 4 billion people without internet access lived in countries with a per capita annual gross domestic product of less than US$3,895, which is the cutoff between lower-middle-income and upper-middle-income countries.[5]

Some additional observations can be made about the B4B. Of the many possible approaches, median values of net worth and per capita income seem to constitute the most logical criteria for analyzing poverty and the B4B

4. Homi Kharas and Kristofer Hamel, "A global tipping point: Half the world is now middle class or wealthier," *Brookings*, September 27, 2018, https://www.brookings.edu/blog/future-development/2018/09/27/a-global-tipping-point-half-the-world-is-now-middle-class-or-wealthier/.

5. Michael Pisa, *Developing countries seek greater control as tech giants woo the "next billion users,"* Center for Global Development, February 5, 2019, https://www.cgdev.org/blog/developing-countries-seek-greater-control-tech-giants-woo-next-billion-users.

Table 1.1. Classification of the world population based on income and the ability to engage in economic exchange

	Rangan et al. (2007)[1]		Kharas and Hamel (2018)[2]		
The year of data on which the classification was based	2001 data from the World Bank		2011 purchasing power parity (PPP) household income and projections by World Data Lab		
The year for which the classification is applicable	2006		2018		2030 (Estimates refined by social enterprise World Data Lab)
	Class (income/day)	Population	Class (household income/day)	Population	Population
	Poverty market (<US$1)	1.2 billion	Poor (<US$1.90)	630 million	450 million
	Submerged market (US$1-US$5)	2.6 billion	Vulnerable to falling into poverty (US$1.90-US$11)	3.16 billion	3.3 billion
	Exchange market (>US$5)	2.1 billion	Middle class (US$11-US$100)	3.59 billion	5.3 billion
			Rich (>US$100)	200 million	300 million

[1]V. Kashturi Rangan, John A. Quelch, Gustavo Herrero, and Brooke Barton, *Business solutions for the global poor: Creating social and economic value* (John Wiley & Sons, 2007).

[2] Homi Kharas and Kristofer Hamel, "A global tipping point: Half the world is now middle class or wealthier," Brookings, September 27, 2018.

population. The 2018 Global Wealth Report, published by Credit Suisse Research Institute, estimated that 50% of the world's population had a net worth of less than US$4,210.[6] Likewise, according to Gallup, in 2013, the median per capita income worldwide was US$2,920 and the median annual

6. Kathleen Elkins, "Here's how much money it takes to be among the richest 50 percent of people worldwide," *CNBC*, November 19, 2018, https://www.cnbc.com/2018/11/19/how-much-money-it-takes-to-be-among-the-richest-50-percent-worldwide.html.

household income was US$9,733.[7] These cutoff points can be used for classifying the B4B.

Some facts and statistics can help us understand the socioeconomic status of the B4B. Among the B4B, according to the World Bank, 3.4 billion people struggle to meet basic needs. For instance, 80% of the population in South Asia lived below US$5.50 per day in 2015.[8] Four billion people lack social security protection,[9] roughly the same number of individuals who belong to the B4B. Four billion people live in places that have no street names or numbers.[10]

Most of the B4B population lives in rural areas. Rural population is 45% of the total world population, but the proportions are much higher in LMICs, for example, 84% in Niger, 87% in Papua New Guinea, 80% in Nepal, and 83% in Rwanda. The proportions are 67% in low-income countries and 19% in high-income countries.[11]

Another approach is to look at ability to engage in market exchanges. In 2006, about 4 billion people were estimated to live on less than US$5 a day. The population that is at the "rock bottom" of the global economic pyramid is referred to as the "poverty market" (living on less than US$1 per day in 2006), which lacks access to sufficient food, clean drinking water, and shelter. The size of this market was about 1.2 billion people in 2006. A widely held assumption is that this population is not in a position to participate in a market economy as consumers or producers and relies highly on nongovernmental organizations (NGOs) and government relief programs.[12]

The population just above the "poverty market"—earning between US$1 and US$5 in 2001—has been referred to as "submerged markets," which

---

7. Glenn Phelps and Steve Crabtree, "Worldwide, median household income about $10,000," *Gallup*, December 16, 2013, https://news.gallup.com/poll/166211/worldwide-median-household-income-000.aspx.

8. World Bank, *Nearly half the world lives on less than $5.50 a day*, October 17, 2018, https://www.worldbank.org/en/news/press-release/2018/10/17/nearly-half-the-world-lives-on-less-than-550-a-day.

9. "Four billion people have no social security protection—UN labour agency," *UN News,* November 29, 2017, https://news.un.org/en/story/2017/11/637771-four-billion-people-have-no-social-security-protection-un-labour-agency.

10. "How 3 words could put 4 billion people without addresses on the map," *CBC*, June 21, 2016, https://www.cbc.ca/radio/thecurrent/the-current-for-june-21-2016-1.3644974/how-3-words-could-put-4-billion-people-without-addresses-on-the-map-1.3645088.

11. "Rural population (% of total population)," World Bank, 2018, https://data.worldbank.org/indicator/SP.RUR.TOTL.ZS.

12. V. Kashturi Rangan, John A. Quelch, Gustavo Herrero, and Brooke Barton, *Business solutions for the global poor: Creating social and economic value* (John Wiley & Sons, 2007).

consisted of about 2.8 billion people that year. They are consumers of wide range of products such as packaged goods, apparel, and appliances, and they participate in the market economy. This population, however, faces challenges related to institutions and markets, which reduce the efficiency with which they can engage in exchange. They lack bank accounts and access to formal credit.[13] The B4B population also faces a number of resource constraints, such as lack of access to clean drinking water and healthcare. A large proportion of the population cannot afford life, health, or agriculture insurance.

This population is also disproportionately affected by adverse social and economic conditions such as corruption and environmental degradation. For instance, the justice needs of about five billion people have not been met, and many live in extreme conditions of injustice. That is, justice is inaccessible for them to deal with everyday problems. They may be excluded from the opportunity that the law provides.[14]

Looking at the regional dimension of poverty, both South Asia and Sub-Saharan Africa (SSA) have high concentrations of poor people. In 2015, half the world's 736 million extremely poor people lived in just five countries in these two regions: India, Nigeria, Democratic Republic of Congo, Ethiopia, and Bangladesh (in descending order).[15] Likewise, in Latin America, 25% of the population lived below the poverty line (US$4 per day) in 2014. Even with a stricter poverty line of US$2.50 per day, 14% of Latin America's population lived below the poverty line in 2014.[16]

Similarly, China's per capita GDP in 2020 was US$10,500, which puts it in the category of upper-middle-income countries. Using the World Bank's poverty line for upper-middle-income countries of US$5.50 a day, about 13% of China's population, or 200 million people, fell below that line in 2021.[17]

---

13. Rangan et al., *Business solutions.*

14. World Justice Project, *Measuring the justice gap*, February 6, 2019, https://worldjusticeproject.org/sites/default/files/documents/Measuring%20the%20Justice%20Gap_WJP%20Update_Feb2019_Final-updated_0.pdf.

15. Roy Katayama and Divyanshi Wadhwa, "Half of the world's poor live in just 5 countries," *Data Blog*, January 10, 2019, https://blogs.worldbank.org/opendata/half-world-s-poor-live-just-5-countries.

16. Santiago Levy, "Poverty in Latin America: Where do we come from, where are we going?" *Brookings*, May 10, 2016, https://www.brookings.edu/opinions/poverty-in-latin-america-where-do-we-come-from-where-are-we-going/.

17. John Ruwitch, "What China's 'total victory' over extreme poverty looks like in actuality," *NPR*, March 5, 2021, https://www.npr.org/2021/03/05/974173482/what-chinas-total-victory-over-extreme-poverty-looks-like-in-actuality.

Key 4R Technologies for the Bottom 4 Billion

Although some 4R technologies can be put to a wide variety of uses across the economy and can create many spillover effects, others have narrower applications in fewer fields. It has been argued that AI especially is rapidly becoming a general-purpose technology (GPT), defined as a technology with the potential to transform a wide range of household and business activities.[18] GPTs such as AI also facilitate complementary innovations and bring transformations and changes in business processes.[19]

Likewise, a survey of 200 applications of blockchain in different functions across various sectors showed that the technology has many characteristics of GPTs. Blockchain can thus be identified as a GPT even though it is still viewed as an emerging technology.[20]

The 4R technologies are diffusing rapidly in LMICs and driving transformative socioeconomic change. For instance, SSA economies are expected to have 300 million IoT connections by 2025.[21] The development of the IoT industry and market in low- and middle-income countries such as those in SSA is facilitated by several factors, including increasing affordability due to falling prices of IoT sensors and hard drives (Figure 1.1).

Among the notable uses of the IoT in LMICs are remote monitoring of water supplies to detect leaks, to identify optimal flow and match supply and demand, to monitor environment and resources such as endangered animals in remote areas, to provide farmers access to irrigation and solar-powered pumping solutions, and to facilitate remote fortification of cereal grains in a cost-effective manner to address nutritional deficiencies in children.[22]

Other 4R technologies are used by international public health, humanitarian, and relief organizations to increase the overall effectiveness of their

---

18. Boyan Jovanovic and Peter L. Rousseau, *General purpose technologies*, National Bureau of Economic Research, January 2005, https://www.nber.org/papers/w11093.

19. Indermit S. Gill, Wolfgang Fengler, and Kenan Karakulah, "The economics of AI-based technologies: A framework and an application to Europe," July 24, 2020, https://papers.ssrn.com/sol3/papers.cfm?abstract_id=3660114#.

20. Ethan Kane, "Is blockchain a general-purpose technology?" *Social Science Research Network (SSRN) Electronic Journal* (2017): 1–27, https://doi.org/10.2139/ssrn.2932585.

21. GSMA, *Scaling the IoT 2018*, https://www.gsma.com/iot/wp-content/uploads/2018/08/GSMA-IoT-Infographic-201.

22. Iris Meijer and Felix Brooks-Church, "The next generation of business: How Sanku is using IoT to fight malnutrition, helping the children of tomorrow have a brighter future," Vodafone, January 17, 2020, https://www.vodafone.com/business/news-and-insights/blog/gigabit-thinking/how-sanku-is-using-iot-to-fight-malnutrition-helping-the-children-of-tomorrow-have-a-brighter-future.

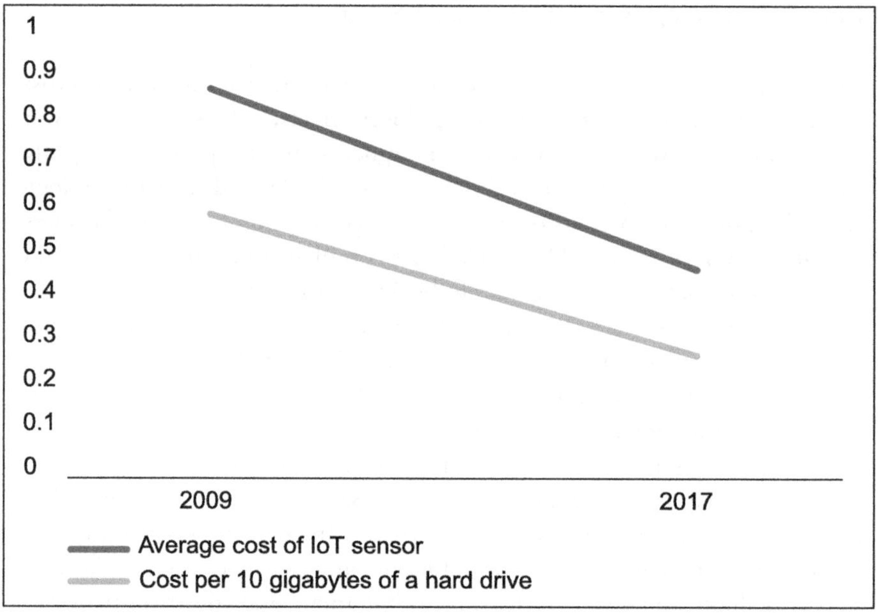

Figure 1.1. Average cost of an IoT sensor (Microsoft, *Manufacturing trends report: Microsoft Dynamics 365*, 2019, https://info.microsoft.com/rs/157-GQE-382/images/EN-US-CNTNT-Report-2019-Manufacturing-Trends.pdf) and cost per 10 gigabytes of a hard drive (US$) (Andy Klein, "The cost of hard drives over time," *Backblaze,* July 11, 2017, https://www.backblaze.com/blog/hard-drive-cost-per-gigabyte/).

programs. For instance, after the 2015 earthquake in Nepal, the UN's World Food Program (WFP) relied on big data to find the locations where people needed food or cash assistance. It analyzed call records to estimate how many people left the capital city of Kathmandu after the earthquake, where they went, and when they returned to the capital.[23]

The 4R technologies can help NGOs and governments more effectively serve the poverty market. For instance, many NGOs in Africa have been fighting malaria, which disproportionately affects poor people. For instance, controlling for other relevant factors, 1995 income levels of countries with intensive malaria were 33% that of countries without malaria.[24] Note that "intensive" malaria means a malaria index greater than 0.5. The index is

23. Amy Maxmen, "Can tracking people through phone-call data improve lives?" *Nature*, May 29, 2019: 614–617.

24. John Luke Gallup and Jeffrey D. Sachs, "The economic burden of malaria," *American Journal of Tropical Medicine and Hygiene*, 64, https://www.ncbi.nlm.nih.gov/books/NBK2624/.

calculated by multiplying the fraction of the population at risk of malaria by the fraction of cases caused by the *Plasmodium falciparum* species of the malaria parasite, which has been a leading cause of death in Africa. Technology startups are developing solutions based on 4R technologies that have focused on key issues such as fighting malaria (In Focus 1.1). This can be a game changer for preventing malaria deaths, a significant humanitarian concern. For instance, the WHO estimated that 409,000 people died due to malaria in 2019, and children aged under 5 years accounted for 67% (274,000) of the deaths worldwide.[25] In 2020, malaria killed six times more people in Africa than COVID-19 did.

In Focus 1.1: Sight Diagnostics' spin-off ZzappMalaria uses AI to eliminate malaria

Israel-based medical device company Sight Diagnostics combines machine vision and AI for blood diagnosis. As of mid-2021, Sight Diagnostics' Parasight AI-based system, which provides lab-grade blood-testing solutions, was used in blood tests to rapidly diagnose malaria in 24 countries.[26]

From Diagnosing to Fighting Malaria

Sight Diagnostics' spin-off ZzappMalaria aims to eliminate malaria using 4R technologies.[27] Zzapp's mission is to use innovative, cutting-edge technologies to eliminate malaria in an efficient and scalable manner.

In 2021, Zzapp won the grand prize of US$3 million in the IBM Watson AI XPRIZE Challenge, a joint venture between nonprofit organization XPrize and IBM's flagship AI platform IBM Watson. It is viewed as one of the toughest technology competitions to win.

The system was first tested in Ghana in 2017. Zzapp collaborated with a local malaria-control company to assess its effectiveness and cost based on number of malaria cases reported in clinics in the

25. World Health Organization, *Malaria*, 2021, https://www.who.int/news-room/fact-sheets/detail/malaria.

26. Mustapha Iderawumi, "Jerusalem start-up Zzapp is using AI to eliminate malaria in Africa," *Space in Africa*, June 25, 2021, https://africanews.space/jerusalem-start-up-zzapp-is-using-ai-to-eliminate-malaria-in-africa/.

27. Iderawumi, "Jerusalem start-up."

area.[28] As of mid-2021, the startup's anti-malaria operations were being conducted in Ghana, Zanzibar, Kenya, and Ethiopia.[29]

Zzapp uses AI to target the breeding sites of malaria-carrying mosquitoes. Its AI-powered, map-based app predicts where stagnant water is likely to form. Zzapp uses AI to analyze satellite imagery and data from various other sources, such as user reports, topography, climate, and local mosquito species. It then guides field workers to manage treatment using pesticides. Field workers use an app to mark water bodies and estimate the need to fumigate in order to free them of mosquitoes and larvae.[30]

Zzapp builds a risk index based on these data and recommends an optimal strategy for every village and town. The recommendations include the areas that need to be scanned for water bodies and houses that need spray. Zzapp also tells when these operations should be conducted and in what order.[31]

Zzapp was reported to be working with IBM Watson's AI and Data Science Elite Team to develop a weather analysis module. Based on weather data, the module predicts the abundance of water bodies, which allows Zzapp to intervene in a timely manner and more accurately determine the resources required to implement them.[32]

A Cost-Effective Solution

Zzapp says its solutions are cost-effective and that operations can be conducted even with limited budgets and in places that face the most challenging environmental conditions. In most SSA econo-

---

28. Abigail Klein Leichman, "AI helps zap mosquito larvae before they become a problem," *Israel21c*, April 7, 2021, https://www.israel21c.org/ai-helps-zap-mosquito-larvae-before-they-become-a-problem/.

29. Heather Landi, "Startups using AI to tackle malaria, mental health and human trafficking take top awards, and cash, from IBM XPrize," *FierceHealthcare*, June 24, 2021, https://www.fiercehealthcare.com/tech/ibm-xprize-startup-winners-using-ai-to-tackle-malaria-mental-health-and-human-trafficking.

30. David Gamba, "Combating malaria with AI and satellite imagery," *Medium*, October 28, 2020, https://medium.com/omdena/combating-malaria-with-ai-and-satellite-imagery-b39528962466.

31. Iderawumi, "Jerusalem start-up."

32. Sai Balasubramanian, "A company that uses AI to fight malaria just won the IBM Watson AI XPrize competition," *Forbes*, June 27, 2021, https://www.forbes.com/sites/saibala/2021/06/27/a-company-that-uses-ai-to-fight-malaria-just-won-the-ibm-watson-ai-xprize-competition/?sh=43c92d294704.

mies, the high costs of installing ICT infrastructure act as a barrier to healthcare.

Given the liberalization of the telecommunication sector, private companies and foreign investors provide information and communications technology (ICT) services. The providers of such services often charge high prices. Most governments have not subsidized ICT services to increase affordability for ordinary citizens.[33] Zzapp's app has low battery consumption, and it does not require continuous internet connectivity. The solution works well even on less advanced smartphones. In Ghana's Obuasi, a project reduced the mosquito population by over 60% in as little as three and a half months for a city with a population of 200,000. According to Zzapp, using the app cost US$0.20 per person protected, compared to US$5 for a basic house-spraying plan. The company reported that it was working on a project to eliminate malaria from São Tomé and Príncipe within two years.

Toward a Perfect Malaria-Fighting Solution

To eliminate malaria, operations need to reach a high level of perfection. They should not miss any house or water body and must tailor the intervention strategy to the particular village or town. Key elements of the intervention strategy are broken down into tasks that include the spraying techniques and materials, the season, and the duration of spray.[34] These tasks are assigned to field workers using the mobile app to guide them in the field and enable them to upload data to a designated dashboard. Its machine learning capabilities improve the system's efficacy from each operation.[35]

Zzapp Malaria's cofounder Arnon Houri Yafin noted: "We see our system as the Waze [Google subsidiary that provides satellite navigation software] of malaria elimination. There's an element that

---

33. Elizabeth Bakibinga-Gaswaga, Stella Bakibinga, David Baxter Mutekanga Bakibinga, and Pauline Bakibinga, "Digital technologies in the COVID-19 responses in sub-Saharan Africa: Policies, problems and promises," *Pan African Medical Journal* 35 (2020), https://www.ncbi.nlm.nih.gov/pmc/articles/PMC7875745/.

34. "Israeli AI company fights malaria with machine learning," *Robotics Business Review*, 2019, https://www.roboticsbusinessreview.com/news/zzapp-malaria-fights-malaria-with-machine-learning.

35. Shoshanna Solomon, "Israeli startup wins IBM top prize to Zzapp out malaria by mapping water sources," June 27, 2021, https://www.timesofisrael.com/israeli-startup-wins-ibm-top-prize-to-zzapp-out-malaria-by-mapping-water-sources.

gives you directions so you don't make mistakes—'Turn right here; spray this house.'" The company's user interface designer worked at Waze before.

Challenge Faced

Zzapp reported various challenges related to the availability and quality of data. First, there was the lack of enough data. Not many organizations were working on data collection for malaria elimination or using AI for that purpose. No annotated dataset existed, which meant that it was not possible to download the data and start. Second, there was the lack of high-resolution data. The dataset used by Zzapp had a three-meter resolution, less detailed than Google Maps imagery. Moreover, satellite imagery cannot detect some water bodies, such as water in a canal covered by a roof.[36]

Solutions based on 4R technologies are being developed to fight climate-related natural disasters such as floods and droughts. The satellite-imagery platform Digital Earth Africa is aggregating data from remote-sensing technologies, which will be used by the nonprofit focusing on the sustainable use of water and land resources in developing countries International Water Management Institute to develop applications for analyzing water levels and supplies on the continent. The data have been collected over many decades. Open Data Cube, an open-source resource for earth observation data, will hold the information and make it publicly available. This will provide an accurate picture of water resources across Africa, which can be used to forecast shortages and water needs. The data can help policy makers assess and monitor water resource availability and use. They can find ways to optimally allocate water across different sectors, such as agriculture and health, and develop strategies to manage water-related climate risks. The data can help scientific research organizations such as the International Water Management Institute develop extreme scenario models. For instance, the institute can use the data to create flood and drought models to develop early-warning systems for districts, counties, and wider regions.[37]

36. "Applying artificial intelligence to identify malaria infected water bodies," 2020, https://omdena.com/blog/artificial-intelligence-malaria/.

37. "Tap big data to fight floods and droughts in Africa," March 22, 2020, https://www.scidev.net/sub-saharan-africa/opinions/tap-big-data-to-fight-floods-and-droughts-in-africa/.

Ability to Engage in Productive Economic Activities and Market Exchange

A widely held belief concerning the "poverty market" is that this population is not in a position to participate in the market economy as consumers or producers. However, many companies have found ways to sell to and profit from this market. While much has been written about "selling to the poor"[38] and making profits at "the Bottom of the Pyramid,"[39] much less has been written about helping the poor start and grow their own ventures, manage risk and vulnerability and sell their produce. Regarding the latter, a challenge this population faces, especially in accessing financial products, concerns informational opacity.[40] Part of the problem is that most LMICs are characterized by the lack of information about or poor performance of credit-rating agencies in providing information about the creditworthiness of small and medium-sized enterprises (SMEs) and low-income population. A national credit bureau would collect and distribute reliable credit information and hence increase transparency and minimize banks' lending risks. This situation puts SMEs in a disadvantaged position in the credit market. This is because SMEs tend to be more informationally opaque than large corporations because the former often lack certified audited financial statements; thus, it is difficult for banks to assess or monitor the financial conditions. Some solutions have been developed to overcome this adverse effect associated with this opacity and help people in the poverty market to engage in productive economic activities and market exchange (In Focus 1.2 and 1.3).

In Focus 1.2: BanQu's "Economic Passport" for the Poverty Market

The U.S. software technology company BanQu has introduced what it calls an "economic passport" for the poverty market. The goal is to establish economic identities and proofs of record, which involves capturing and storing detailed records of economic transactions, for unbanked persons in extreme poverty zones or the poverty market. The solution utilizes the Ethereum network's blockchain.[41]

38. Allen L. Hammond and C. K. Prahalad, "Selling to the poor," *Foreign Policy* 142 (2009): 30.

39. Erik Simanis and Duncan Duke, "Profits at the bottom of the pyramid," *Harvard Business Review*, October 2014, https://hbr.org/2014/10/profits-at-the-bottom-of-the-pyramid.

40. Joseph E. Stiglitz and Andrew Weiss, "Credit rationing in markets with imperfect information," *American Economic Review* 71 (1981): 393–410.

41. Aaron Stanley, "Microlending startups look to blockchain for loans," *Coindesk*, December 8, 2017, www.coindesk.com/microlending-trends-startups-look-blockchain-loans.

Among the first projects, BanQu created a secure identity for displaced people in Kenya's Dadaab refugee camp. By 2018, it had connected 25,000 farmers, displaced people, and refugees to the platform across eight countries. BanQu users maintains ownership of their personal information and decide which information to share with whom.[42]

BanQu defines economic identity as "the marriage of identity and commerce, resulting in a global, vetted, and manageable asset."[43] A blockchain-based verifiable digital identity can help marginalized groups establish ownership, business assets, and production values. Such an identity would thus help them engage in economic transactions and participate in the global economy.

BanQu aggregates information from a number of sources, such as financial history, land records, trust networks documenting relationships with others, business registrations, vaccination records, and remittance income. Identity-related information sources include selfies, biometrics, and key physical attributes. Blockchain's decentralized, secure ledger also provides "know your customer" (KYC) and other information to partners that can potentially offer products and services to these marginalized individuals.[44]

Individuals and organizations that engage in transaction with unbanked persons—such as financial institutions and supply chain partners—can also join BanQu's platform. In this way, unbanked individuals are connected with parties that might be banked or unbanked to build a "mini financial network."[45] This allows the creation of history of transactions on the platform and the formation of economic identity.

Theoretically potential borrowers with blockchain-based economic ID can thus more easily receive loans by showing such infor-

42. Banqu, *Blockchain for supply chain transparency*, 2018, https://banqu.co/case-study/.

43. Vanessa Bates Ramirez, *How blockchain can make identification borderless and immutable*, Singularity Hub, June 11, 2017, https://singularityhub.com/2017/06/11/how-blockchain-is-helping-in-the-battle-against-extreme-poverty/#sm.001dojtf51cf3dbxr292hdqg31bks.

44. George Baker White, "BanQu seeks to transform the lives of millions by empowering them with an economic identity," *Social Fintech*, 2018, https://socialfintech.org/banqu-seeks-to-transform-the-lives-of-millions-by-empowering-them-with-an-economic-identity/.

45. A. Krishnakumar, "BanQu—Financial network on blockchain creating economic identity for the unbanked," *Daily Fintech*, 2017, https://dailyfintech.com/2017/05/12/banquapp-blockchain-driving-economic-identity-for-the-unbanked/.

mation to lenders.[46] For instance, farmers can access their records using a mobile phone to present proof of identity. A potential borrower can also ask the bank to do so.

A case study provided by one of the founders of BanQu is that of a poor Rwandan female farmer near the border of Tanzania. She lacks any identity but owns a piece of land. She can produce about 100 kilograms of corn from the land. She needs to feed a family of eight to ten. When she harvests her crops, she is often forced to sell them to a broker at a low price. If she can show that she owns the land that produces 100 kilograms of corn annually, she can sell her crops at higher prices to contract buyers that may also be willing to enter into long-term agreements with her.

BanQu solution can also greatly reduce post-harvest deterioration, which affects tradability of some crops.[47] For instance, if the Rwandan farmer is able to dry her corn to 13% moisture content, the corn's price may increase by 100%. With her economic identity, which includes the land she owns, a production forecast, and a buyer, she can purchase a dryer, which also becomes a part of her economic history for future transactions. After her BanQu identity has gone through a few farming cycles, the lender charges her lower interest rates for future loans.

In Focus 1.3: Afreximbank's Initiative to Facilitate SMEs' Market Access

Intra-African trade was around 2% of the continent's total trade between 2015 and 2017, compared to 67% in Europe.[48] Moreover, up to 50% of intra-African trade is conducted on an informal basis.[49]

46. Stanley, "Microlending."

47. Andrew Dorward, Jonathan Kydd, Jamie Morrison, and Ian Urey, "A policy agenda for pro-poor agricultural growth," *World Development* 32, no. 1 (2004): 73–89.

48. Omoniyi Kolade, "Why paytech is the key to unlocking Africa's new free trade zone," August 30, 2021, https://europeansting.com/2021/08/30/why-paytech-is-the-key-to-unlocking-africas-new-free-trade-zone/.

49. "AFCFTA critical to Nigeria's economic potential—AFREXIM Bank," *Vanguard News,* March 28, 2019, https://www.vanguardngr.com/2019/03/afcfta-critical-to-nigerias-economic-potential-afrexim-bank/.

African Continental Free Trade Agreement and Facilitation of Digitalization

The 55-member African Continental Free Trade Agreement (AfCFTA), which officially went into force on May 30, 2019, is likely to play a key role in facilitating digitalization and promoting Africa's economic development. AfCFTA has the potential to lift Africa's 30 million people out of extreme poverty and boost the incomes of nearly 68 million others living on less than US$5.50 a day.[50]

The goal of AfCFTA is to create a single African market for goods and services and to promote freer movement of capital and people.[51] As of February 9, 2020, 28 of the 55 African Union member states had ratified and deposited the ratification of AfCFTA with the African Union (tvcnews.tv, 2020). Trading activities under AfCFTA started on January 1, 2021.

Following the establishment of AfCFTA, SSA's financial institutions are developing solutions to facilitate trade between countries with diverse financial systems. They are developing new technologies, such as AI systems, advanced analytics and digital trade finance platforms.[52]

African Export-Import Bank's Roles

African Export-Import Bank (Afreximbank) is a pan-African multilateral trade finance institution created by the African Development Bank. Afreximbank is creating tools to eliminate bottlenecks and facilitate the full realization of AfCFTA's potential. The bank aims to use AI and other technologies to break trade barriers in Africa caused by fragmentation.[53] For instance, it has developed a Trade

50. World Bank, *The African Continental Free Trade Area*, 2020, https://www.worldbank.org/en/topic/trade/publication/the-african-continental-free-trade-area.

51. Trudi Hartzenberg, *The African Continental Free Trade Area Agreement—What is expected of LDCs in terms of trade liberalisation?*, United Nations, 2019, https://www.un.org/ldcportal/afcfta-what-is-expected-of-ldcs-in-terms-of-trade-liberalisation-by-trudi-hartzenberg/.

52. Baker McKenzie, "Massive increase in M&A deal value in Sub-Saharan Africa in the first half of 2021," *Lexology*, August 28, 2021, https://www.lexology.com/library/detail.aspx?g=e3f54dc6-bd5e-4a67-834b-ee943cb0f52e.

53. "Afreximbank supports intra-regional trade with USD 25 bln," 2019, https://thepaypers.com/e-invoicing-supply-chain-finance/afreximbank-supports-intra-regional-trade-with-usd-25-bln-779393.

Information Portal (TIP), which will utilize AI to predict regional supply chains.[54]

Afreximbank's African Trade Gateway uses multiple interrelated digital platforms to eliminate some of the main challenges to intra-African trade. These are related to the complexities of payments transfers within Africa, low access to trade and investment information, and costs associated with conducting KYC checks on African entities. Afreximbank invested over US$1.7 billion in supporting the development of regional value chains to accelerate the process of industrialization and lift the supply-side constraints that have acted as a barrier to the growth of African trade.[55]

Afreximbank is effectively utilizing technology-led innovations to address the continent's structural issues. Afreximbank's director of information technology Olaleye Babatunde noted that the "ABCD of Tech—Artificial Intelligence; Blockchain; Cloud; and Big Data" has been the driving principles at Afreximbank. For instance, more than 90% the company's system solutions run from secure cloud platforms. The bank migrated to the cloud in the early 2010s.[56]

Afreximbank found that around 38% of local banks in Africa reported rising rejection rates for trade finance instruments such as letters of credit in the first four months of 2020. More correspondent banking relationships were canceled.[57] To address this challenge, Afreximbank's goal was to hold over 500,000 datasets in its trade finance platform by 2020, which would make it the largest repository of customer due diligence information on African entities.[58]

Two flagship projects of Afreximbank are Pan African Payment and Settlement System (PAPSS) and MANSA.

54. Benedict Oramah and Kanayo, "Afreximbank: Access to finance for SMEs would play a key role in intra-regional trade," *ANA*, March 5, 2021, https://www.africanewsagency.fr/afreximbank-lacces-au-financement-pour-les-pme-jouerait-un-role-cle-dans-le-commerce-intra-regional/?lang=en.

55. Benedict Okey Oramah, "Facilitating the transformational AfCFTA: Tools for eliminating bottlenecks," *Brookings*, March 11, 2021, https://www.brookings.edu/blog/africa-in-focus/2021/03/11/facilitating-the-transformational-afcfta-tools-for-eliminating-bottlenecks/.

56. "Afreximbank's love affair with cutting-edge technology," *African Business*, 2021, https://african.business/2021/07/finance-services/afreximbanks-love-affair-with-cutting-edge-technology/.

57. Graphic Online, *Narrowing the trade finance gap for LDCs*, 2021, https://www.graphic.com.gh/entertainment/narrowing-the-trade-finance-gap-for-ldcs.html.

58. "Afreximbank supports."

PAPSS

PAPSS will help reduce the turnaround time for settlement of payment for intra-African trade. It is also expected to reduce cost, duration, and time variability of cross-border payments. It can decrease liquidity requirements of commercial banks and central banks. It will strengthen central banks' oversight of cross-border payment systems.[59]

The pilot phase of PAPSS was launched in Nigeria, Gambia, Ghana, Guinea, Liberia, and Sierra Leone. Following the pilot, the PAPSS was commercially launched on January 13, 2022.[60]

In July 2020, Afreximbank advertised for several positions for PAPSS, whose responsibilities included developing a data services and research framework combining artificial intelligence and natural language processing (NLP), advanced data modeling, predictive modeling, and other data analytics techniques to draw insights, discover hidden opportunities, and develop ideas and concepts to guide strategic decisions

MANSA

In many countries, enforcing contracts may involve collecting and disseminating information on noncompliant farmers and buyers, which increases incentives for farmers and firms to comply and help each party avoid high-risk partners. Benin has a clearinghouse for information on farmers who produce cotton and receive inputs on credit. This information can be used to punish farmers who violate the terms of their contract, and buyers knowingly purchase cotton

59. World Economic Forum, "Why paytech is the key to unlocking Africa's new free trade zone," *European Sting*, August 30, 2021, https://europeansting.com/2021/08/30/why-paytech-is-the-key-to-unlocking-africas-new-free-trade-zone/.

60. "AfCFTA gathers 40 ratifications," *New Times*, July 12, 2021, https://www.newtimes.co.rw/news/afcfta-gathers-40-ratifications; Zainab Usman and Alexander Csanadi "African Continental Free Trade Area: The Pan-African Payment and Settlement System," Carnegie Endowment for International Peace, February 7, 2022, https://carnegieendowment.org/2022/02/07/latest-milestone-for-african-continental-free-trade-area-pan-african-payment-and-settlement-system-pub-86376; African Export Import Bank "Data Services and Research (DSR) Manager (PAPSS) at the African Export Import Bank (Afreximbank)," 2020, https://hotjobsng.com/jobs/data-services-and-research-dsr-manager-papss-at-the-african-export-import-bank-afreximbank/.

from growers who have contracts with other companies.[61] Many other countries lack such information.

Afreximbank's data repository platform MANSA aims to address such challenges and lower the perceived risk of African entities. Mansa launched in 2018 and had onboarded about 200 African financial institutions by August 2020.[62] MANSA is expected to reduce the costs of compliance in African trade and facilitate the participation of the continent's 450 million SMEs in international trade.[63]

MANSA's centralized digital platform provide a single source of primary data required to store and perform customer due diligence checks on counterparties in Africa and provide investment information.[64] African corporates, including SMEs and financial institutions, voluntarily contribute information.[65] It can reduce the costs of compliance and increase the availability of trade finance.

MANSA connects key value-chain players in intra-African trade—such as farmers, processors, financiers, logistic providers, manufacturers, policy makers, traders, and regulators. Each player accesses the platform as a contributor, subscriber, or both. This is expected to result in standardization of information and improvement of corporate governance standards in Africa, particularly among SMEs and family businesses. Informal businesses are expected to formalize.[66]

61. Nicholas Minot, *Contract farming in sub-Saharan Africa: Opportunities and challenges*, International Food Policy Research Institute, 2011, https://www.ifpri.org/publication/contract-farming-sub-saharan-africa-opportunities-and-challenges.

62. *Global Trade Review*, "Afreximbank provides update on Mansa due diligence platform, launches trial for international FIs," 2020, https://www.gtreview.com/news/africa/afreximbank-provides-update-on-mansa-due-diligence-platform-launches-trial-for-international-fis/.

63. Boubacar Diallo, "Mansa, the new digital platform to boost intra-African trade," Afrika Tech, July 23, 2018, https://www.afrikatech.com/economy/mansa-the-new-digital-platform-to-boost-intra-african-trade/.

64. International Finance Corporation, "*IFC navigating essential anti-money laundering and combating the financing of terrorism requirements in trade finance: A guide for respondent banks public disclosure,* International Finance Corporation," 2018, https://documents1.worldbank.org/curated/en/761701540793262064/pdf/131400-WP-Navigating-Essential-anti-money-laundering-Guide-for-Respondent-Banks-PUBLIC.pdf.

65. The Payers, *Digital onboarding and KYC report 2020*, 2020, https://www.keesingtechnologies.com/wp-content/uploads/2020/12/PAYPERS_-Digital-onboarding-and-kyc-report-2020.pdf.

66. ITU*: The Mansa Repository Platform Document Classification*, 2021, https://figi.itu.int/wp-content/uploads/2021/06/GamuchiraiMoyo_AfreximBank-.pdf.

Regarding the ability of the B4B population (e.g., the poverty market and the submerged market earning between US$1 and US$5 in 2001) to engage in market exchange, four major variables have been proposed that influence the costs of exchange and hence the rate of growth of economies.[67] The first is the cost of measuring. A key component of transaction costs involves measuring the dimensions and attributes of the goods and services being exchanged or the performance of agents.[68] 4R technologies such as IoT, machine vision, and AI can provide new methods for accurate measurements to describe precisely what the parties engaged in a transaction are exchanging and which performance characteristics can be expected.

The second variable is the exchange process, which results from the nature and size of the market. It is argued that reputations based on previous experience and the history of behavior form the basis of economic exchanges in industrial societies. However, such exchanges are based on social networks in preindustrial societies.[69] 4R technologies can provide an opportunity to extend reach to markets beyond their social network. For instance, costly specification and enforcement are not needed in markets in preindustrial societies that are characterized by informal personal exchange, kinship ties, and friendship. Repeat dealings and personal knowledge of participants in the relationship constrain behaviors. In impersonal exchanges, there is the absence of factors that constrain the parties to gain from taking advantage of each other. All this increases the importance of precisely specified contracts and safeguards to enforce compliance.

The third variable is enforcement. In a society characterized by perfect enforcement, a neutral third party impartially evaluates disputes and awards compensation to the party affected by a contract violation. In such a situation, opportunism, shirking, and cheating are not attractive options for transacting parties. However, the real world is far from ideal. The high costs of measurement often make it difficult to determine whether a contract has been violated and, if so, who violated it. In countries with strong rule of law, well-developed court systems, and the state's coercive power to enforce judgements, complex contracting is an important mode of formal governance.

---

67. D. C. North, "Institutions, transaction costs, and economic growth," *Economic Inquiry* 25 (1987): 419–428.

68. D. C. North, "Dealing with a nonergodic world: Institutional economics, property rights, and the global environment," *Duke Environment, Law, and Policy Forum* 10, no. 1 (1999): 1–12.

69. Marcel Fafchamps, *Market institutions in Sub-Saharan Africa: Theory and evidence* (Cambridge, MA: MIT Press, 2003).

The final variable is ideology, which is related to one's convictions and beliefs about the justice of the rules and the society's contractual arrangements.[70] Specialization and division of labor have produced different perceptions of reality across communities, resulting in divergent and conflicting views on the fairness and justice of institutional arrangements.[71] Justice and fairness are key motivational forces in exchange relationships.[72] A challenge is that the ideology of profit maximization has been the main goal for many managers.[73] This means that the private sector has been doing very little to genuinely help marginalized groups (In Focus 1.4). Worse still, many companies have engaged in unjust and unfair activities that hurt the poor. Overall, however, 4R technologies have potential to expose injustice and unfair dealings. It is thus possible for marginalized groups such as farmers to get fair price of their produce.

In Focus 1.4: Alibaba's 4R Solutions for Small Business

China's online commerce company Alibaba has attracted vendors to its e-commerce websites Taobao Marketplace and Tmall by offering advertising and other services based in big data and AI. Theoretically, Taobao is a fee-free marketplace, which means that neither sellers nor buyers pay fees for completed transactions. In contrast, vendors that want to sell on Tmall, the platform for Chinese consumers to buy international brands, are required to pay annual technical support fees and security deposits. In 2012, depending on the size of the business-to-consumer stores, the fees amounted 60,000 RMB and 150,000 RMB respectively.[74]

Big-data solutions provide deep insights into shoppers' preferences.[75] Sellers that want to rank higher on Taobao's search engine are required to pay a fee for better placement of their products and advertising.[76] The active sellers on Taobao, which numbered about 7 million in 2016, paid

70. North, "Institutions."

71. North, "Institutions."

72. North, "Institutions."

73. Sumantra Ghoshal, "Bad management theories are destroying good management practices," *Academy of Management Learning and Education* 4, no. 1 (2005): 75–91.

74. Adina-Laura Achim, "Amazon vs. Alibaba: Everything you need to know about the two biggest e-tailers," *Jing Daily*, September 12, 2019.

75. L. Lorenzetti, "Alibaba's first public earnings reveal major revenue growth," *Fortune*, 2014, https://fortune.com/2014/11/04/alibabas.

76. Achim, "Amazon vs. Alibaba."

to rank higher on the site's internal search engine, which generated huge advertising revenue for Alibaba.[77]

AI services are also being given to merchants. For instance, Alibaba launched the customer-service chatbot Alime Shop Assistant in March 2017 to support merchants on its Taobao and Tmall platforms, which was offered free to merchants using its platforms. As of August 2018, about 600,000 merchants were using the chatbot.[78] While independent surveys conducted to assess the effectiveness of Alibaba chatbots have been found, the company claimed that online retailers can cut half of their call-center costs by using Alime Shop Assistant.[79]

Ling Shou Tong for Small Physical Stores

Alibaba also developed a big-data-based retail-management platform known as Ling Shou Tong for small physical stores in China. The solution aims to help store owners in making decisions on product procurement and sales. In 2017, Alibaba started providing the platform to Chinese retail shops. The shops get the platform for "free," but they are required to use their storefronts as Alibaba's fulfillment and delivery centers. They also need to provide data on their customers' shopping habits and patterns.[80]

Unclear Cost-Benefit Trade-Offs

Having access to big data and AI solutions is no guarantee that small businesses will improve their improve efficiency and profitability. For instance, it was reported that, due primarily to Alibaba's high advertising rates, most vendors on Taobao were making losses.[81] According to an article published in the Chinese-language newspaper *Enterprise Observer* in August 2013, over 80% of sellers

77. Matteo Talmassons "Taobao Marketplace" *Hackernoon*, May 5, 2016, https://hackernoon.com/taobao-marketplace-c7797581af6a.

78. Eileen Yu, "Alibaba smartens up chatbot ahead of 11.11 shopping festival," Zdnet, August 28, 2018, https://www.zdnet.com/article/alibaba-smartens-up-chatbot-ahead-of-11-11-shopping-festival/.

79. Yu, "Alibaba smartens."

80. Karen Hao, "Alibaba is trying to reinvent China's mom-and-pop stores," *Quartz*, January 5, 2018, https://qz.com/1171743/alibaba-is-trying-to-reinvent-chinas-mom-and-pop-stores/.

81. C. Clover, "Alibaba has almost single-handedly brought ecommerce to China," *Financial Times*, 2014, https://www.ft.com/content/11022ce8-a61a-11e3-8a2a-00144feab7de#axzz3V2tjJstS.

on Taobao did not make a profit. It was also reported that thousands of shops on Taobao close down every day.[82] Ling Shou Tong's own survey found that only one in five small shops deploying the technology reported a positive revenue growth.[83]

In many cases the benefits to small businesses of so-called free solutions provided by technology giants are not clear. Indeed, free solutions may hurt some firms rather than help them. The providers of such solutions tend to use them to promote their own interests rather than those of small businesses. For instance, while the use of Ling Shou Tong may make it easier to run stores, many small stores worry about unfair competition from Alibaba's online marketplace, which has a huge selection of products to choose from. This means that these small stores' customers may decide to take advantage of the convenience of online shopping on Alibaba's online marketplace and pick up the products from these stores.[84] Furthermore, Alibaba is in a position to make a better utilization of data on these stores' customers that they are required to provide.

A key benefit for Alibaba is also that merchants could be locked in its platform. In general, due to fierce competition in the Chinese e-commerce market, the country's e-commerce platform owners, such as Alibaba and JD.com, attempt to lock in the merchants by offering deals, which include technology tools.[85]

Different Value Propositions of 4R Technologies

4R technologies have different value propositions for the B4B than for middle-income and rich consumers. In this section, I illustrate this by considering two GPTs: AI and blockchain.

---

82. Rachel Lu, "Tea leaf nation: For Alibaba's small business army, a narrowing path," *Foreign Policy*, September 10, 2014, https://foreignpolicy.com/2014/09/10/for-alibabas-small-business-army-a-narrowing-path/.

83. He Wei, "Alibaba bets on Plan W for retailers," *China Daily*, August 29, 2020, https://global.chinadaily.com.cn/a/202008/29/WS5f49b17aa310675eafc563b6.html.

84. Hao, "Alibaba is trying."

85. Leo Sun, "Alibaba and JD.com want to lock in merchants with big data deals," *Fool*, June 22, 2019, https://www.fool.com/investing/2019/06/22/alibaba-jdcom-lock-in-merchants-data-deal.aspx.

Artificial Intelligence

In many LMICs, credit bureaus do not track large proportions of the populations. Consequently, a large proportion of consumers are excluded from many financial products and face obstacles to getting loans. Mainstream banks lack incentives to assess creditworthiness of low-income customers. One of the main focus areas of AI has been on expanding inclusion. For instance, AI is being used to find new ways of credit scoring.

In richer economies, however, the focus of the financial service industry has been on providing a high-quality customer experience using solutions such as chatbot, robo-advisory platforms, and customer service robot. For instance, the UAE's National Bank of Dubai (NBD) has developed AI-based personal banking assistant Eva. The NBD's lifestyle banking arm Liv also offers a conversational chatbot as well as a customer service robot "Pepper." AI-based robo-advisory platforms are also being used by financial services players such as the Dubai-based startup Sarwa in wealth management. They offer advanced investment services at low cost by using AI to allocate and manage funds.[86]

Blockchain

An analysis of Blockdata, the Netherlands-based market intelligence platform for blockchain and distributed ledger technology, found that the leading blockchain use case for Blockchain 50 companies—the world's biggest brands with over US$1 billion in annual revenue that are using blockchain[87]—was traceability and provenance.[88] Blockdata's analysis indicted that fifteen companies used blockchain solutions in traceability and provenance.[89] While other technologies make it possible to trace and track products, blockchain will lead to confidence and trust in wide range of

86. "The global AI agenda: The Middle East and Africa," *MIT Technology Review,* 2020, https://www.technologyreview.com/2020/06/19/1004121/the-global-ai-agenda-the-middle-east-and-africa/.

87. Michael del Castillo, "Blockchain 50: Billion dollar babies," *Forbes*, https://www.forbes.com/sites/michaeldelcastillo/2019/04/16/blockchain-50-billion-dollar-babies/#145541e757cc.

88. Andrew Fenton, "Blockchain traceability overtakes payments among major corporations," *Cointelegraph*, April 8, 2020, https://cointelegraph.com/news/blockchain-traceability-overtakes-payments-among-major-corporations.

89. Jonathan Knegtel, "Forbes blockchain 50—products data deep dive" *Blockdata*, August 16, 2021, https://blog.blockdata.tech/2020/04/forbes-blockchain-50-products-data-deep-dive/.

products such as fresh produce, raw materials, and diamonds. When goods change hands, relevant records can be added. Reports related to inspections and deliveries can be uploaded, and payments can be released automatically when conditions are fulfilled. PwC has identified provenance as the leading use of blockchain and estimated that by helping organizations verify the sources of their goods and track their movement and enhancing supply chain transparency, the technology has the potential to increase the global GDP by US$962 billion by 2030.[90]

For the B4B, one of the biggest value propositions of blockchain is the ability to make payments and settlement more efficient. There is especially a great need to make cross-border payments and settlement more efficient. For instance, many families living in refugee camps in Africa rely on remittances from friends and relatives overseas to survive.[91]

Crypto remittances are being used to address challenges related to high costs and inefficiency in the international remittances market. In August 2021, the bitcoin exchange LocalBitcoins announced that there would be no deposit fees and transaction fees between wallets in the platform. The platform does not deal with fiat currency itself. Users can use the platform to transact with one another.[92] For instance, to send money from Venezuela to a family member in Colombia, the sender buys bitcoin with Venezuelan bolivares using a bank transfer on LocalBitcoins. The sender then can search sell offers of bitcoin in Colombia and choose the offer with the best exchange rate. After the seller of bitcoin in Colombia transfers Colombian pesos to the family member's bank account, the sender transfers the bitcoin. A user reported that the whole process takes less than an hour. The platform charges a 1% fee to the user who offered to sell bitcoins.[93] Other platforms, such as Binance P2P and LocalCryptos, offer similar services.[94]

New regulations related to cryptocurrency assets are being put in place in some countries, which can reduce remittance costs. One example is the

---

90. PwC, *Time for trust: The trillion-dollar reasons to rethink blockchain 2020*, https://image.uk.info.pwc.com/lib/fe31117075640475701c74/m/2/434c46d2-a889-4fed-a030-c52964c71a64.pdf.

91. ReliefWeb, "Global remittances to suffer a blow from COVID-19—World," June 17, 2020, https://reliefweb.int/report/world/global-remittances-suffer-blow-covid-19,

92. Namcios, "LocalBitcoins cuts deposit, transaction fees between wallets in platform to zero," *Bitcoin Magazine*, 2021, https://bitcoinmagazine.com/business/localbitcoins-reduce-fees-to-zero.

93. "LocalBitcoins.com: Fastest and easiest way to buy and sell bitcoins," https://localbitcoins.com/fees.

94. "Crypto remittances prove their worth in Latin America," *CoinDesk*, 2020, https://www.coindesk.com/crypto-remittances-latin-america-geopolitical-tension.

UAE. In 2020, migrant workers in the Gulf region sent US$43 billion in remittances to their home countries.[95] The high cost of remittance transfer fees has been a concern. In the third quarter of 2020, the global average cost of sending US$200 was 6.75% of the amount.[96]

In 2020, the UAE's Securities and Commodities Authority specified conditions for offering crypto assets in the Emirates. Any entity providing such services must be formally licensed and comply with a number of laws on anti–money laundering, cybersecurity, and data protection. As of August 2021, six companies had satisfied the requirements of the regulations to create crypto exchanges. Two had reached the first stages of going live. One is the crypto asset-trading platform MidChains, which will be open to anyone. To use the services of crypto assets providers in the UAE, clients are required to provide a number of documents, such as proof of residence, income, and secure assets. Such requirements may prevent migrant workers from using the services. In the future, remittances are expected to be a regular feature of the UAE's cryptocurrency services.[97]

Internet of Things

We can see some sophisticated deployments of the IoT in industrialized countries. For instance, the Edge in Amsterdam was named by Bloomberg as the smartest buildings in the world.[98] It has about 28,000 sensors. The accounting and consulting company Deloitte, the building's main tenant, uses data from the sensors and other sources to adapt and lower energy use. For instance, sections of the building not in use are shut down to conserve energy. The building also knows which areas need cleaning.

The building recognizes employees' cars when they arrive and directs them to a parking spot. At the Edge, no one has a permanent desk. A smartphone app finds a desk for employees when they arrive. A variety of desk options or room types, such as sitting desk, standing desks, meeting room,

95. Salim Essaid, "Cryptocurrency promise for UAE's unbanked migrants—but not yet," *Reuters*, August 3, 2021, https://www.reuters.com/article/emirates-tech-migrants/update-1-cryptocurrency-promise-for-uaes-unbanked-migrants-but-not-yet-idUSL8N2PA7UG.

96. "How Covid-19 transformed the UAE's remittance sector in 2020," *The National*, 2020, https://www.thenationalnews.com/business/money/how-covid-19-transformed-the-uae-s-remittance-sector-in-2020-1.1131325.

97. Essaid, "Cryptocurrency promise."

98. Tom Randall, "The smartest building in the world: Inside the connected future of architecture," *Bloomberg*, September 23, 2015, https://www.bloomberg.com/features/2015-the-edge-the-worlds-greenest-building/.

and balcony seat are assigned depending on an employee's schedule. The app knows the employee's preferences for environmental conditions such as light and temperature.[99] With the smartphone app, workers can locate colleagues or schedule their workout in the gym.[100]

In some LMICs, relatively simple IoT solutions have helped address common challenges, such as the lack of grid infrastructure. To take an example, Kenya-based pay-as-you-go energy company M-KOPA (*kopa* is Swahili for "borrow") Solar uses IoT-based solutions to offer solar home systems on credit to poor households. The household can make small daily payments using mobile money, which allows storing and managing money in an account linked to a mobile phone.[101]

In 2016, M-KOPA's solar package cost US$200, including US$25 in interest, which could be purchased with a down payment of about US$30. Users were required to pay a daily instalment of US$0.45 for one year and then own an 8-watt solar panel, two LED lights, a USB phone charger and a portable solar-powered radio. M-KOPA solar panels can be used to power services such as lighting, cell phone charging, and radios. Most off-grid people in SSA are reported to spend US$0.50 daily on kerosene. GSM sensors placed inside the solar systems can monitor and regulate usage based on payments.[102] Each solar panel has a SIM card powered by Safaricom. The households can use the mobile payment system M-PESA to pay bills. Sensors have enabled this pay-as-you-go model to serve remote communities. When a customer makes a payment on a mobile phone using the M-PESA system, the SIM card sends a signal to activate the battery. The battery is powered by the panels.[103] M-KOPA also utilizes Microsoft Azure's data and analytic platform tools Cortana Intelligence Suite to predict customer repayment behavior.[104]

---

99. Randall, "Smartest building."

100. Brian Buntz, "Here now: The Edge in Amsterdam," *IoT World Today*, October 25, 2018, https://www.iotworldtoday.com/slides/here-now-the-edge-in-amsterdam/.

101. Anmar Frangoul "Pay-as-you-go solar power takes off in Africa," *CNBC*, February 25, 2015, https://www.cnbc.com/2015/02/25/pay-as-you-go-solar-power-takes-off-in-africa.html.

102. Frangoul, "Pay-as-you-go."

103. Stephan Faris, "The solar company making a profit on poor Africans: M-Kopa plans to be a $1 billion company by selling solar panels to rural residents—and providing them with credit," *Bloomberg*, December 2, 2015, https://www.bloomberg.com/features/2015-mkopa-solar-in-africa/.

104. Microsoft Airband Initiative, *Using the cloud and IoT to deliver safe, affordable energy in Africa*, 2018, https://download.microsoft.com/download/6/C/9/6C955541-5053-4A1C-BF0E-22F3BA34CE0F/Microsoft_Airband_M-KOPA_Casestudy.pdf.

Chapter Summary and Conclusion

In light of various resource constraints and challenges, 4R technologies can deliver a higher value proposition to the B4B population than they can for rich consumers. These technologies have already brough socioeconomic changes to the B4B population, and the potential for more change is even greater. This chapter has explicated a number of mechanisms by which 4R technologies can help the B4B population to become significant producers and consumers of goods and services in the global economy. These technologies can help increase access to resources to engage in productive activities. They provide alternative ways of measuring the dimensions and attributes of the goods and services being exchanged or the performance of agents, which can benefit the B4B. For smallholder farmers, the current reliance on intermediaries to measure the quality of their produce can be reduced. These technologies can thus provide new possibilities to participate in the market economy, which can help billions of poor people move out of poverty and surface from their current "submerged" state into the market.

4R technologies can also help create more effective welfare development policies to target the poor or those belonging to the poverty market. A significant proportion of those eligible for receiving welfare benefits fail to receive such benefits. The main reasons include lack of information (e.g., potential beneficiaries do not know that they are eligible),[105] administrative hassles, accessibility problems (e.g., inadequate transportation or childcare, limited staffing hours in welfare offices, problems filling out the forms).[106] With these technologies, relief programs can be better designed to achieve their immediate objectives. For example, the beneficiaries of Ethiopia's Productive Safety Net Program, a regular cash transfer program for poor households, increased from 6.5 million to 9.6 million beneficiaries within two months after the drought there in 2011. Data from satellite-based technology helped quickly evaluate the drought situation and effects on the communities.[107]

There currently is a big gap between the potential of the 4R technolo-

105. Richard D. Coe, "Nonparticipation in welfare programs by eligible households: The case of the food stamp program," *Journal of Economic Issues* 17 (1983): 1035–1056.

106. Marlene Kim and Thanos Mergoupis, "The working poor and welfare recipiency: Participation, evidence, and policy directions," *Journal of Economic Issues* 31 (1997): 707–728.

107. Keiko Saito and Felix Lung, "Using satellite technologies to protect African farmers from climate shocks," World Bank, October 29, 2015, https://blogs.worldbank.org/digital-development/using-satellite-technologies-protect-african-farmers-climate-shocks.

gies to improve the lives of the B4B and the reality of applications of these technologies on the ground. The 4R technologies' transformative potential can be enhanced through outside interventions to create high-quality, relevant data and make it available to relevant actors. Policy measures need to be adopted to facilitate the use of new tools to increase the productivity and reach new markets.

A point that needs to be elucidated is that the B4B populations have significantly different usage patterns of 4R technologies compared to more developed economies. Nonetheless, the B4B population cannot be treated as a single market. The populations that are above "rock bottom" on the global economic pyramid have generally been viewed as markets for big companies' products. With some effort, these populations can make use of 4R technologies to find markets for their products.

Finally, the private sector needs a clear pathway to profit in order to be convinced to serve the B4B.[108] The 4R may provide such pathways.

This Book's Organization

Following this introductory chapter, this book has three additional parts and eleven chapters. The four chapters in part 2 give an overview of the major fourth revolution technologies. Chapters 2 and 3 will cover two major 4R technologies—AI and blockchain—that are viewed as GPTs. These two chapters will explain how these technologies work and discuss their impacts and some notable applications to overcome challenges facing the B4B.

Chapter 4 highlights the roles of remote sensing and satellite imagery tools in addressing economic, healthcare, environmental, and other concerns facing the B4B. It provides a perspective on the amplified benefits of remote sensing and satellite imagery tools by combining them with other technologies. A special emphasis of this chapter is also on the usefulness of these tools in dealing with disasters such as COVID-19 pandemic.

Chapter 5 provides a perspective on how the falling costs of IoT sensors and hard drives to store data is driving the internet of things in low- and middle-income countries and how the technology has helped overcome barriers related to deficient infrastructures and markets and increase the welfare of vulnerable groups. This chapter also delves into the entrepreneurial activities of IoT firms to develop new markets that serve the unmet needs of B4B populations.

108. Rangan et al., *Business solutions.*

The three chapters in part 3 analyze the effects of the 4R on economic and social developments by looking at three key sectors and issues. Chapter 6 provides glimpse into the use of the 4R technologies to improve healthcare outcomes and deal with pandemic preparedness, especially in the context of the COVID-19 pandemic. This chapter gives special consideration to the roles of 4R technologies in achieving key goals of a healthcare system—to keep people healthy, to treat sick people, and to provide affordable, efficient, and cost-effective healthcare services.

Chapter 7 focuses on roles of the 4R technologies in the agricultural sector. A key emphasis of the book is on the roles of 4R technologies in enhancing the welfare of smallholder farmers, who comprise the majority of poor people in LMICs. Key topics covered include smallholder farmers' participation in agricultural markets, access to agricultural finance, and threshold costs in economic exchange. This chapter also touches on 4R technologies' roles in promoting sustainability issues in the agricultural value chain.

Chapter 8 highlights how these technologies are transforming finance, banking, and insurance activities in LMICs. It explains 4R technologies' roles in addressing information asymmetry problems and reducing transaction costs in the financial system, which can potentially increase poor people's access to finance. This chapter argues that LMICs can gain more than rich countries from decentralized networks. It also offers an analysis of 4R technologies' roles in financing and delivering welfare programs.

Part 4's final four chapters discuss opportunities, challenges, and implications, and suggest the way forward. In chapter 9, I highlight the opportunities, barriers, and enablers of the 4R from the perspective of the B4B. Key enablers facilitating digital transformation of LMICs discussed in this chapter include freely available data, local capacity-building initiatives, efforts to develop skills and exchange knowledge, and rapidly falling costs of 4R technologies. Major opportunities created by 4R technologies discussed in the chapter include south-south cooperation (SSC) and marginalized groups' participation in economic activities. The chapter also takes a look at 4R technologies' potential malicious uses by corporations and governments, and various other challenges that need to be addressed in order for the benefits of these technologies to be realized for the B4B.

Chapter 10 discusses economic developmental implications of the 4R. Several theories of economic development and concepts such as poverty trap, middle-income trap, leapfrogging theory, and the flying geese paradigm (FGP) are discussed in relation to the 4R in the context of the B4B. Potential roles of the 4R to pull poor people out of poverty are analyzed, as are conditions that can facilitate leapfrogging in the 4R era.

Chapter 11 delves into social, policy, legal, and ethical implications. It gives an overview of regulations in several key realms to benefit from the 4R. It emphasizes the importance of data protection in LMICs. It presents examples of questionable ethical practices that negatively affect low-income people. It discusses digital colonialism in the context of the 4R. Discussion, conclusion, and recommendations are provided in chapter 12.

PART 2

Major 4R Technologies

CHAPTER 2

Artificial Intelligence

Artificial intelligence is a major fundamental 4R technology and among the most transformative and powerful technological innovations.[1] Economists view AI as among the four most important GPTs,[2] the other three being the steam engine, electric power, and information technology. GPTs possess the potential to transform household and business activities.[3] GPTs such as AI also facilitate complementary innovations and bring about transformations and changes in business processes.[4]

AI has already started having a powerful impact on LMICs. For instance, AI's business value in Africa is expected to surpass US$45 billion by 2025.[5] AI applications, though in a nascent stage of development, are already having considerable economic, health, and social outcomes for the B4B.[6] The first of AI's direct benefits to LMICs stems from the democratization and decentralization of medical knowledge and excellence. Medical AI is already achieving things that even the world's best medical professionals are not

1. Celine Herweijer and Dominic Waughray, *Fourth industrial revolution for the Earth harnessing artificial intelligence for the Earth,* PWC, January 2018.

2. Indermit Gill, "Whoever leads in artificial intelligence in 2030 will rule the world until 2100," *Brookings,* January 17, 2020, https://www.brookings.edu/blog/future-development/2020/01/17/whoever-leads-in-artificial-intelligence-in-2030-will-rule-the-world-until-2100/.

3. Boyan Jovanovic and Peter L. Rousseau, "General purpose technologies," National Bureau of Economic Research, January 2005, https://www.nber.org/papers/w11093.

4. Gill, "Whoever leads."

5. J. Preston, "The blockchain and AI Africa conference 2020 is moving beyond the hype," *Cointelegraph,* November 30, 2019, https://cointelegraph.com/press-releases/the-blockchain-and-ai-africa-conference-2020-is-moving-beyond-the-hype.

6. Nir Kshetri, "Artificial intelligence in developing countries," *IEEE IT Professional* 22, no. 1 (2020): 63–68.

capable of doing. AI is especially likely to have a powerful impact on the fields of ophthalmology and radiology. Several AI apps use images of human eyes that practitioners, technicians, and even patients can use to diagnose possible diseases without the expertise of ophthalmologists (In Focus 2.1). This is especially important in LMICs that have severe shortage of health specialists.[7]

In Focus 2.1: Aravind Eye Hospital's AI-Based Solutions to Help the Visually Impaired

Aravind Eye Hospital, based in Madurai, India, uses AI to diagnose blindness-causing diabetic retinopathy, among the leading causes of blindness worldwide. According to the World Health Organization (WHO), about 70 million Indians suffer from diabetes, which carries elevated risk of blindness. But the country has a severe shortage of eye doctors. There are only 11 eye doctors for every million people in India. Aravind Eye Hospital uses neural networks as a new screening method; this involves the use of complex mathematical systems to learn tasks by analyzing large amounts of data. Neural networks have been used to improve tasks such as face recognition services, digital assistants, self-driving cars, and instant translation services (e.g., Google Translate). For instance, by analyzing millions of retinal scans that show signs of diabetic blindness, a neural network learns to identify the blindness conditions.[8]

In a pilot project sponsored by Google, nurses in Aravind Eye Hospital's 70 satellite clinics in India, known as rural teleconsultation centers, captured images of patients' retinas. They upload the photos to the cloud, where Google's machine learning (ML) algorithm works in combination with the specialists to detect and diagnose the disease.[9] The algorithm can diagnose the problem in a few seconds.

7. W. Nicholson Price II, "Risks and remedies for artificial intelligence in health care," *Brookings*, November 14, 2019, https://www.brookings.edu/research/risks-and-remedies-for-artificial-intelligence-in-health-care/.

8. Cade Metz, "India fights diabetic blindness with help from A.I.," *New York Times*, March 10, 2019, https://www.nytimes.com/2019/03/10/technology/artificial-intelligence-eye-hospital-india.html.

9. Shweta Ganjoo, "Google teams up with Indian doctors to use AI to help diabetes patients in India," *India Today*, July 11, 2019, https://www.indiatoday.in/technology/news/story/google-teams-up-with-indian-doctors-to-use-ai-to-help-diabetes-patients-in-india-1566874-2019-07-11.

Starting 2014, Google and Aravind trained an algorithm to recognize the signs of diabetic retinopathy, which include distinctive spots and bleeding in the retina. The ML algorithm can detect problems that eye doctors could not.[10] As of 2018, about 2,000 patients had benefited from the service.

In 2021, Aravind Eye Hospital introduced AI-based smart vision spectacles (SVS), to help people with visual impairments move around with more confidence. Bengaluru-based software company Smart Global Technology manufactured the spectacles by teaming up with the Boston-based nonprofit Vision Aid. Each SVS pair costs ₹22,000 (US$292). Wearers can listen to texts in English and 73 Indian regional languages. The SVS is fitted with two devices that look like USB pen drive—one has a camera that can capture images within two meters and warns of obstacles, then stores images and details of people known to the wearer. If a known person is in the line of sight, it can identify the person's name and other details of the person.[11]

AI tools are also useful for locating people who live in poverty, as accurate and up-to-date poverty datasets do not exist. These tools' usefulness for this purpose was especially apparent during the COVID-19 pandemic. For instance, Meta platform's Data for Good developed microestimates of wealth and poverty for LMICs based on nontraditional data such as satellite imagery, topographic maps, and Facebook and Instagram user data. Although the Data for Good project began in 2017, COVID-19 brought a new urgency and importance to it, accelerating the project.[12] To predict where poor populations live, ML algorithms were trained using demographic and health survey data from 56 countries. The resulting dataset—the Relative Wealth Index—is published as a publicly available dataset that covers 135 LMICs.[13]

10. Anthony Lydgate, "To an AI, every eye tells a story," *Wired*, September 18, 2018, https://www.wired.com/story/wired25-sundar-pichai-r-kim-artificial-intelligence-vision/.

11. "Good news for those with low or no vision," *The Hindu*, August 18, 2021, https://www.thehindu.com/news/cities/Madurai/good-news-for-those-with-low-or-no-vision/article35964002.ece.

12. Talib Visram, "These new poverty maps could reshape how we deliver international aid," *Fast Company*, April 16, 2021, https://www.fastcompany.com/90625436/these-new-poverty-maps-could-reshape-how-we-deliver-humanitarian-aid.

13. Laura McGorman, Guanghua Chi, and Han Fang "How AI-powered poverty maps can increase equity in the COVID-19 response," *Brookings*, May 7, 2021, https://www.brookings.edu/blog/future-development/2021/05/07/how-ai-powered-poverty-maps-can-increase-equity-in-the-covid-19-response/.

The detailed, fine-grained maps generated by the program are being used by the governments of Nigeria and Togo. In Nigeria, poverty levels for 100% of the country's wards are known. In Togo, the data is reported at the canton level.[14]

An important use of the dataset is likely the timely and equitable administration of COVID-19 vaccines, which are currently in short supply in LMICs. As of mid-April 2022, only 16% of the population in low-income countries had received at least one dose of a COVID-19 vaccine, compared to 80% in high- and upper-middle-income countries.[15] The Relative Wealth Index is expected to help show where people live, especially the poorest among the world's poor.[16]

Unsurprisingly, many LMICs have made a serious commitment to developing AI capabilities. They are accelerating the process of establishing new policies, regulations, and practices for using AI. Moreover, LMICs such as China, India, Kenya, Mexico, and Russia have released AI strategy documents.[17] Others have developed broader visions to transform their economies using 4R technologies.[18]

There is, however, wide variation among LMICs in developing their AI industry. China is a clear global leader. Most other countries perform poorly in producing or having the main ingredients needed for building high-quality AI applications, such as big data, computing power, and workforce. Nonetheless, the AI ecosystem is gradually emerging in these countries.

Organizational impacts of AI are far from clear at this point. Thus, there are some well-founded rationales for the adoption of AI as well as a number of misinformed and ill-guided viewpoints. AI has not yet reached the level of handling complex tasks.[19] AI may create a false sense of hope among the

---

14. Visram, "These new poverty maps."

15. Josh Holder, "Tracking coronavirus vaccinations around the world," *New York Times*, April 16, 2021, https://www.nytimes.com/interactive/2021/world/covid-vaccinations-tracker.html.

16. Visram, "These new poverty maps."

17. Anand Rao, "Is AI the next frontier for national competitive advantage?," *strategy+business*, January 22, 2019, https://www.strategy-business.com/blog/Is-AI-the-Next-Frontier-for-National-Competitive-Advantage.

18. Helmo Preuss, "WEF Africa: Botswana tells the world it wants to reduce dependence on diamonds," *Independent Online (IOL)*, September 5, 2019, https://www.iol.co.za/business-report/economy/wef-africa-botswana-tells-the-world-it-wants-to-reduce-dependence-on-diamonds-31998283.

19. Phil Britt, "Where AI customer experience investments are paying off," *Digital Asset Management Buyer's Guide*, November 15, 2018, https://www.cmswire.com/customer-experience/where-ai-customer-experience-investments-are-paying-off/.

leadership of many companies. AI vendors tend to present the technology as a means to replace humans. This situation is especially problematic and paradoxical for the B4B, as a large proportion of this population lacks formal jobs and is un- or underemployed.

In this chapter, I provided a perspective on the applications, and impacts of AI on the B4B. I also provide an overview of key enablers as well as major opportunities and barriers in the development and implementation of AI for the B4B.

How Artificial Intelligence Works

AI involves machines simulating human intelligence by machines. The key processes involved in AI are learning (acquiring information and understanding the rules for using the acquired information), reasoning (applying the rules to reach conclusions), and self-correcting. Machine learning—or ML—is a type of AI that helps increase accuracy of software applications in predicting outcomes without explicit programming. The basic idea of ML is simple: algorithms receive input data and, using statistical analysis, predict output values within acceptable ranges. ML processes are similar to those involved in data mining and predictive modeling, which also look for patterns in data to adjust program actions (Figure 2.1 and In Focus 2.2).

In Focus 2.2: TechnoServe Uses Machine Learning to Identify Cashew Trees in Benin

The U.S. nonprofit TechnoServe uses remote sensing, drone mapping, machine learning, and satellite data to boost cashew-nut production in Benin. Cashews account for 8% of the country's export earnings. TechnoServe helps farmers increase both the quantity and the quality of yields and providing supports to plant trees in the best possible way. The organization has plans to replicate the project in other West African countries and in Mozambique.[20]

The challenge for Benin has been low per-hectare yield of cashew nuts. The government's goal is to increase production from 120,000MT to 300,000MT. Drones map wide cashew-producing areas and identify cashew production areas that have not adopted climate-smart agricultural practices. Cropland, forest, and other

20. B. Debusmann Jr., "Farms are going to need different kinds of robots," March 1, 2021, *BBC News*, https://www.bbc.com/news/business-56195288.

landscape elements in such areas perform poorly in addressing the challenges of food security and climate change. This allows policymakers and service providers to adjust services to efficiently direct resources in the areas that are in urgent needs.[21]

Remote sensing—the process of acquiring information from a distance—often involves drones, airplanes, and satellites, all of which provide important aerial imagery. While drones provide high-resolution imagery, they cover a small area. Satellites, however, cover a large area, but their effectiveness is hampered by cloud cover and other issues. To get a complete picture, information from several sources is used, including ground data collected by survey teams.

The large amount of agricultural information in the imagery is analyzed using machine learning, which teaches a computer to make intelligent assessments about the imagery. For instance, cashew trees are often planted with other types of trees and crops. In overhead images of farmland, it takes a lot of time to identify cashew trees to analyze how they are maintained. TechnoServe teamed up with the University of Minnesota to develop a machine-learning algorithm to analyze satellite images to identify cashew trees. To train the algorithm, a set of high-resolution 50 cm/pixel data from Airbus Pleiades satellites for a 1,000-square-kilometer area was used. Images from part of the area were used to train the algorithm on what cashew farms look like and how they differ from forest, bare land, or urban areas (Figure 2.1). Images from another part were used to validate the algorithm's accuracy. Some examples of relevant variables for training and validating the ML models include tree diameter and height, trunk length and shape, leaf shape and texture, and flower shape, size, and color. Real-world data from outside the dataset (e.g., from elsewhere in Benin or another country) is used to test the model in order to ensure that the model works with a satisfactory level of accuracy.

As of October 2020, the algorithm used satellite imagery to identify cashew plantations with over 85% accuracy. Over time, the algorithms are expected to become even more accurate. Program staff and policymakers can then determine which farms need more support or where deforestation is taking place.[22]

---

21. Wehubit, *Drone-assisted land mapping for climate smart cashew production—Cajulab*, https://www.wehubit.be/en/node/48.

22. TechnoServe Technology, Innovation, and Climate-Smart Cashew Production,

As of 2021, through the BeninCajù program, TechnoServe had trained over 71,000 farmers in Benin in best agricultural practices, such as intercropping and agroforestry management.[23] Over 145,000 hectares of cashew orchards were under sustainable forestry practices. Cashew production also provides an opportunity for climate mitigation through tree planting.[24]

In a type of ML known as supervised machine learning, computers are "trained" or "taught" about what to look for. The training is based on reference or training data, which may consist of examples on the ground. Data from existing classifications or from other places can also be used. The ML algorithms then automatically detect and classify different objects and generate insights, analytics and visualizations.[25]

Neural network models developed in the 1940s by McCullough and Pitts,[26] which are modeled on the human brain, are the foundation that has given rise to machine learning.[27] Neural networks use complex mathematical systems and learn tasks by analyzing large amounts of data. Neural networks have been used to improve tasks such as face recognition, digital assistance, self-driving cars, and instant translation (e.g., Google Translate). For instance, by analyzing large amounts of data involving manufacturing activities, a neural network learns to identify such activities. Put simply, a neural network is an extension of linear regression. The goal in a neural network model is to capture complex nonlinear relationships that exist between

---

2020, https://www.technoserve.org/blog/technology-innovation-and-climate-smart-cashew-production/.

23. James Obarowski, "Remote sensing and satellite imagery providing data for African cashews," *Nutfruit Magazine*, July 19, 2021, https://www.nutfruit.org/industry/publications/inc-magazine/articles/detail/remote-sensing-and-satellite-imagery-providing-data-for-african-cashews.

24. "TechnoServe and Alteia partner to deploy AI solutions for climate-smart cashew program in Benin," TechnoServe, January 20, 2021, https://www.technoserve.org/news/technoserve-and-alteia-partner-to-deploy-ai-solutions-for-climate-smart-cashew-program-in-benin/.

25. Ran Goldblatt, Kai Kaiser, Huong Thi Lan Tran, and Kien Vu, "Artificial intelligence for smart cities: Insights from Ho Chi Minh City's spatial development," *World Bank Blogs*, March 8, 2018, https://blogs.worldbank.org/opendata/artificial-intelligence-smart-cities-insights-ho-chi-minh-city-s-spatial-development.

26. Warren S. McCulloch and Walter Pitts, "A logical calculus of the ideas immanent in nervous activity," *Bulletin of Mathematical Biophysics* 5 (1943): 115–133.

27. Larry Hardesty, "Explained: Neural networks: Ballyhooed artificial-intelligence technique known as 'deep learning' revives 70-year-old idea," *MIT News Office*, April 14, 2017, http://news.mit.edu/2017/explained-neural-networks-deep-learning-0414.

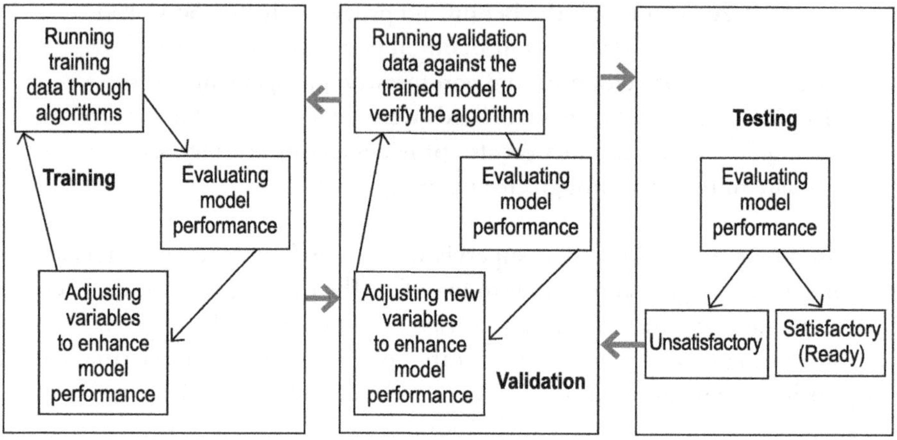

Figure 2.1. How machine learning works

input variables (e.g., satellite image) and an outcome variable (e.g., correct identification of a road). In neural network models, the associations between input variables and the outcome are represented through a large number of combinations of hidden layers. The associations are analyzed according to prespecified functions. A main goal is to estimate the weights of various associations in the input and outcome data so that the error between the outcome and the prediction is minimized.[28]

A type of ML known as deep learning uses algorithms, which teach computers to learn by example and perform tasks based on classifying structured and unstructured data such as images, sound, and text. A simple way to view deep learning is to consider it as a neural network model involving many layers. The exponential growth of computing power enables deep learning models to build up neural networks with a large number of layers, which was not possible in classic neural networks. As a result, deep learning can analyze complex nonlinear patterns in high-dimensional data that involves a large number of traits. Traditional ML algorithms, however, work when the number of traits is small.[29]

Deep learning for visual tasks is advancing at a breathtaking pace. In 2017, an article in *Nature* described a system that had the same performance

28. Fei Jiang, Yong Jiang, Hui Zhi, Yi Dong, Hao Li, Sufeng Ma, Yilong Wang, Qiang Dong, Haipeng Shen, and Yongjun Wang, "Artificial intelligence in healthcare: past, present and future," *Stroke and Vascular Neurology* (2017), https://svn.bmj.com/content/2/4/230.full.

29. Jiang et al., "Artificial intelligence."

rate as dermatologists in classifying cancer.[30] Similar systems are being developed to assess other diseases, such as diabetic retinopathy, stroke, bone fracture, and Alzheimer's disease. Such a program can be installed on smartphones to provide a low-cost universal access to diagnose diseases.

Some Notable AI Applications for the B4B

AI applications are becoming ubiquitous and used to improve economic, social and environmental performance in LMICs. Table 2.1 provides illustrative examples of how AI is being used in diverse economic sectors that can benefit the B4B.

Farming and Agriculture

AI has gained popularity in the farming and agricultural sectors. Especially large-scale agriculture firms use AI to improve the speed and accuracy of planting and crop management. Brazil's sugar and ethanol producer Raízen teamed up with the startup Space Time Analytics to develop an AI-based solution to forecast the size of the sugarcane harvest one year in advance.[31]

AI-based apps are diffusing rapidly among smaller firms to benefit the B4B. The Brazilian startup Solinftec's AI assistant Alice integrates and processes data from machines, people, climate stations, and other sources.[32] To use Alice, farmers embed smart black boxes in their machinery and deploy IoT devices in fields. Alice calculates farmers' needs and provides real-time recommendations.[33]

The initial Alice solution was developed for the sugarcane industry but now has offerings for other crops. As of August 2019, Solinftec was being used on more than 6.5 million hectares, monitoring 20,000 pieces of equipment and managing 100,000 active daily users.[34] Solinftec opened North

30. Apurv Mishra, "Deep-learning networks rival human vision," *Scientific American*, June 26, 2017.

31. Omar Abbosh, "Artificial intelligence could help reverse Latin America's economic slowdown," *World Economic Forum*, March 30, 2017, https://www.weforum.org/agenda/2017/03/artificial-intelligence-could-help-reverse-latin-america-s-economic-slowdown/.

32. "7 Brazilian agriculture technology startups," *Nanalyze*, August 19, 2019, https://www.nanalyze.com/2019/08/brazil-agriculture-technology.

33. CropLife, "Growmark teams up with Solinftec to increase farm efficiency," *CropLife*, August 23, 2019.

34. Nanalyze, "7 Brazilian."

American offices in West Lafayette, Indiana,[35] and is expanding to other South American countries as well as in Russia and Ukraine.[36]

Simpler and more user-friendly AI tools are available to help predict near-term crop productivity for smallholder farmers. For instance, PlantVillage is a free smartphone app in use in Kenya and other African countries to diagnose crop diseases such as cassava and potato. The app provides a highly accurate diagnosis from talking with an AI assistant Nuru (which means "light" in Swahili).[37] In a test of ML models in the typically high light and temperature of an African farm, the app was found to perform twice as better than human experts in making accurate diagnoses. The UN Water Productivity through Open Access of Remotely Sensed Derived Data portal is the main data source for the PlantVillage Nuru.[38] The portal's database uses the National Aeronautics and Space Administration's (NASA) satellite-derived data and computes relevant metrics for crop productivity. Other databases incorporated in PlantVillage Nuru include weather forecast data, a soil dataset for Africa, and the UN Crop Calendar, a series of algorithms on adaptive measures for particular conditions.[39] By integrating diverse data, the AI assistant can provide information about crops' drought tolerance and suitability of crops in different areas.[40]

The app uses TensorFlow, Google's open-sourced AI.[41] The app's early application has been to help farmers in Africa identify fall armyworm infections.[42] Not much technical knowledge or literacy is needed to use the app: farmers

---

35. "Brazil's Solinftec sets up shop in the U.S.," *Indianapolis Business Journal*, June 3, 2019, https://www.precisionag.com/market-watch/brazils-solinftec-sets-up-shop-in-the-u-s/.

36. Rob Leclerc, "The road to automated agriculture begins in Brazil," *AgFunderNews*, July 10, 2019, https://agfundernews.com/the-road-to-automated-agriculture-begins-in-brazil.html.

37. Jan Kreuze, "The bots are here—and they're protecting our crops," Scientific American, September 2, 2019, https://blogs.scientificamerican.com/observations/the-bots-are-here-and-theyre-protecting-our-crops/.

38. FAO Water Productivity Open-Access Portal (WaPOR), *ReliefWeb*, 2018, https://reliefweb.int/report/world/fao-water-productivity-open-access-portal-wapor.

39. Penn State, "New AI app predicts climate change stress for farmers in Africa," *EurekAlert*, 2019, https://www.eurekalert.org/pub_releases/2019-09/ps-naa091919.php.

40. "AI-based tool can predict climate change stress for farmers," *The Quint*, September 23, 2019, https://www.thequint.com/news/hot-news/ai-based-tool-can-predict-climate-change-stress-for-farmers.

41. Nnamdi Oranye and Ryan Peter, "A.I. agribusiness is good *business*," *CNBC Africa*, April 24, 2019, https://www.cnbcafrica.com/news/special-report/2019/04/24/a-i-agribusiness-is-good-business/.

42. Katherine Walla, "21 projects democratizing data for farmers," *Food Tank*, September 2019, https://foodtank.com/news/2019/09/21-projects-democratizing-data-for-farmers/.

imply point a phone at the diseased crop. PlantVillage has been adopted by the UN's Food and Agriculture Organization (FAO) and is now used to monitor the spread of the voracious armyworm in about 70 countries. The Nuru uses ML and AI to tell a farmer whether a worm is damaging crops and provides information to eradicate the pest. Nuru runs on standard Android phones.[43]

When farmers' devices are online, the Fall Armyworm Monitoring and Early Warning System (FAMEWS) mobile app uploads the collected data. As of early 2019, there were about 10,000 registered users of the system, which was available in 13 languages.[44]

The data collected is validated by national fall armyworm focal points and subsequently added to a global database.[45] Anyone can access the database from the FAO Fall Armyworm home page.[46] In this way, the app provides a real-time view of infestations across maps of Africa to help other app subscribers.[47] In Kenya, the PlantVillage AI tool is used to send SMS messages to farmers across the country.[48]

Mining and Energy

In the mining and energy sector, AI-based tools are helping to enhance health and safety of mine workers and improve short- and long-term operational performance. Chile's copper producer Codelco uses autonomous trucks at its mines. The vehicles reduce costs by operating for longer hours, and they reduce accidents.[49] It uses AI to monitor mining equipment to predict maintenance needs and make sure that operations run efficiently.[50]

---

43. Mariana López, "AI startup, Olivia, raises funds to aid Brazil Gill, 2018' finances," *Contxto*, January 16, 2020, www.contxto.com/en/brazil/ai-startup-olivia-raises-funds-aid-brazilians-finances/Lu.

44. Food and Agriculture Organization of the United Nations, "A mobile app and a global platform for managing fall armyworm," February 2019.

45. Chuck Gill, "'Nuru' becomes African farmers' newest ally against fall armyworm," *phys.org*, June 27, 2018.

46. Food and Agriculture Organization of the United Nations, "Fall armyworm," http://www.fao.org/fall-armyworm.

47. Kreuze, "The bots."

48. Penn State, "New AI."

49. Cody McFarlane, "Chile: Mining technology—where it is and where it is going in Chile?," June 14, 2018, https://www.mondaq.com/mining/710550/mining-technology-where-it-is-and-where-it-is-going-in-chile.

50. "Codelco to deploy AI solution," *Mining Journal*, March 26, 2019, https://www.mining-journal.com/innovation/news/1359598/codelco-to-deploy-ai-solution.

As another example, an autonomous four-wheeled robot developed by Peru's National Engineering University explores mines to detect levels and trends of dangerous substances such as methane, carbon dioxide, and ammonium. Using sensors, it detects location and generates actions to be taken such as the best routes for the workers to escape.[51] In this way, AI promotes the health and safety of mine workers.

IBM's AI-based adviser used in the oil and gas industry in Brazil provides information for better interpretation and explanation of seismic information to increase the performance of the reservoir-modeling process. The process entails developing computer models of a petroleum reservoir in order to improve the estimation of reserves and help decision making on developing fields and adding new wells. Such models can also predict future production and evaluate the effectiveness of reservoir management scenarios. The tool thus can be used to improve the performance of geological models and provide a better risk assessment of new oil field projects.[52]

Fighting Environmental Risk Factors

AI-based systems are also being developed to help understand factors that affect air pollution levels. Such systems can predict air pollution levels in advance, which can help individuals take actions to improve their health and reduce risks; for instance, people with serious health problems may avoid going outside if air pollution levels are predicted to increase. Researchers at Loughborough University reported in early 2020 that they tested such a system with historical air pollution data from Beijing to train and test the algorithms. The plan was to test the developed system with live data captured by sensors in Shenzhen, China.[53]

The health effects of such systems are especially important for the B4B: most of the world's polluted cities are in the developing world. For instance, of the 161 major Chinese cities monitored for air quality, 145 failed to meet the national standard for clean air in 2014.[54]

51. "Robot minero también tendría utilidad en agricultura y desastres naturales," *Andina*, July 13, 2016, http://www.andina.com.pe/Agencia/noticia-robot-minero-tambien-tendria-utilidad-agricultura-y-desastres-naturales-621069.aspx.

52. Ken Milam, "IBM's AI adviser promises 'transformation' of oil exploration," *AAPG Explorer*, August 2018, https://explorer.aapg.org/story/articleid/49008/ibms-ai-adviser-promises-transformation-of-oil-exploration.

53. Institution of Engineering and Technology, *AI system can predict air pollution before it happens*, March 19, 2020, https://eandt.theiet.org/content/articles/2020/03/ai-system-can-predict-air-pollution-before-it-happens/.

54. "China environment communique exposes poor air, water quality," *Xinhua*, June 5, 2015, http://en.people.cn/n/2015/0605/c90882-8902445.html.

China's mobile-phone-based ride-sharing service Didi uses algorithms to help predict traffic jams to minimize the impact of traffic congestion on drivers.[55] The analysis combines past traffic data with real-time data to predict traffic congestion; the company claims that it can forecast demand 15 minutes in advance at 85% accuracy. This information is used to build predictive dispatching models and send cars earlier to areas with high degrees of congestion (https://technode.com/2017/08/28/three-interesting-facts-just-learned-didis-big-data/).

Chinese bike-sharing platforms are using AI and big data to identify locations that have the highest demand for bikes. Using such data, bikes are provided only in locations that need them. This reduces their overall number and leads to less cluttered sidewalks and discarded bikes.[56]

Healthcare and Medical

A wide range of innovative healthcare applications have been developed and used in LMICs to address issues related to the lack of healthcare access and to improve healthcare delivery. For instance, the Chinese company Ping An's AI-powered system can predict chronic illnesses and infectious diseases with high accuracy. The systems can predict the likelihood that a patient may suffer from an illness even before physical symptoms appear. It can also quickly scan large number of medical images for abnormalities.[57] In 2018, an AI-based system developed by Ping An detected a pulmonary nodule known as "coin lesion," a small ovoid growth, with 95.1% accuracy.[58]

To take another example, Vietnam's Viettel Military Industry and Telecoms Group, which has expanded its activities to include several business lines including IT and digital services, uses AI in endoscopy. Its AI-based solutions can identify, locate and assess damages in digestive systems. It was reported that AI diagnosed problems related to digestive systems five times

55. Stephany Zoo, "What Africa can learn from China about data privacy," *World Economic Forum*, June 26, 2019, https://www.weforum.org/agenda/2019/06/what-africa-can-learn-from-china-about-data-privacy/.

56. Winston Ma Wenyan, "Here are 4 major bike-sharing trends from China after lockdown," *World Economic Forum*, July 22, 2020, https://www.weforum.org/agenda/2020/07/4-big-bike-sharing-trends-from-china-that-could-outlast-covid-19/.

57. Saheli Roy Choudhury, "A.I. can improve health care in China, says Ping An Technology CEO," *CNBC*, July 9, 2019, https://www.cnbc.com/2019/07/09/ping-an-tech-ceo-says-ai-can-improve-health-care-in-china.html.

58. "Ping An Technology applies ai to healthcare, delivering solutions that benefit the general public," *PR Newswire*, 2018, https://www.prnewswire.com/news-releases/ping-an-technology-applies-ai-to-healthcare-delivering-solutions-that-benefit-the-general-public-300597721.html.

as fast compared to traditional methods with an accuracy level of 90%.[59] While AI's use in endoscopy is a growing trend in Northern countries,[60] what is encouraging here is that LMICs such as Vietnam are developing such solutions locally.

E-Commerce

A number of innovative AI applications have appeared in the e-commerce sector. For instance, in 2015, Alibaba's affiliate Ant Group launched an AI-based automated customer calls system known as "smart customer service."[61] Ant Group's customers can use the voice-recognition system to buy airline tickets and book hotels.[62] The system reportedly outperformed human customer service agents in delivering customer satisfaction.[63]

Alibaba has been using AI-powered chatbots on the company's Taobao site. In 2017, Its customer-service chatbot Alime Shop Assistant handled more than 93% of customer queries, which would require 83,000 human customer service agents.[64] The proportion of such queries handled by chatbots increased to 97% on its e-commerce platforms Taobao and Tmall on Singles Day in 2019. Alibaba also launched the AI-enabled smart speaker Tmall Genie. Over 1 million orders were placed and processed through voice commands using Tmall Genie.[65]

The Thai startup Pomelo plans to use big data and AI for pricing, design, e-commerce personalization, and supply chain automation.[66] Likewise, in

---

59. "Việt Nam puts priority on artificial intelligence development," Việt *Nam News*, 2019, https://vietnamnews.vn/society/534918/viet-nam-puts-priority-on-artificial-intellig ence-development.html.#867UukgQvAkd1KST.97.

60. Kazutomo Togashi, "Applications of artificial intelligence to endoscopy practice: The view from Japan Digestive Disease Week 2018,"*Digestive Endoscopy* 31, no. 3 (2019): 270–272, https://onlinelibrary.wiley.com/doi/10.1111/den.13354.

61. Chris Kapfer, "Whither digital transformation? The emergence of global digital finance institutions," *Asian Banker: Financial Technology Working Group*, 2018, http://tech .theasianbanker.com/updates?&docid=0008226985027197164408442O.

62. Lia Zhu, "Alibaba's Ant Financial using artificial intelligence to advance," *China Daily*, January 24, 2018.

63. Will Knight, "Meet the Chinese finance giant that's secretly an AI company," *MIT Technology Review*, June 16, 2017, https://www.technologyreview.com/s/608103/ant-finan cial-chinas-giant-of-mobile-payments-is-rethinking-finance-with-ai.

64. Britt, "Where AI."

65. Dan Berthiaume, "Alibaba reveals secret to its Singles Day mega-sale success," *Chain Storeage*, November 20, 2019, https://chainstoreage.com/alibaba-reveals-secret-its-singles -day-mega-sale-success,

66. Ng Ren Jye, "Thai fashion e-commerce startup Pomelo raises US$52m in Series C

2018, the Southeast Asian e-commerce firm Lazada launched an AI-driven app that uses ML algorithms to show products to users based on purchase and viewing history.[67] Lazada's mobile app also has an AI-powered image search. When users take pictures of items they want, Lazada suggests similar items available. Users can tap the camera icon labeled "Scan" to take a photo or select an existing image from the gallery. The AI-powered image search recognizes the scanned item, and the app displays suggested items.[68]

While the most significant transformations of AI systems are those that involve redesigning and augmenting workplace tasks, organizations tend to use them to replace workers.[69] Most customers value interactions with live human agents. AI providers often present AI as an opportunity to reduce or eliminate the workforce. When companies understand the importance of interactions with human agents, AI is likely to be used for augmenting human intelligence.[70]

Finance, Banking, and Insurance

Firms in the developing world have successfully incorporated AI in financial and insurance services, which appeal to consumers and have increased the performance and efficiency of such services. China's Ant Group uses big data and AI to manage credit risk and lower loan delinquency rates and detect fraud.[71] Banks in Latin America also use chatbots to provide customer service. A survey found that 83% of Brazilian financial services consumers would trust banking advice entirely generated by a computer.[72]

In 2017, Ant Financial launched an AI-driven, image-recognition system

---

round," *Business Times*, September 12, 2019, https://www.businesstimes.com.sg/garage/thai-fashion-e-commerce-startup-pomelo-raises-us52m-in-series-c-round.

67. Charlotte Trueman and Cristina Lago, "How is AI benefiting industries throughout Southeast Asia?," *CIO*, January 10, 2020, https://www.cio.com/article/3311756/how-is-artificial-intelligence-benefiting-industries-throughout-southeast-asia.html.

68. Vernon, "Lazada adds AI Image Search to app," *Venturebeat*, September 23, 2018, https://vernonchan.com/lazada-ai-image-search-mobile-app/.

69. Khari Johnson, "AI ethics is all about power," *Venturebeat*, November 11, 2019, https://venturebeat.com/2019/11/11/ai-ethics-is-all-about-power/.

70. Britt, "Where AI."

71. Bien Perez and Zen Soo, "China a fast learner when it comes to artificial intelligence-powered fintech, experts say," *South China Morning Post*, October 28, 2017, https://www.scmp.com/tech/innovation/article/2117298/china-fast-learner-when-it-comes-artificial-intelligence-powered.

72. Accenture, *2020 global banking consumer study*, 2020, https://www.accenture.com/us-en/insight-financial-services-distribution-marketing-consumer-study.

Table 2.1. AI use in diverse industries in developing economies

Sector/industry/ activities	Examples of companies/apps	Use of AI	Outcome
Farming and agriculture	Brazil's Solinftec's AI assistant Alice	Integrates and processes data from machines, people, network of climate stations, and other sources	Provides real-time actionable recommendations
Mining and energy	Chile's Codelco	Monitorw mining equipment to predict maintenance needs, detect anomalies and help prevent problems	Efficiency in operations.
Fighting environmental risk factors such as air pollution	China's Didi	Predicts traffic jams	Minimize the impact of traffic congestions
Healthcare and medical	Viettel Military Industry and Telecoms Group	Diagnoses problems related to digestive systems	5 times as fast compared to traditional methods with an accuracy level of 90%
E-commerce	Alibaba's Alime Shop Assistant	Offers customer-service chatbot	97% of queries on Taobao and Tmall handled on the 2019 Singles Day
Finance, banking and insurance	China's Ant Group	Uses deep-learning technology to detect fraud.	Reduction of losses related to fraud
Human resources management and development	Chile's AIRA	Publishes vacancy announcements, reads and ranks résumés, uses psychometric tests, conducts video interviews, and measures performance of applicants.	Reduces time human recruiters need to spend.

to process vehicle insurance claims. Ant Group's algorithm assessed damages in 12 different cases in six seconds. Human investigators take longer than six minutes to assess a claim.[73] Ant Group's accident-processing system uses deep learning. The system is trained by feeding thousands of examples of images into a neural network. Compared to the human cognitive system's limited information-processing capabilities, AI-based systems can process necessary and relevant information about damages.[74] AI is especially suitable for handling routine and clear tasks. In China, exterior damage claims account for 60% of private-vehicle insurance claims.[75]

Likewise, Kenya's Kenindia Assurance plans to use AI to detect fraudulent motor insurance claims.[76] In addition to the industry's Integrated Motor Insurance Data System (IMIDS) through the Association of Kenya Insurers, it plans to establish a data center of customers' insurance history.

Human Resources Management and Development

The Chilean company Artificial Intelligence Recruitment Assistant's (AIRA) system publishes vacancy announcements on job recruitment websites. It reads and ranks résumés and uses psychometric tests. It also conducts video interviews with applicants and measures an applicant's performance with indicators related to analyses of emotions, converting factors such as attention levels and facial expressions into numbers. After all these processes are completed, human recruiters conduct in-depth interviews with the highest-ranked candidates.[77]

LMICs face a large deficit of human capital: the key issue is a surplus of unskilled labor and a significant shortage of highly skilled human capital.[78]

73. Yujing Liu, "Chinese insurance giant Ping An launches cloud service aimed at banks, hospitals," *South China Morning Post*, October 18, 2018, https://www.scmp.com/business/companies/article/2169023/chinese-insurance-giant-ping-launches-cloud-service-aimed-banks.

74. Knight, "Meet the."

75. Zhu, "Alibaba's Ant."

76. Beth Nyaga, "Kenyan insurers utilizing artificial intelligence to curb fraud cases," October 24, 2019, https://www.kbc.co.ke/kenyan-insurers-utilizing-artificial-intelligence-to-curb-fraud-cases/.

77. Armen Ovanessoff and Eduardo Plastino, *How artificial intelligence can drive South America's growth*, Accenture, 2017, https://f.etica.ai/cplp/2017_accenture_how-ai-can-drive-south-america-growth(original).pdf.

78. Mariassunta Giannetti, Guanmin Liao, and Xiaoyun Yu, "The brain gain of corporate boards: Evidence from China," *Journal of Finance* 70 (2015): 1629–1682.

A related problem is lack of information on human resources. A key step to address this challenge is to ensure the availability of reliable and comprehensive workforce information.[79]

AI tools can tackle these problems by enhancing the skills of people in LMICs and supporting empowerment, through training and job placement. In this way, social inclusion can be achieved. One example of this is the South African social enterprise Harambee's AI tool to help young people find jobs. Harambee uses Google's open-sourced AI TensorFlow to interact with more than 1 million young people. It uses precise geographical attributes and preferential behavioral metrics to achieve its goals. Harambee uses ML to more effectively use the data it collects.[80] Harambee's youth employment accelerator CEO, Tamera Campbell, noted that 2,600 jobs for young people were found in the first two years after the organization was established in 2011. By early 2019, 50,000 young people had benefited.[81]

Harambee's young recruiters, referred to as "feet on the streets," visit townships and villages to collect the contact information of unemployed people. Some are invited into its offices to have their interests and skills assessed and their analytical capabilities tested. Harambee helps them create email accounts and CVs and facilitates the interview process. It provides advice and information on preparing for interviews, such as how to dress and questions that might be asked. Potential candidates are also offered free clothes for the interview. The company provides work-readiness interventions to address risks that employers have identified.[82] Promising candidates can also get additional assessment and vocational training in call centers or similar facilities.

Harambee plans to expand its service to over 7 million unemployed South African youths. It has also expanded into Rwanda.

Harambee has developed strong partnerships with companies. It first learns the skills needed by employers. It then works to identify candidates who are likely to be a good match. Harambee's corporate partners provide information about the number of candidates they need and the target hiring

79. Krishna D. Rao, Aarushi Bhatnagar, and Peter Berman, "So many, yet few: Human resources for health in India," August 13, 2012, https://human-resources-health.biomedcentral.com/articles/10.1186/1478-4491-10-19.

80. Amy Paul, Craig Jolley, and Aubra Anthony "Reflecting the past, shaping the future: making AI work for international development," USAID, 2019, www.usaid.gov/sites/default/files/documents/15396/AI-ML-in-Development.pdf.

81. John Kennedy, "The future of AI: Why Google is betting big on Africa," *Silicon Republic*, February 19, 2019, www.siliconrepublic.com/machines/africa-ai-google.

82. Paul et al., "Reflecting the past."

date. This process provides them with larger pools of potential candidates by including demographic groups that were overlooked.[83]

Harambee has reduced the cost barriers for employers to hire unemployed youth. These marginalized demographic groups now have access to opportunities that were unavailable to them before—and unthinkable.

Economic Effects of AI

Some have argued that AI will have a negative impact on LMICs' export-led growth model. AI and automation in developed countries will arguably lead to a decline in the outsourcing of manufacturing and other jobs.[84] However, by using AI in diverse industries, LMICs might be able to compensate for such losses. Even more impressive, technology firms from LMICs such as Brazil, China, Nigeria, and Russia are selling AI solutions in foreign markets. The entry of young AI firms, such as Brazil's Solinftec, into developed countries is an important trend that can shape AI's global competitive landscape.

Regarding the economic growth mechanisms, traditionally economists viewed new technologies as economic growth drivers through their effects on total factor productivity (TFP). The factors of production are capital (e.g., machines, buildings) and labor that drive economic growth. Broadly, TFP is a function of technological, economic, and other factors. TFP measures how efficiently the factors of production are being used. Technologies of the past, such as electricity and IT, boosted productivity. For example, the World Bank's *World Development Report 2016* found that the internet and e-commerce increased TFP in Vietnam by 1.9% and 3.6%, respectively, during the period 2007–2012.

The global professional services company Accenture argues that AI, in addition to its role in driving TFP, also is a new factor of production. Various uses of AI illustrate this possibility. First, AI functions as a new workforce by replicating labor activities at a higher scale and speed. AI performs tasks that humans cannot, and it learns faster. AI-enabled robots and intelligent machines can also function as a physical capital. Unlike conventional forms of capital such as machines and buildings, AI's self-learning capabilities help improve TFP over time.

83. Paul et al., "Reflecting the past."

84. Robert A. Manning and Peter Engelke, "A quest to win the future the global innovation sweepstakes," June 2018, https://www.atlanticcouncil.org/wp-content/uploads/2018/06/The-Global-Innovation-Sweepstakes.pdf.

Accenture has used case studies from Latin American countries to illustrate AI as a new factor of production. The company forecasts that Brazil's gross value added (GVA) in 2035 will be US$3.5 trillion without AI. If AI's impact is limited to TFP, its projected GVA will be US$3.5trillion. If AI does become viewed as a factor of production, the GVA will further increase to US$3.9 trillion.[85]

Value Chain of the Global AI Industry and Effects on the B4B

According to the professional services firm PwC, AI's contribution to the global economy will reach about US$16 trillion by 2030.[86] Accurately labeling data for training ML models is integral to the creation of this value. Data-labeling activities, however, are extremely time and labor intensive. According to a 2018 McKinsey report, data labeling is the biggest obstacle to AI adoption.[87]

Data need to be cleaned, categorized into appropriate groups, and labeled so that AI algorithms can recognize patterns.[88] For instance, for ML algorithms to accurately recognize a car, the algorithms may need to be trained with a large number of car pictures.[89] In the most common form of AI—supervised learning—algorithms need to be fed with millions of tagged examples of car pictures until they correctly identify the pictures. These activities need a large amount of human labor. For instance, one hour of video data related to autonomous driving may need as much as 800 hours of data-labeling work.[90] Data-labeling activities are estimated to account for as much as 80% of the time needed to build AI systems.[91]

85. Ovanessoff and Plastino, "How artificial."

86. "Sizing the prize: PwC's global artificial intelligence study: Exploiting the AI revolution," PWC, 2017, https://www.pwc.com/gx/en/issues/data-and-analytics/publications/artificial-intelligence-study.html.

87. Michael Chui, James Manyika, and Mehdi Miremadi "What AI can and can't do (yet) for your business," January 11, 2018, https://www.mckinsey.com/business-functions/mckinsey-analytics/our-insights/what-ai-can-and-cant-do-yet-for-your-business.

88. Kate Kaye, "These companies claim to provide 'fair-trade' data work—Do they?," August 7, 2019, https://www.technologyreview.com/s/614070/cloudfactory-ddd-samasource-imerit-impact-sourcing-companies-for-data-annotation/.

89. Huizhong Wu, "China is achieving AI dominance by relying on young blue-collar workers," *Vice*, December 21, 2018, https://www.vice.com/en_us/article/7xyabb/china-ai-dominance-relies-on-young-data-labelers.

90. Kaveh Waddell, "Data labeling for AI is set to become a billion-dollar market by 2023," *Axios*, March 25, 2019, https://www.axios.com/ai-data-labeling-billion-dollar-market-409704bc-e63c-4af0-b0d0-44424abcd561.html.

91. Cade Metz, "A.I. is learning from humans, many humans," *New York Times*, August 16, 2019, https://www.nytimes.com/2019/08/16/technology/ai-humans.html.?auth=linked-google.

A fascinating aspect of the development of the global AI industry is that a high proportion of jobs that require relatively low skills are being performed in LMICs, where most of the B4B population lives. The multibillion-dollar data-labeling industry is an interesting illustration of this trend. According to the analyst firm Cognilytica, the third-party data-labeling solutions market was US$150 million in 2018.[92] The global market for data labeling, also known as content labeling or data annotation, is expected to reach US$5 billion by 2023.[93]

LMICs are in a position to take advantage of the opportunities provided by the global AI industry. This phenomenon has created a whole new industry of data labeling in LMICs, which is described as "a new type of blue-collar industry."[94] Data labelers are referred to as the blue-collar workers of the AI age.

Labeling data for some AI apps may require high levels of skills and knowledge. For instance, to develop an AI app to detect cancer on images from medical CT scans, experienced radiologists may have to train the algorithms.[95] However, there are many tasks that computers lack the capability to perform as well as human beings, and less-skilled workers can easily be trained to perform such tasks. Most data-labeling trainings can be completed in a short time. For instance, to learn their tasks, data labelers at iMerit typically take a one-week online training course via video calls with U.S.-based trainers

Global companies are undertaking major initiatives to perfect ML training, which has led the emergence of the data-labeling industry. This industry employs hundreds of thousands of workers in India and other developing countries. Some examples of data labeling include teaching self-driving cars the meanings of road signs or the difference between a child and a fox.

The India- and U.S.-based data annotation company iMerit had 2,200 employees in India to label data generated by manufacturing, medical imaging, autonomous driving, retail, insurance, and agriculture. Its Kolkata operation employs 460 women to train computer vision algorithms used in

---

92. Madhumita Murgia, "AI's new workforce: the data-labelling industry spreads globally," *Financial Times*, 2019, https://www.ft.com/content/56dde36c-aa40-11e9-984c-fac8325aaa04.

93. "Data-labelling startups want to help improve corporate AI," *The Economist,* October 17, 2019, https://www.economist.com/business/2019/10/17/data-labelling-startups-want-to-help-improve-corporate-ai.

94. Kori Hale, "Google & Microsoft banking on Africa's AI labeling workforce," *Forbes*, May 28, 2019, https://www.forbes.com/sites/korihale/2019/05/28/google-microsoft-banking-on-africas-ai-labeling-workforce/#34b0fd0d541c; Maximilian Gahntz, "The invisible workers of the AI era," December 12, 2018 https://towardsdatascience.com/the-invisible-workers-of-the-ai-era-c83735481ba.

95. Gahntz, "The invisible workers."

Table 2.2. Examples of data-labeling companies operating in LMICs

Company	Operations and workforce	Profiles of clients
iMerit	Based in India and the U.S. 2,500 in data labeling centers in India such as Ranchi, Shillong, Vizag and Kolkata.[1]	90% are U.S.-based.[2]
Samasource	Offices in San Francisco, New York, the Hague, Kenya and Uganda. Global staff of 2,900: East Africa's largest AI and data annotation employer.[3]	Include 25% of the Fortune 50 companies including major automakers and U.S.-based technology companies such as Google, Microsoft, and IBM.[4]
MBH	300,000 in China's backward provinces.[5]	Mainly Chinese companies such as the Beijing-based video-sharing social networking platform TikTok.
Playment	Based in India and the U.S. More than 300,000 crowdsourced data labelers.[6]	Over 100 customers in more than 12 countries. Some include Samsung, Didi Chuxing Technology, Alibaba, Drive.ai and Continental AG. Most clients are in the autonomous vehicle industry.[7]

[1]Sonam Joshi, "*How artificial intelligence is creating jobs in India, not just stealing them*" *Times of India* September 9, 2019, [online]. https://timesofindia.indiatimes.com/india/how-artificial-intelligence-is-creating-jobs-in-india-not-just-stealing-them/articleshow/71030863.cms.

[2]Inc42 Media, "*Facebook responds to reports of Indian workers labelling user data.*" *Inc42 Media*, May 8, 2019, https://inc42.com/buzz/facebook-responds-to-reports-of-indian-workers-labelling-private-user-data/.

[3]Jake Bright, "*Google & Microsoft banking on Africa's AI labeling workforce.*" *Techcrunch*, November 20, 2019, https://techcrunch.com/2019/11/20/samasource-raises-14-8m-for-global-ai-data-biz-driven-from-africa/.

[4]Jason D. Rowley, "*Last week in venture: scads of scoots, sourcing ai data in Africa, and games with friends.*" *Crunchbase News*, November 22, 2019, https://news.crunchbase.com/startups/last-week-in-venture-scads-of-scoots-sourcing-ai-data-in-africa-and-games-with-friends/.

[5]*The Economist*, "*China's success at AI has relied on good data.*" January 4, 2020, https://www.economist.com/technology-quarterly/2020/01/02/chinas-success-at-ai-has-relied-on-good-data?utm_medium=affiliates.offer.pd&utm_source=partnerize-viglink&utm_campaign=a.io&utm_content=conversion.direct-response.anonymous&utm_term=1100lwbk5nQx&partnerize_clickref=1100lwbk5nQx/.

[6]Anand Murali, "*How India's data labellers are powering the global AI race.*" *FactorDaily*, March 21, 2019, https://factordaily.com/indian-data-labellers-powering-the-global-ai-race/.

[7]Murali, "How India's data."

autonomous vehicles and augmented reality systems for companies such as Amazon, Microsoft, eBay, and TripAdvisor.[96]

Western companies are also undertaking initiatives to perfect training in ML. The data-labeling industry employs hundreds of thousands of workers in developing countries such as Kenya, India, and the Philippines. Samasource, based in Nairobi, Kenya, labels data for Walmart, Google, Microsoft, Glassdoor, Continental, and General Motors. It employs more than 2,800 people.[97]

Some economic sectors are highly digitized, which makes it easier to develop AI-based solutions. For instance, in Africa, financial services, telecommunications, and retail have relatively large amounts of data. Likewise, banking and retail industries are expected to dominate AI expenditure in the Middle East.[98]

Table 2.2 provides some examples of data-labeling companies operating in developing economies. Most of these firms mainly serve foreign clients. Chinese data-labeling firms such as MBH, however, mainly support the domestic AI industry. China has gained global prominence in recent years in the AI field, research and development activities to support the growth of the AI industry are conducted in wealthy cities such as Beijing, Shanghai, Hangzhou, and Shenzhen. Most of the data-labeling work, though, is performed in the country's disadvantaged regions, such as smaller towns and rural northern areas in Shandong, Henan, Hebei, and Shanxi provinces.[99]

The growth in data labeling also reflects the influences of trends in the manufacturing sector. First, manufacturing activities have become increasingly automatized. For instance, between 2012 and 2016, Foxconn was reported to deploy tens of thousands of robots to replace more than 400,000 jobs. The company's plan includes full automation of 30% of production activities by 2020.[100] Second, manufacturing jobs in China are declin-

96. Kshetri, "Artificial intelligence."

97. Bianca Wright, "What Africa's approach to AI can teach the world," *CIO*, August 20, 2019, https://www.cio.com/article/3431656/what-africas-approach-to-ai-can-teach-the-world.html.

98. Alkesh Sharma, "AI spending to grow 43% in Middle East and Africa in 2019," *The National*, October 1, 2019, https://www.thenational.ae/business/technology/ai-spending-to-grow-43-in-middle-east-and-africa-in-2019-1.917352.

99. Sarah Dai, "AI promises jobs revolution but first it needs old-fashioned manual labour—from China," *South China Morning Post*, October 8, 2018, https://www.scmp.com/tech/article/2166655/ai-promises-jobs-revolution-first-it-needs-old-fashioned-manual-labour-china.

100. Cissy Zhou, "Could robotic automation replace China's 100 million workers in its manufacturing industry?," *South China Morning Post*, February 14, 2019, https://www.sc

ing along with a decline in the country's economic growth and demand for products, which is increasing costs and competition from other economies.[101] A Boston Consulting Group study conducted in 2015 found that manufacturing costs in some of China's major manufacturing hubs were almost at the same levels as in the U.S. Unsurprisingly, many Western companies have been moving and relocating manufacturing activities into other developing countries in Asia and also to the U.S., Canada, and Mexico.[102] For instance, factory workers in Bangladesh and Vietnam can be hired for 25% and 50% less, respectively, than a Chinese worker. Finally, there has been a generational shift in preference for work. An increasing number of Chinese millennials are not choosing the tedium of factory jobs.[103]

Henan Province is the epitome of AI-led transformation. The province's city Zhengzhou is known for being home to the manufacturing plants of the Taiwanese electronics company Foxconn. It was estimated that Foxconn employed about 350,000 people and produced about half the world's iPhones in Zhengzhou in 2016.[104] The number of people employed by Foxconn decreased to about 100,000 in early 2019. Henan's mobile phone exports reduced nearly 24% in January 2019 compared to the previous year.[105]

Partly in response to the decline in manufacturing activities, in recent years, data-labeling companies are mushrooming in the towns and villages of rural regions such as Henan.[106] In the province's Pingdingshan village, for example, some large data-labeling projects employ tens of thousands of people.[107] Likewise, northern China's landlocked Shanxi Province, among the

---

mp.com/economy/china-economy/article/2185993/man-vs-machine-chinas-workforce-starting-feel-strain-threat.

101. Michael Schuman, "Is China stealing jobs? It may be losing them, instead," *New York Times*, July 6, 2016, https://www.nytimes.com/2016/07/23/business/international/china-jobs-donald-trump.html.

102. Schuman, "Is China stealing."

103. Robyn Dixon, "Chinese millennials are rejecting dull factory jobs—and transforming the economy," *Los Angeles Times*, 2019, https://www.latimes.com/world/la-fg-china-millennials-jobs-20190512-story.html.

104. Harrison Jacobs, "Inside 'iPhone City,' the massive Chinese factory town where half of the world's iPhones are produced," *Business Insider*, May 7, 2018, https://www.businessinsider.com/apple-iphone-factory-foxconn-china-photos-tour-2018-5.

105. Cissy Zhou, "What is it like inside Foxconn: The world's largest iPhone assembly plant in China," *South China Morning Post*, March 1, 2019, https://www.scmp.com/economy/china-economy/article/2188162/foxconn-tale-slashed-salaries-disappearing-benefits-and-mass.

106. Wu, "China is achieving."

107. Cate Cadell, "Faces for cookware: Data collection industry flourishes as China pursues AI ambitions," *Reuters*, June 27, 2019, https://www.reuters.com/article/us-china-ai-data-insight/faces-for-cookware-data-collection-industry-flourishes-as-china-pursues-ai-ambitions-idUSKCN1TS3EA.

poorest in China, plans to attract more than 100 data-labeling companies and train more than 10,000 workers by 2022. The province's goal is to have RMB 5 billion (US$726 million) in industry revenue by 2025.[108] Workers who were employed in assembly lines and construction sites before are thus finding new jobs in the Chinese data-labeling industry.[109]

According to China's official data, 1 million people in Henan lived below the poverty line at the end of 2018.[110] It is possible that data-labeling jobs can help lift some of these people out of poverty. Many data workers previously worked on assembly lines and on construction sites in big cities. But it is becoming difficult to find work, and wage growth has slowed. Many Chinese prefer to live closer to home.[111]

The Development and Deployment of AI

Innovative AI Activities

Innovative activities have stimulated the development of AI industry. For instance, in 2017, Chinese firms accounted for 99 of the 314 blockchain patents and 473 of the 649 AI patents filed with the World Intellectual Property Organization (WIPO).[112] According to WIPO, China and India ranked first and fourth in the world, respectively, in ML-related scientific publications and in overall AI-related scientific publications (the U.S. and the U.K. ranked second and third, respectively). A 2017 report by the *Economist* noted that China could be a "close second" to or "even ahead of" the U.S. in AI.[113] In a 2016 report, the U.S. White House noted that

108. "Data labeling jobs are coming to underdeveloped regions in China, but can they stay?," KrASIA, 2019, https://kr-asia.com/data-labeling-jobs-are-coming-to-underdeveloped-regions-in-china-but-can-they-stay.

109. Li Yuan, "How cheap labor drives China's A.I. ambition," *New York Times*, November 25, 2018, https://www.nytimes.com/2018/11/25/business/china-artificial-intelligence-labeling.html.

110. "Henan lifts over one million out of poverty in 2018," *Xinhua,* http://www.xinhuanet.com/english/2019-01/16/c_137748581.htm.

111. Yuan, "How cheap labor."

112. Antony Peyton, "China bosses blockchain and AI patents," *Fintech Futures*, January 21, 2019, https://www.fintechfutures.com/2019/01/china-bosses-blockchain-and-ai-patents/.

113. "China may match or beat America in AI: Its deep pool of data may let it lead in artificial intelligence," *The Economist*, July 15, 2017, https://www.economist.com/news/business/21725018-its-deep-pool-data-may-let-it-lead-artificial-intelligence-china-may-match-or-beat-america.

China had overtaken the U.S. in the number of published journal articles on deep learning.[114]

Local technology hubs are rapidly evolving in many developing countries, which are creating solutions to solve local problems. As of October 2019, Africa had 618 tech hubs, together serving as the foundation for the AI industry. Countries such as Ethiopia have launched high-profile AI initiatives. Most of Ethiopia's more than 30 official universities and 130 polytechnics emphasize technology. In 2012, the Ministry of Science and Technology established its own university and developed a US$250 million technology park.[115] The country's AI and robotics research company iCog has produced several apps.[116] One such project involves developing software for AI tablets for children, who could then learn coding, mathematics, and English.

Multinational Corporations

Some multinational corporations (MNCs) are taking advantage of various resources in developing countries to develop cutting-edge AI solutions. Such activities are likely to create positive externalities and spillover effects of AI-related knowledge for the local economy. Unilever, for example, developed an autonomous forklift in Brazil and launched its manufacturing units in the country before any other markets.[117]

Stimulation of New Economic Activities

The development in the global AI industry is also likely to lead to stimulating new economic activities in LMICs. One such example is the data-labeling industry, a notable feature of which is that it does not favor a specific cultural context. The data-labeling market thus is characterized by a low entry barrier for most developing countries. Whereas the outsourcing of call-center jobs gravitated to countries with sizable English-speaking populations, such as India and the Philippines, English skill is less a factor in data-labeling jobs.

114. "China may match or beat America in AI."

115. Christina Galbraith, "Artificial intelligence catches fire in Ethiopia," Techonomy, August 25, 2015, https://techonomy.com/artificial-intelligence-catches-fire-in-ethiopia/.

116. Robert A. Manning, "Will AI cripple or leapfrog developing nations' growth? A world-class expert gives us his opinion," *National Interest*, October 14, 2019, https://nationalinterest.org/blog/buzz/will-ai-cripple-or-leapfrog-developing-nations-growth-87716.

117. Armen Ovanessoff and Omar Abbosh, "Artificial intelligence could help reverse Latin America's economic slowdown," *World Economy Forum*, March 30, 2017, https://www.weforum.org/agenda/2017/03/artificial-intelligence-could-help-reverse-latin-america-s-economic-slowdown/.

Digital literacy is sufficient to participate in most data-labeling tasks, such as image classification.

Among LMICs, China has emerged as a key global AI player. The country's wealthy regions and big cities are not attractive to the data-labeling services industry. Such services are mostly performed in the poorer, rural regions, which are providing economic incentives for data-labeling firms.

While finding high-quality AI talent such as ML engineers has been big challenge for companies in developing economies,[118] there is an abundant supply of low-skilled, low-wage laborers in India and other developing countries. For instance, Indian high schools graduated 20 million students in 2017,[119] but there are not nearly as many job opportunities available to absorb these graduates. To take an example, in a Mumbai city police's job advertisement for 1,137 constable positions with a salary of US$357 per month, which required only a high school education, over 200,000 people applied. Many candidates had been trained in highly skilled professions such as doctors, lawyers, and engineers.[120]

Relatively Low Data Privacy Barriers

The availability of data at more granular levels is important for developing better algorithms, which is especially critical in personalizing user experience. However, doing so might compromise users' privacy or data confidentiality.[121]

Firms in some developing countries face relatively lower data privacy barriers than developed countries. Whereas ethical and data privacy issues hinder development of the AI industry, such issues are less of a concern in developing countries. For instance, due to China's almost nonexistent privacy controls, Chinese companies have easy access to the data of over 1 billion users. Such large datasets help algorithms produce more accurate results and predictions.[122]

---

118. Sunny Sen, "India moves to address AI talent supply gap, gets a leg-up from Google, Microsoft, Intel," *Factor Daily*, January 18, 2018, https://factordaily.com/india-ai-talent-gap-google-microsoft/.

119. Kelsey Sheehy, "High school grads in China, India are better prepared for college," *U.S. News*, August 27, 2012, https://www.usnews.com/education/blogs/high-school-notes/2012/08/27/high-school-grads-in-china-india-are-better-prepared-for-college.

120. Neha Thirani Bagri, "India is trapping its young people," *Foreign Policy*, May 14, 2019, https://foreignpolicy.com/2019/05/14/india-is-trapping-its-young-people/.

121. Ovanessoff and Plastino, "How artificial."

122. Steven Feldstein, "We need to get smart about how governments use AI," Carnegie Endowment for International Peace, January 22, 2019, https://carnegieendowment.org/2019/01/22/we-need-to-get-smart-about-how-governments-use-ai-pub-78179.

Researchers have noted that the data divide—the gap in the availability of data needed for scientific research and decision-making—adversely affects LMICs.[123] While LDCs lack data and other resources to develop a native AI industry, big middle-income countries such as China have an advantage over small economies with strict data privacy laws (Table 2.3).

Negative Effects of AI on Vulnerable Groups

A big worry among consumers and activists has been AI's malicious use by corporations and governments.[124] Data and algorithms can be used to manipulate individuals, groups, and organizations. These actors can be misled into making decisions that promote the interests of corporations and states.[125] In such situations, a prediction of the median voter theory is that major adverse social and welfare effects of AI are more likely to be experienced by consumers who are poorer,[126] have less formal education, and are less likely to vote.[127] These descriptions fit the profile of the B4B population. As an example of the negative social effects of AI, fintech companies that use AI algorithms to lend money to low-income people are reported to engage in publicly shaming borrowers into paying back loans.

Likewise, government agencies in many countries use AI in mass surveillance. The negative effects of mass surveillance are more likely to be experienced by vulnerable groups that are disadvantaged because of their poverty, race, religion, and ethnicity.[128] To take an example, in China, the Uyghurs, an economically disadvantaged ethnic minority group that is mostly Muslim and lives in the Xinjiang Autonomous Region, have become a major target of the government's surveillance program (In Focus 2.3). For instance, in Xinjiang, the annual rural income in 2012 was less than 6,400 yuan (¥)

---

123. Jonathan Cinnamon and Nadine Schuurman, "Confronting the data-divide in a time of spatial turns and volunteered geographic information," *GeoJournal* 78 (2013): 657–674.

124. François Chollet, "What worries me about AI," *Medium*, March 28, 2018, https://medium.com/@francois.chollet/what-worries-me-about-ai-ed9df072b704.

125. El Adl, "Debunking."

126. B. A. Abrams and A. L. Kenneth, "A median-voter model of economic regulation," *Public Choice* 52 (1987): 125–142.

127. L. Strahilevitz "Toward a positive theory of privacy law," *Harvard Law Review* 126 (2013): 2010–2042.

128. Barton Gellman and Sam Adler-Bell, *The disparate impact of surveillance*, Century Foundation, December 21, 2017, https://tcf.org/content/report/disparate-impact-surveillance/?agreed=1.

Table 2.3. Classification of countries with respect to data availability and strictness of privacy laws

Data availability Strictness of privacy laws	High	Low
Low	Big middle-income countries such as China	Small LDCs such as small island states (e.g., Samoa and Vanuatu).
High	Big industrialized countries such as the U.S.	Small economies recognized by the European Commission as providing adequate data protection such as Faroe Islands and Isle of Man[1]

[1]European Commission. *Adequacy decisions*. 2019, https://ec.europa.eu/info/law/law-topic/data-protection/international-dimension-data-protection/adequacy-decisions_en.

($1,000), which was ¥1,500 less than the national average and more than ¥11,000 less than Shanghai's rural residents.[129]

In Focus 2.3: China's Techno-Utilitarian Approach Targeting the Uyghur Population

AI solutions have advanced considerably, such that it is easy to identify from images and label people based on sociopolitical and demographic factors such as race and ethnicity. Technology companies such as IBM, for instance, advertise AI software that can sort people on such factors.[130] The Chinese government and its technology companies have elevated the notion of AI-based surveillance to new heights.

In general, 4R technologies in China are being developed and deployed without sufficient consideration of ethical issues.[131] China's approach to privacy has been described as "techno-utilitarian." The utilitarian approach is a fundamental ethics principle that bases decisions upon achieving the greatest good for the greatest number

129. Michael Martina, "In China's Xinjiang, poverty, exclusion are greater threat than Islam" *Reuters*, November 3, 2013, https://www.reuters.com/article/us-china-xinjiang/in-chinas-xinjiang-poverty-exclusion-are-greater-threat-than-islam-idUKBRE9A20GS20131103.

130. IBM, *Attribute detection with body camera analytics*, March 3, 2021, https://www.ibm.com/support/knowledgecenter/SS88XH_2.0.0/iva/attribute_detectors_ranked_search.html.

131. David Meyer, "A.I. regulation is coming soon: Here's what the future may hold," *Fortune*, October 24, 2019.

of people. Thus, the protection of personal privacy and individual rights receives less emphasis. This approach is also referred to as techno-authoritarianism, which involves the government's use of internet networks, surveillance systems, and algorithms to monitor people.

Some critics have noted that such arguments have been used to justify the use of AI for surveillance and to suppress political opponents and minorities, such as Uyghers, who are Muslims living in Xinjiang.[132] China has made it mandatory for Xinjiang residents to install surveillance software on their cellphones.[133] The software allegedly scans pictures and texts in the phone. A *Financial Times*' analysis of the software found that it scans for digital fingerprints matching content objectionable to the Chinese Communist Party. When such content is found, authorities are informed. Possession of sensitive digital content may lead to detention. Some Uyghurs do not use smartphones as a result.[134]

In addition to scanning Uyghers' cellphones, China allegedly uses a secret system of advanced facial recognition technology to track them. An article published in the *New York Times* described the use of this facial recognition technology as "automated racism" integrated into the country's surveillance camera networks. The system looks exclusively for Uyghurs on the basis of their appearance and keeps records of their movements. Such records can be used for search and review.[135]

While some law enforcement agencies and AI providers described the practice as "minority identification," the tools are exclusively used to identify Uyghurs, who have a phenotype similar to people from Central Asia. As a result, it is easy for facial recognition software to identify them.[136]

In cities outside of Xinjiang, police run facial recognition systems

---

132. John Thornhill, "Formulating values for AI is hard when humans do not agree," *Financial Times*, July 22, 2019, https://www.ft.com/content/6c8854de-ac59-11e9-8030-530adfa879c2.

133. Emily Feng, "In China, a new call to protect data privacy," *NPR*. January 5, 2020, https://www.npr.org/2020/01/05/793014617/in-china-a-new-call-to-protect-data-privacy.

134. Feng, "In China."

135. Paul Mozur, "One month, 500,000 face scans: How China is using A.I. to profile a minority," *New York Times*, April 14, 2019.

136. Mozur, "One month."

to investigate people who may be Uyghurs.[137] Chinese police were reported to be using such systems to target Uyghurs in cities such as Hangzhou and Wenzhou and in the coastal province of Fujian. Likewise, law enforcement agencies in central China's city Sanmenxia reportedly ran a system 500,000 times in one month in early 2019 to determine whether certain residents were Uyghur. Likewise, in 2018, law enforcement agencies from Shaanxi Province wanted to acquire a smart camera system with functionalities to "support facial recognition to identify Uighur/non-Uighur attributes."[138]

Ethics Dumping and Ethics Shirking in the Global North

To test the predictive abilities of AI, organizations in the Global North use countries in the Global South, which tend to lack data safeguards and regulations. This practice is ethics dumping, and it often is illegal in the Global North.[139] Unethical practices are exported to marginalized and vulnerable populations in LMICs, a striking resemblance to "the old fault lines of colonialism."[140]

An example of ethics dumping in AI is beta testing in LMICs. Software developers perform a beta test on commercial off-the-shelf software to test external market acceptance, acquire feedback from the market, and create interest among potential customers.[141] This helps identify issues when the software is launched for "real" users and cases.[142]

Some AI developers try out AI algorithms they develop on vulnerable groups in LMICs. Such practices often result in adverse social and political effects (In Focus 2.4).

---

137. Adrian Zenz, "Xinjiang's new slavery, *Foreign Policy*, December 11, 2019, https://foreignpolicy.com/2019/12/11/cotton-china-uighur-labor-xinjiang-new-slavery/.

138. Mozur, "One month."

139. UN Conference on Trade and Development, *Information economy report 2013: The cloud economy and developing countries* (Geneva: UNCTAD).

140. D. Schroeder, J. Cook Lucas, F. Hirsch, S. Fenet, S., and V. Muthuswamy, *Ethics dumping case studies from north-south research collaborations* (Cham: Springer International Publishing, 2018).

141. Niklas Leicht, Ivo Blohm, and Jan Marco Leimeister, "Leveraging the power of the crowd for software testing," *IEEE Software*, March 28, 2017, https://ieeexplore.ieee.org/abstract/document/7888424.

142. Shakir Mohamed, Marie-Therese Png, and William Isaac, "Decolonial AI: Decolonial theory as sociotechnical foresight in artificial intelligence," *Philosophy & Technology* 33, no. 4 (2020): 659–684, https://link.springer.com/article/10.1007/s13347-020-00405-8#ref-link-section-d2630e1034.

In Focus: 2.4: Cambridge Analytica's Beta Testing in Africa before U.S. and U.K. Elections

The political consulting firm Cambridge Analytica executed beta testing of its algorithms on the 2015 Nigerian and 2017 Kenyan elections before using them in the U.S. and U.K. This testing has been viewed as a reinforcement of Western companies' historical tendency to treat former colonies in Africa and other parts of the world as laboratories for new medicines and technologies.[143] Studies found that these experiments disrupted the Kenyan election process and eroded social cohesion. Cambridge Analytica was found to have manipulated Kenyan voters during the elections and to have undermined democracy in Kenya.[144]

President Uhuru Muigai Kenyatta's party had used Cambridge Analytica for his reelection campaign. Cambridge Analytica conducted a 47,000-person survey in 2013 to identify Kenyan voters' "real needs (jobs)" and "fears (tribal violence)," as well as their "preferred information channels." An AI-based system was used to send targeted text messages to voters during the campaign season. Gacheke Gachihi of the Mathare Social Justice Center in Nairobi noted that Cambridge Analytica relied on "divisive propaganda" and raised ethnic enmity to help Kenyatta's party win.[145]

Following that test, Cambridge Analytica is alleged to have subsequently manipulated U.S. voters in the 2016 presential election. In early 2014, Cambridge Analytica obtained the private information of about 87 million Facebook users. The data was collected illegally, without users' knowledge. A subset of the data was used to build a system that profiled individual U.S. voters and to pre-

143. Karen Hao, "The problems AI has today go back centuries," *MIT Technology Review*, July 31, 2020, https://www.technologyreview.com/2020/07/31/1005824/decolonial-ai-for-everyone.

144. Nanjala Nyabola, "Digital democracy, analogue politics: how the internet era is transforming politics in Kenya," Foreign Affairs, October, 2019, https://www.foreignaffairs.com/reviews/capsule-review/2019-08-12/digital-democracy-analogue-politics-how-internet-era-transforming.

145. Jina Moore, "Cambridge Analytica had a role in Kenya election, too," *New York Times*, March 20, 2018, https://www.nytimes.com/2018/03/20/world/africa/kenya-cambridge-analytica-election.html.

dict and influence their choices on Election Day.[146] Depending on their profile, U.S. voters were targeted with personalized political advertisements.[147] According to a former employee, the company specifically targeted users who were "more prone to impulsive anger or conspiratorial thinking than average citizens."[148] Methods used included Facebook group posts, ads, and article sharing to provoke voters to vote for or against a party. To do so, Cambridge Analytica also created fake Facebook pages such as "I Love My Country."[149]

A related phenomenon is ethics shirking—when harms emerge and firms do nothing to protect people beyond what is legally required by the law.[150] Because regulations do not yet exist for many areas of the 4R, those who are harmed are less likely to be compensated.

Conclusion

AI applications are tackling economic and social challenges facing LMICs. AI holds great promise and potential to fight key sources of poverty. Economically speaking, AI possesses mechanisms that allow it to have significant impacts on productivity.

The true potential of AI comes from its ability to complement and enhance traditional factors of production. Some AI applications already perform activities that humans or other technologies are not capable of doing. Other applications perform many tasks efficiently and more accurately than humans. However, sufficient evidence has not yet accumulated to establish claims related to some measures of AI-led efficiency gains. While some surveys conducted by the providers of AI systems such as Alibaba have reported

146. Carole Cadwalladr and Emma Graham-Harrison, "Revealed: 50 million Facebook profiles harvested for Cambridge Analytica in major data breach," *The Guardian*, March 17, 2018, https://www.theguardian.com/news/2018/mar/17/cambridge-analytica-facebook-influence-us-election.

147. Cadwalladr and Graham-Harrison, "Revealed."

148. Rosalie Chan, "Cambridge Analytica whistleblower on how the firm used Facebook data," *Business Insider*, October 5, 2019, https://www.businessinsider.com/cambridge-analytica-whistleblower-christopher-wylie-facebook-data-2019-10.

149. Chan, "Cambridge Analytica."

150. L. Floridi, "Translating principles into practices of digital ethics: five risks of being unethical," *Philosophy & Technology* 32, no. 2 (2019): 185–193.

positive economic effects, most findings have not been supported by independent surveys.

A related point is that overreliance on AI may lead to overall low organizational performance, such as customer service and loyalty. A survey of the global provider of business applications, enterprise learning, and outsourcing services CGS found that many consumers prefer to interact with human agents over chatbots. Chatbots perform relatively well when consumers have straightforward questions, but they are less helpful and provide less detailed answers. For nonstandard questions or complex issues, customers prefer to speak with a human specialist[151] While customer satisfaction with AI is undoubtedly improving,[152] there are still many areas in which human agents perform better.

AI in developing countries depends on many factors, including technology entrepreneurship, knowledge and expertise, data availability, and government policy. The success of an AI project is heavily dependent on the quality of data and algorithms. The lack of data availability is a big obstacle to develop good-quality AI systems. AI performance is influenced by a lower level of training in AI algorithms.

The B4B is disproportionately affected by the negative consequences of AI. Among the negative outcomes associated with AI, oppressive regimes have identified AI as the most effective tool for repressing and defeating dissidents and opposition groups. Marginalized groups are more likely to be the victims of government surveillance programs.

In terms of AI workforce, developing countries experience a much more severe shortage than their developed counterparts. While challenges associated with the transferability and adaptability of foreign technologies to developing countries are well recognized, AI-based systems face issues not encountered by past technologies. The challenges range from lack of training in ML algorithms to recognize physical traits to underdeveloped language skills.

The rapidly growing global AI industry has created demands for highly skilled jobs such as ML engineers and data scientists, and for less-skilled jobs such as data labeling. Most AI systems heavily rely on human-powered data-labeling activities. Developing countries provide a very large workforce

151. Christopher Elliott, "Chatbots are killing customer service: Here's why," *Forbes*, 2018, https://www.forbes.com/sites/christopherelliott/2018/08/27/chatbots-are-killing-customer-service-heres-why/#3be59fd913c5.

152. Britt, "Where AI."

to support these activities and boost the global AI industry. Data labelers in these countries are playing a key role in curating the data that powers AI systems around the globe. While developing countries may experience a decline in outsourcing jobs from developed countries, the potential negative impact of such a decline can be minimized by appropriate policies around the deployment of AI solutions.

CHAPTER 3

Blockchain

Blockchain is a major 4R technology that is considered to have the potential to cause significant economic, political, and social transformations in LMICs. Blockchain affects economic, social, and political outcomes in the developing world through many direct and indirect pathways. The first of blockchain's direct benefits is the potential reduction of corruption and fraud. For instance, blockchain can empower donors. It can ensure that donations reach intended recipients. To take an example, donors can buy electricity for a South African school using bitcoin. A blockchain-enabled smart meter makes it possible to send money directly to the meter. There are no organizations involved in distributing funds. Donors can also track electricity being consumed by the school to calculate the power of their donations.[1] This program was launched by the South African bitcoin startup Bankymoon via the crowdfunding platform Usizo, to allow public schools to use blockchain to crowdsource utility credits.[2]

Increased efficiency and reduced transaction costs constitute a second benefit: there is no third party or central body involved. That is, blockchain transactions are conducted by the concerned parties themselves. There are already some signs of blockchain-led disintermediation in international remittances and international trade.

To be sure, blockchain is in infancy. Some compare the current level of development to "the World Wide Web in the early 1990s."[3] Nonetheless,

1. Stan Higgins, "How bitcoin brought electricity to a South African school," *CoinDesk*, 2016, http://www.coindesk.com/south-african-primary-school-blockchain/.

2. Gabriella Mulligan, "5 African crowdfunding startups to match," *Disrupt Africa*, 2015, http://disrupt-africa.com/2015/11/5-african-crowdfunding-startups-to-watch/.

3. Willis Towers Watson, "Want to get an insurer's attention? Just say blockchain," June

MNCs, local companies, and policymakers have devoted considerable attention to blockchain.

In many ways, blockchain has a much higher value proposition for LMICs than for the Global North. This technology has the potential to make up for the lack of effective formal institutions—rules, laws, regulations, and enforcement—in developing economies. These economies are also in desperate need of improving administrative aspects, such as maintaining standards, monitoring, and enforcing compliance. Blockchain technology is perfectly suited to address these issues.

There are different mechanisms that lead to such benefits, which can be better understood by looking at blockchain-led reduction in the costs of verification and networking. Regarding the cost of verification, blockchain makes it possible to verify information about past transactions and the current ownership of a digital asset. As to blockchain's effect on reducing the cost of networking, various parties can start a self-sustaining process and operate a marketplace. It is not necessary to assign control to a centralized body because blockchain can verify the state at a low cost. Economic incentives can be targeted to reward valuable activities from a network perspective, including contribution of resources to operate and scale up the network and to secure a decentralized stage. The digital marketplaces that result from such collaborations allow participants to make joint investments to create shared digital assets.[4]

To illustrate these points, consider the example of blockchain-based solutions to fight global slavery. According to studies conducted by the International Labour Organization, the Walk Free Foundation, and the International Organization for Migration, in 2016, 40.3 million people were estimated to be living in modern slavery—forced to work under threat against their will or living in a forced marriage—70% of whom were women and girls.[5] Many of the products that these people produce are then exported to rich countries. In 2018, the G20 countries imported US$354 billion worth of products at risk of having been produced by forced labor. This is an extremely sad situation because it continues despite Western brands' efforts over three decades to address issues related to forced labor, bondage, sweatshops, and other abuses in their supply chains.[6]

2016, https://www.the-digital-insurer.com/wp-content/uploads/2016/12/793-want-to-get-an-insurers-attention-just-say-blockchain-wtw.pdf.

4. Christian Catalini and Joshua S. Gans, "Some simple economics of the blockchain," Rotman School of Management Working Paper No. 2874598, 2019.

5. Walk Free Foundation, "The 2018 global slavery index," https://downloads.globalslaveryindex.org/ephemeral/GSI-2018_FNL_190828_CO_DIGITAL_P-1596333264.pdf.

6. Peter Bengtsen, "Why are monitory democracies not monitoring supply chain slav-

The huge marine fishing industry exhibits a high propensity to use "slave" or grossly underpaid labor because of the lack of clear regulations and enforcement mechanisms. Migrant workers especially are exploited. For instance, in 2014, 82% of 172,430 fishermen employed by the Thai fishing industry were migrant workers, mainly from Cambodia and Myanmar. Most workers in seafood-processing plants are also migrants who often arrive there after falling prey to recruiters promising well-paying jobs in Thailand. However, they are paid about 25% less than Thailand's minimum wage. The migrant workers often sign a contract in their home country, but their contracts change when they arrive in the host country to begin work.[7] Unlike Thai workers, they cannot join unions and do not have other protections that Thai workers are entitled to.[8]

Some initiatives are expected to improve this issue. The blockchain solutions provider Diginex has been working with the International Organization for Migration and the antislavery nongovernmental organization Mekong Club to ensure ethical recruitment of migrant workers by increasing transparency of workers' contracts. The British embassy in Bangkok partly funded the pilot phase of the project.[9]

Blockchain-based mobile app eMin (https://www.eminproject.com) is used to store copies of employment contracts for workers in this sector. Employment contracts and related data are stored on the Ethereum blockchain. Workers can access their contracts, which allows them a basis for claiming the rights and benefits they were offered at the time of recruitment.[10]

The eMin pilot started in February 2019 with Verifik8, a data intelligence and analytics provider for agribusiness suppliers, at a shrimp farm in Phuket, Thailand.[11] In October 2019, Diginex signed an agreement with Verifik8 to integrate eMin into the latter's existing monitoring tools. As of October 2019, Verifik8's farming monitoring tools called Blue 8/Green 8

---

ery?," *Global Policy Journal* August 28, 2020, https://www.globalpolicyjournal.com/blog/28/08/2020/why-are-monitory-democracies-not-monitoring-supply-chain-slavery.

7. "Detecting modern slavery in the supply chain," *Business Fights Poverty*, January 21, 2020, https://wordonthestreets.net/Articles/560874/Detecting_modern_slavery.aspx.

8. Kate Nicholl, Miriam Wilhelm, and Vikram Bhakoo, "Almost every brand of tuna on supermarket shelves shows why modern slavery laws are needed," *The Conversation*, 2019, https://theconversation.com/almost-every-brand-of-tuna-on-supermarket-shelves-shows-why-modern-slavery-laws-are-needed-108421.

9. Toan Dao, "Companies ink deal to use blockchain for protecting Thai aquaculture sector workers," *Seafood Source*, 2019, https://www.seafoodsource.com/news/aquaculture/companies-ink-deal-to-use-blockchain-for-protecting-thai-aquaculture-sector-workers.

10. Business Fights Poverty, "Detecting modern."

11. Business Fights Poverty, "Detecting modern."

were used by 5,000 workers in Thailand.[12] Diginex plans to expand into different sectors in other Southeast Asian nations, as well as in Bangladesh and Bahrain.[13] The following sections focus on how blockchain can address some of the key challenges facing the B4B. I also offer an overview of enablers and barriers in implementing blockchain in LMICs.

How Blockchain Works

Blockchain can be viewed as a decentralized ledger that maintains digital records of a transaction on multiple computers simultaneously. After a block of records is entered into the ledger, the information in the block is mathematically connected to other blocks. In this way, a chain of immutable records is formed.[14] With this mathematical relationship, the information in a block cannot be changed without changing all blocks. Any change would create a discrepancy, which others are then likely to notice.[15]

To take an example, facing pressures from various stakeholders to demonstrate ethical practice, South Africa's De Beers Group has launched a GemFair program to log diamonds produced by artisanal and small-scale miners (ASMs), who are among the most vulnerable of the B4B (Figure 3.1). In the first phase of the program, De Beers trained ASMs at 16 mine sites in Sierra Leone. The training program focuses on digitally tracking diamonds throughout the supply chain. The goal is to make sure that diamonds that originated in conflict zones do not enter the supply chain. ASMs are required to identify and manage key risks defined in a due diligence guide by the Organisation for Economic Co-operation and Development to participate in the program.[16] Among the major requirements is that ASMs identify the worst forms of child labor and address them. Compliance is ensured through first-party (e.g., a member completes a self-assessment workbook provided by GemFair), second-party (GemFair's biannual site monitoring), and third-party (assessment of a sample of sites twice a year) verifications.

---

12. Dao, "Companies ink."
13. Business Fights Poverty, "Detecting modern."
14. Dylan Yaga, Peter Mell, Nik Roby, and Karen Scarfone, *Blockchain technology overview,* National Institute of Standards and Technology Internal Report (NISTIR) 8202, October 2018.
15. Nir Kshetri, "Blockchain could be the answer to cybersecurity: Maybe," *Wall Street Journal,* May 29 2018, https://www.wsj.com/articles/blockchain-could-be-the-answer-to-cybersecurity-maybe-1527645960.
16. GemFair, *Artisanal and small-scale mining standard,* 2019, https://gemfair.com/static/files/GemFair_ASM_Requirements_2019_v2.pdf.

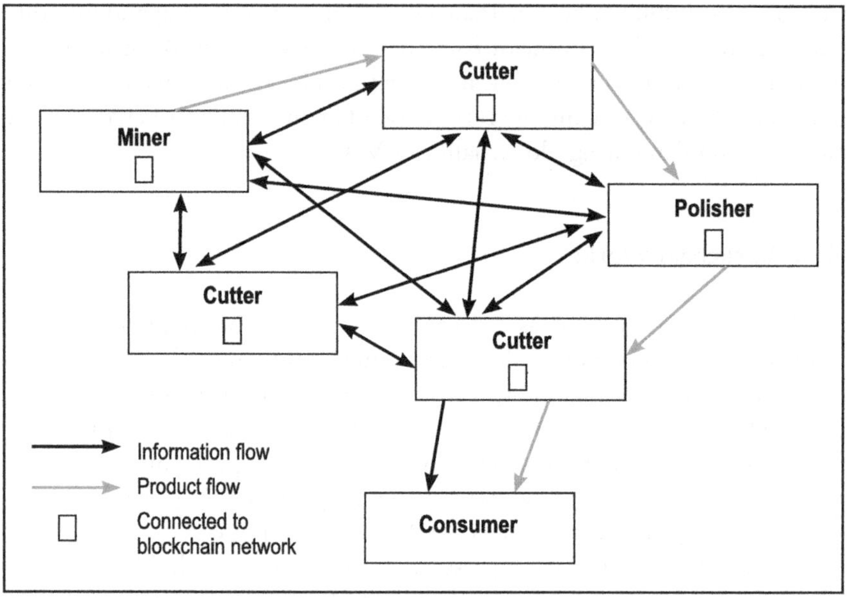

Figure 3.1. An illustration of the use of blockchain to trace a diamond in a supply chain

De Beers' blockchain solutions track diamonds as they move from the mine to cutter and polisher and then jeweler. Each organization involved in this traceability can use a smartphone or other device to sign into a blockchain platform (Figure 3.1). De Beers' program records the GPS locations for each diamond found. The diamond is then placed in a tamper-proof bag with a QR code.[17] GemFair provides a tablet for a participating mine to log into the GemFair app, which can also function offline. However, the tablet must be connected to the internet to store production records in the GemFair system. After this step, the raw diamonds move on to the supply chain's next stage. All relevant participants receive information about all transactions.

Economic Prosperity and Poverty

There are many and varied sources of underdevelopment, which include colonialism, dependence on commodities,[18] ethnic tension, corruption, exploita-

17. Padraig Belton, "How do you know your diamond isn't fake?," *BBC News*, July 2, 2019, https://www.bbc.com/news/business-48824290.

18. Alice Sindzingre, "The impact of the 2008–2009 crisis on commodity-dependent

tion, lawlessness, and political violence.[19] In this chapter, I focus on institutional environments. Poor countries mostly lack good institutions that ensure strict enforcement of property rights, have the ability to deal with corrupt practices effectively and provide equal opportunity to all members of society.[20]

The Lack and Poor Enforcement of Property Rights

According to a 2011 report of the FAO and Transparency International, in over 61 countries, weak governance led to corruption in land occupancy and administration. Corruption varied from small-scale bribes to the abuse of government power at the national, state, and local levels.[21]

Enforcement of property rights increases incentives to invest and provides resources for individuals to get out of the poverty trap. Clear property rights would allow entrepreneurs to use the assets as collateral and thus increase their access to capital. A large proportion of poor people in the developing world lack property rights. For instance, about 90% of land in rural Africa is undocumented or unregistered. Likewise in India, the lack of land ownership remains among the most important barriers to entrepreneurship and economic development.[22] One estimate is that over 20 million rural families in India did not own land and millions more lacked legal ownership to the lands where they built houses, lived, and worked.[23] Indeed, lack of land ownership or tenure is arguably a more powerful predictor of poverty in India than caste or illiteracy.[24]

Disregard of the Rule of Law

In some developing economies, the rule of law is disregarded and not respected by corrupt politicians, government officials, and other powerful

low-income African countries: Confirming the relevance of the concept of poverty trap?," *Journal of International Development* 24, no. 8 (2012): 989–1007.

19. S. Mansoob Murshed, "Conflict, civil war and underdevelopment: An introduction," *Journal of Peace Research* 39, no. 4 (2002): 387–393.

20. Daron Acemoglu, "Root causes: A historical approach to assessing the role of institutions in economic development," *Finance & Development*, 2003, 27–30, http://isites.harvard.edu/fs/docs/icb.topic637539.files/Acemoglu.pdf.

21. UN News Center, "Corruption leading to unequal access."

22. Nir Kshetri, "Fostering startup ecosystems in India," *Asian Research Policy* 7, no. 1 (2016): 94–103.

23. Tim Hanstad, "The case for land reform in India," *Foreign Affairs*, 2013, https://www.foreignaffairs.com/articles/india/2013-02-19/untitled?cid=soc-twitter-in-snapshots-untitled-022013.

24. Hanstad, "The case."

groups. These groups sometimes expropriate the incomes and investments of poor people or create an uneven playing field in other ways.

Less Opportunity for Marginalized Groups

Economically and socially disadvantaged groups have less opportunity to access finance, credit, insurance, and education. Thus, they cannot make investments and participate in productive economic activities. Consider insurance, for instance. In India, 86% of rural populations and 82% urban populations lack health insurance.[25]

Regarding access to finance, in China, SMEs account for 70% of GDP but have access to 20% of financial resources.[26] However, 89% of SMEs face difficulty satisfying banks' loan requirements.[27] Small borrowers often lack sufficient collateral required by most traditional banks.[28]

Unavailable financing is a critical barrier faced by most entrepreneurs. For instance, despite high interest rates, demand for credit exists in most developing economies. Banks in the Democratic Republic of Congo (DRC) reject over one-third of credit and loan applications. The fact that they cannot enforce their legal rights as lenders has led to the industry's risk-averse behavior, and this is a manifestation of a broader structural problem in LMICs where a large proportion of the population lacks access to formal banking institutions.[29] The situation is not much different in other economies. For instance, only 10% of people in Kenya have bank accounts, 5% in Tanzania, and 15% in Liberia.[30]

Barriers

Poor-quality institutions lead to transaction cost barriers. To make this statement meaningful requires a more detailed discussion of transaction costs. In the context of business transactions involving two or more parties, for

---

25. Samarth Bansal, "Health cover: Too little, too scarce," *The Hindu*, 2016, http://www.thehindu.com/sci-tech/health/policy-and-issues/health-insurance-in-india-too-little-too-scarce-reveal-national-sample-survey-data/article8462747.ece.

26. Brian P. Klein and Kenneth Neil Cukier, "Tamed tigers, distressed dragon," *Foreign Affairs*, 2009, https://www.foreignaffairs.com/articles/asia/2009-07-01/tamed-tigers-distressed-dragon.

27. Jing, "Alibaba, lenders team up for SME financing."

28. Nir Kshetri, "Big data's role in expanding access to financial services in China," *International Journal of Information Management* 36, no. 3 (2016): 297–308.

29. Nir Kshetri, *Global entrepreneurship: Environment and strategy*, 2nd ed. (New York: Routledge, 2019).

30. Efami Dovi, "Boosting domestic savings in Africa," 2011, http://www.un.org/africarenewal/magazine/october-2008/boosting-domestic-savings-africa.

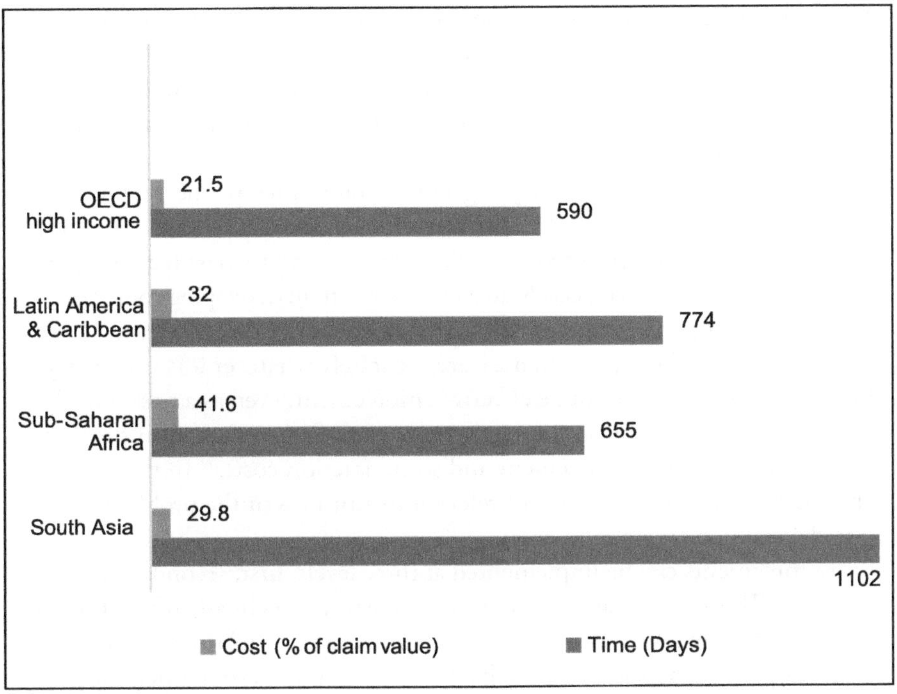

Figure 3.2. Time and cost required to enforce contracts in different geographic regions and groups of countries. Data source: "Doing business measuring business regulations," World Bank, 2020, https://www.doingbusiness.org/en/data/exploretopics/enforcing-contracts#.

Douglas North, "transaction costs are . . . two things: (1) the costs of measuring the dimensions of whatever it is that is being produced or exchanged and (2) the costs of enforcement."[31] He goes on to say that "a lot of what we need to do is to try to measure the dimensions of what we are talking about in such a way that we can define them precisely."[32]

Many developing economies are faced with challenges in enforcing commercial contracts, social and economic rights, laws and regulations (e.g., agro-environmental) and standards (e.g., pollution). Put differently, these economies lack effective contract enforcement mechanisms (Figure 3.2). A main goal of contract laws and their enforcement is to increase the value of contracting in order to facilitate the organization of economic activities.[33]

31. D. C. North, "Dealing with a nonergodic world: Institutional economics, property rights, and the global environment," *Duke Environment, Law, and Policy Forum* 10, no. 1 (1999): 1–12.

32. North, "Dealing with."

33. Richard A. Posner, "The law and economics of contract interpretation," *Texas Law Review* 83 (2005): 1581–1614.

When there is a legal remedy for breach of contract, a party to a contract is less likely to engage in opportunistic behavior.[34] In developing economies, laws in general and contract laws in particular are not adequate to meet the needs of the population.[35] Enforcing a contract takes longer and is more costly in LMICs than in developed countries.

Detailed contracts can protect against a contracting partner's opportunism through the threat of legal enforcement.[36] Most parties, however, rarely use detailed contracts in practice because of the high costs associated with drafting and enforcing them.[37] This is especially an issue of concern in developing countries.

Emphasizing the importance of measurement to enforcement, North argues: "Without being able to measure accurately whatever it is you are trying to enforce, there cannot be effective enforcement, even as a possibility."[38] The technology available is among the important factors that affect the costs of measurement and enforcement and so transaction costs.[39] In this, blockchain can make up for the lack of relevant institutions or the problems associated with high transaction costs.

Enforcements can be implemented at three levels: first, second, and third parties.[40] Third-party enforcement mechanisms, which are often formal coercive enforcement measures by the state, have been relatively ineffective in developing economies.[41] Blockchain has a potential to strengthen the governments' enforcement powers and sanctions against individuals or organizations that breach regulations.

Notable Blockchain Applications for the B4B

Some key current applications and future prospects of blockchain are presented in Table 3.1. As it is clear there, various barriers and challenges faced by the B4B can be addressed through blockchain.

---

34. Oliver E. Williamson, *The economic institutions of capitalism* (New York: Free Press, 1985).

35. L. Michael Hager, "The role of lawyers in developing countries," *American Bar Association Journal* 58 (1972): 33–38.

36. Paul L. Joskow, "Contract duration and relationship-specific investments: Empirical evidence from coal markets," *American Economic Review* 77, no. 1 (1987): 168–185.

37. Stewart Macaulay, "Non-contractual relations in business: A preliminary study," *American Sociological Review* 28, no. 1 (1963): 55–67.

38. North, "Dealing with."

39. North, "Dealing with."

40. North, "Dealing with."

41. Nir Kshetri, "The economics of the internet of things in the Global South," *Third World Quarterly* 38, no. 2 (2016): 311–339.

Promoting Transparency and Reducing Risk

Blockchain can help achieve transparency in various settings. In mid-2016, Ant Financial, Alibaba's online payments affiliate, announced the launch of blockchain payment technology. Blockchain was first applied to Alipay's donation platform. Donors on the "Ant Love" charity platform can track

Table 3.1. Blockchain in LMICs: Some applications in use or being developed

Blockchain use	Example
Promoting transparency and reducing frauds and corruption	• South Africa's Bankymoon allows public schools in Africa to use blockchain to crowd-source utility credits (ODR). • Standard Chartered and DBS Group's platform detects falsification and frauds in trade transactions (SRL)[1] • MonetaGo: creates an invoice's hash—if a trader submits the same invoice to more than one trade finance platforms, the hash will match, which raises a red flag.
Reducing friction and costs of property registration	• Bitland's land registry system based in Ghana (EPR). • BitFury and Republic of Georgia agree to develop a system for registering land titles using blockchain (EPR). • India's Andhra Pradesh implements land recording project in Amaravati (EPR).
Promoting efficiency in international B2B trade and increasing access to trade and supply chain finance	• Chained Finance is used by Foxconn to pay electronics suppliers on cryptocurrencies based on ethereum (ODG).)
Reducing costs and increasing efficiency in international payment systems	• Ripple's On-Demand Liquidity: XRP to send money faster at a lower fee (ODG). • AlipayHK and GCash: real-time money transfer between Hong Kong and the Philippines at lower fees (ODG). • Crypto-denominated international commerce (ODG).
Finance, banking, and Insurance	• Saldo.mx's microinsurance service (ODG).
Blockchain-based digital identity.	• Sierra Leone's blockchain-based National Digital Identity Platform developed by Kiva (ODG).

Note: In parentheses are indications of how the use cases have potential to address various causes of poverty by strengthening the rule of law (SRL), helping to enforce property rights (EPR) and creating opportunity for disadvantaged groups (ODG).

[1]Chanjaroen and Boey, "Fraud in $4 trillion."

transaction histories and understand where their funds go and how they are used.[42] The goal is to increase transparency and provide a trust mechanism by recording each payment and spend of donations on the blockchain.

Blockchain solutions can help reduce fraudulent activities. To take an example, the use of fake export invoices to disguise cross-border capital flows is pervasive in China. Since China has maintained strict capital controls, some importers and exporters falsify transactions to move capital in and out of the country. Many banks do not check the authenticity of trade documents.[43] From April to September 2014, China found US$10 billion in fake transactions.[44] Some major fraud cases were in Qingdao, the world's seventh-busiest port, where firms had used fake receipts to secure multiple loans against a single cargo of metal.[45]

The Qingdao fraud involved 300,000 tons of aluminum, 20,000 tons of copper, and 80,000 tons of aluminum ingots.[46] As a result of this scandal, Chinese banks charge higher interest rates and are less likely to provide collateral financing.[47] Blockchain, though, can arguably stop such scandals.

Recent high-profile fraud has increased blockchain's attractiveness. The British multinational banking and financial services company Standard Chartered lost about US$200 million from Qingdao fraud. Standard Chartered teamed up with DBS Group and Singapore's Infocomm Development Authority to develop a blockchain-based platform.[48] Standard Chartered is a participant in blockchain-based trade finance platforms such as eTradeConnect and Bay Area Trade Finance Blockchain Platform.

To take another example, the Reserve Bank of India licensed three entities—RXIL, A.TReDS, and M1xhange—to provide receivable financing to microbusinesses and small businesses. These three platforms wanted to share information to prevent fraud but also keep data private. Blockchain can help achieve this. Using blockchain it is possible to create a cryptographic representation of an invoice, known as a hash. A hash is an indecipherable text and does not reveal anything about the invoice. It is nearly impossible to

---

42. Chen, "Jack Ma."

43. Song Shengxia, "China uncovers $10b worth of falsified trade," *People's Daily Online*, September 26, 2014.

44. Shengxia, "China uncovers."

45. Paul Smith, "7 ways blockchain technology could disrupt the post-trade ecosystem," Kynetix White Paper, 2015, http://www.the-blockchain.com/docs/Seven%20ways%20the%20Blockchain%20can%20change%20the%20trade%20system.pdf.

46. Shengxia, "China uncovers."

47. Smith, "7 ways."

48. Chanjaroen and Boey, "Fraud in $4 trillion."

convert a hash back to original data. If a trader submits the same invoice to more than one trade finance platform, the hash will be flagged. New York–based MonetaGo also hashes some invoice elements to prevent traders from modifying an invoice. An invoice with a high degree of similarity to another invoice already submitted to a different platform will not be rejected, but it will trigger an amber flag; then the trader is asked to explain.[49] As of April 2022, MonetaGo made over 2.5 million transactions in India with a total value of INR 692.7793 billion (US$8.7 billion).[50]

Detecting fake receipts has been a challenge for Chinese tax collectors, businesses, and state-run enterprises. In 2009, Chinese authorities detained 5,134 people and closed 1,045 fake invoice production sites. Likewise, in 2010, Chinese authorities took action against 1,593 criminal gangs and 74,833 enterprises that filed false invoices. From 2007 to 2012, the pharmaceutical company GlaxoSmithKline's four senior executives at its China operation submitted fake receipts and defrauded the company and shareholders of millions of dollars.[51]

Some financial technology companies aim to address this problem by providing blockchain-based systems for the reimbursement process (known as *fapiao* in Chinese). In August 2018, Tencent piloted a blockchain-based feature using WeChat Pay data to inform employers of employees' purchases. Employees can use the system to automatically send transaction data to employers for reimbursement. The feature is expected to bring efficiency to the corporate expense reimbursement process and to reduce fraud and tax evasion.[52]

Currently, reimbursement requires that merchants issue different receipts for purchases with the employer's taxpayer ID and other information. Merchants manually enter this information to generate receipts and process

49. Nicky Morris, "MonetaGo's blockchain solution for trade finance fraud," *Ledger Insights*, 2018, https://www.ledgerinsights.com/monetago-blockchain-trade-finance-fraud/.

50. "MonetaGo delivers major milestones in strategy for trade finance de-duplication in India, sets the stage for global adoption in 2022," *Businesswire*, April 28, 2022, https://www.businesswire.com/news/home/20220428005053/en/MonetaGo-Delivers-Major-Milestones-in-Strategy-for-Trade-Finance-De-duplication-in-India-Sets-the-Stage-for-Global-Adoption-in-2022/.

51. David Barboza, "Coin of realm in China graft: phony receipts," *New York Times*, August 3, 2013, https://www.nytimes.com/2013/08/04/business/global/coin-of-realm-in-china-graft-phony-receipts.html.

52. Jide Idowu, "Tencent to use blockchain on WeChat for faster refunds of company expenses," *BTCNN*, July 22, 2019, https://www.btcnn.com/tencent-to-use-blockchain-on-wechat-for-faster-refunds-of-company-expenses/.

additional paperwork.[53] In December 2018, Tencent announced that qualified merchants can use WeChat's blockchain-backed e-invoices.[54]

Blockchain-based effective risk assessment tools have also been developed for banks. OneConnect, a fintech subsidiary of Ping An Group, has developed a blockchain-based platform for small and medium-sized banks.[55] The solution can evaluate potential customers' creditworthiness by extracting a wide range of company data at low cost and with less need for documentation and labor to process transactions.

Reducing Frictions in Property Registration

Blockchain can play a major role in improving the protection of property rights.[56] Some potentially high-impact uses of blockchain are likely to be in improving property registry,[57] and also securely managing land records and land tenure.[58] A blockchain-based property registry system can reduce title fraud and guarantee title protection.[59] Blockchain can reduce friction, conflict, and costs in property registration. Some have suggested that land-titling systems could be low-hanging fruit for blockchain applications,[60] carried out in an inexpensive way.[61]

Various benefits of blockchain in developing a national system for property management have been suggested.[62] They include elimination of paper-

---

53. Connie Wang, "Tencent integrates blockchain e-invoicing with WeChat," *8BTC*, December 13, 2018, https://news.8btc.com/tencent-integrates-blockchain-e-invoicing-with-wechat.

54. Wang, "Tencent."

55. "OneConnect is a financial technology services company that provides financial technology solutions for small and medium-sized banks," *Crunchbase*, 2018, https://www.crunchbase.com/organization/oneconnect#section-overview.

56. Kshetri, *Global entrepreneurship*.

57. M. Swan, *Blockchain: Blueprint for a new economy* (Sebastopol, CA: O'Reilly Media, 2015).

58. A. Zwitter and M. Boisse-Despiaux, "Blockchain for humanitarian action and development aid," *Journal of International Humanitarian Action* 3, no. 1 (2018): 16, https://doi.org/10.1186/s41018-018-0044-5.

59. M. Themistocleous, "Blockchain technology and land registry," *Cyprus Review* 30, no. 2 (2018): 195–202, http://cyprusreview.org/index.php/cr/article/view/579.

60. M. Swan, "Anticipating the economic benefits of blockchain," *Technology Innovation Management* 7 (2017): 6–13.

61. G. Dwyer, "Blockchain: A primer," MPRA Paper 76562, University Library of Munich, Germany, 2016.

62. G. Gabison, "Policy considerations for the blockchain technology public and private applications," *SMU Science and Technology Law Review* 19 (2016): 327–350, https://heinonline.org/HOL/LandingPage?handle=hein.journals/comlrtj19&div=19&id=&page=.

work, reduction of fraud, and increase in speed with which transactions can be conducted.[63]

Both actual and proposed implementations of blockchain for land registry systems in Honduras,[64] Ghana,[65] Georgia, India, and elsewhere have been given as examples of the technology's benefits.[66] With a blockchain-based land titling project, the government of Georgia aims to enable landowners to borrow against their land and engage in entrepreneurial activities.[67] Economist Hernando de Soto was also involved in the development of a blockchain-based platform for property records in the country.

In 2017, India's Andhra Pradesh State collaborated with a Swedish startup, ChromaWay, to implement a blockchain-based land-recording project in the capital Amaravati. As of January 2018, about 100,000 land records were put in blockchain.[68] A typical land record in blockchain includes 58 attributes:[69] static attributes that describe the property such as unique ID, plot code, geo-coordinates, survey number, boundary information (e.g., neighboring plots, location in relation to roads or landmarks), and land classification, as well as dynamic attributes, such as owner (e.g., Aadhaar number) and mortgage information, right of first refusal, and litigation status. Events such as mutation, court case filing, court-issued stays, sale, approval of buildings, conversion of lands (e.g., from agricultural to commercial), mortgage, and owner's death are also recorded. The system also provides flexibility to add new attributes in the future.[70]

---

63. Lantmateriet, *The land registry in the blockchain: Blockchain land registry report*, 2016, http://icait.org/pdf/Blockchain_Landregistry_Report.pdf.

64. V. Lemieux, "Trusting records: Is blockchain technology the answer?," *Records Management Journal* 26, no. 2 (2016): 110–139, https://doi.org/10.1108/RMJ-12-2015-0042.

65. Nir Kshetri, "Will blockchain emerge as a tool to break the poverty chain in the Global South?," *Third World Quarterly* 38, no. 8 (2017): 1710–1732, http://www.tandfonline.com/doi/full/10.1080/01436597.2017.1298438.

66. Nir Kshetri, "Blockchain's roles in strengthening cybersecurity and protecting privacy," *Telecommunications Policy* 41, no. 10 (2017): 1027–1038.

67. S. Manski, "Building the blockchain world: Technological commonwealth or just more of the same?," *Strategic Change* 26, no. 5 (2017): 511–522, https://doi.org/10.1002/jsc.2151.

68. K. V. Kurmanath, "In AP Capital, blockchain technology secures land," *Businessline*, January 8, 2018, https://www. thehindubusinessline.com/info-tech/in-ap-capital-blockchain-technology-secures-land-records/article10020465.ece#.

69. Nir Kshetri, "Blockchain as a tool to facilitate property rights protection in the Global South: Lessons from India's Andhra Pradesh state," *Third World Quarterly* 43, no. 2 (2022): 371–392, https://doi.org/10.1080/01436597.2021.2013116.

70. N. S. S. Sai Baba, *Securing land records through blockchain*, Andhra Pradesh Human Resource Development Institute, February 2, 2020, http://www.aphrdi.ap.gov.in/docum

A U.S.-based platform for real estate registration, Bitland, announced the introduction of a blockchain-based land registry in Ghana, where 78% of land is unregistered.[71] There is a long backlog of land-dispute cases in Ghanaian courts.[72] Bitland records transactions securely with Global Positioning System (GPS) coordinates, written descriptions, and satellite photos. The process is expected to guarantee property rights and reduce corrupt practices. As of mid-2016, 24 communities in Ghana had expressed interest in the project.[73]

Bitcoin company BitFury and the government of Georgia signed a deal to develop a system for registering land titles using the blockchain.[74] To buy or sell land in the Republic of Georgia, currently the buyer and the seller go to a public registry house. They are required to pay between US$50 and US$200, which depends on the speed with which they want the transaction to be notarized. The pilot project will move this process onto the blockchain. The costs for buyer and seller are expected to be in the range of US$0.05-US$0.10.[75]

Blockchain-based land registry is likely to be a major force in fighting poverty. For instance, due to poor land tenure regularization institutions, landlessness is a major predictor of poverty in India over other socioeconomic factors such as caste or illiteracy.[76] Poor people can access financing more easily by showing their land ownership documents.

Promoting Efficiency and Access to International Trade and Financing

The global trade finance market, valued at US$18 trillion, is likely to be transformed by blockchain's disintermediation and other efficiency measures. A current challenge is that there is a big gap between the demand

---

ents/Trainings@APHRDI/2020/feb_2/Citizen%20Centric%20Services/Block%20Chain%20Technology.pdf.

71. O. Ogundej, "Land registry based on blockchain for Africa," *IT Web*, May 24, 2016, https://itweb.africa/content/raYAyModrd4qJ38N.

72. A. Jones, "How blockchain is impacting industry," *International Banker*, 2016, http://internationalbanker.com/finance/ blockchain-impacting-industry/.

73. Ogundeji, "Land registry."

74. S. Higgins, "Republic of Georgia to develop blockchain land registry," *Coindesk*, 2016, http://www.coindesk.com/bitfury-working-with-georgian-government-on-blockchain-land-registry/.

75. S. Higgins, "Survey: Blockchain capital markets spending to reach $1 billion in 2016," 2016, http://www.coindesk.com/capital-markets-1-billion-2016-blockchain/.

76. Hanstad, "The case."

and supply of trade financing. According to the Asian Development Bank (ADB), the global trade finance gap was US$1.5 trillion, or 10% of merchandise trade volume in 2018. This gap is expected to increase to US$2.4 trillion by 2025.[77]

First, the global trade finance market relies on paper documentation for most processes. A typical cross-border transaction involves many parties. A letter of credit (LC)—a promise to pay for goods if certain conditions are fulfilled—is sent to the exporter by the importer's bank. After receiving the LC, the exporter ships the goods. The bank faces the risk that the importer may be unable or unwilling to pay. The exporter then presents proof of shipping to get financing from its bank, which is paid directly by the importer's bank. Estimates suggest that, on average, a single cross-border trade transaction involves the exchange of 36 documents[78] to 40 documents[79] As many as 240 copies of documents need to be exchanged among various parties, such as financiers, logistics providers, customs officers, and warehouse managers.[80]

Paper-based methods such as letters of credit and factoring account for about US$5 trillion of annual trade worldwide.[81] It costs 1%–3% of a trade's value to buy a LC. Paper documents need to be physically exchanged, an extremely slow process that became especially apparent during the COVID-19 pandemic. Documents such as letters of credit, bills of lading, and invoices are normally carried in the cargo holds of passenger aircraft. Most passenger flights, though, could not operate during the COVID-19 pandemic. Millions of documents related to cross-border trade transactions were forced to move to alternative means, such as ships, to reach their destinations. However, many could not be delivered to banks, which were closed. Due to the emergency nature of the situation, many banks started accepting scanned signatures and documents. While electronic documents show

77. Peter Vanham, "Blockchain could enable $1 trillion in trade, mostly for SMEs and emerging markets," World Economic Forum, 2018, https://www.weforum.org/press/2018/09/blockchain-could-enable-1-trillion-in-trade-mostly-for-smes-and-emerging-markets/.

78. Laurence Fletcher, "Forget the paper trail—blockchain set to shake up trade finance," *Financial Times*, 2019, https://www.ft.com/content/04a4fcde-dfb5-11e9-b8e0-026e07cbe5b4.

79. J. Lamoureux and T. Evans, "Supply chain finance: A new means to support the competitiveness and resilience of global value chains," *Social Science Research Network (SSRN)*, 2011, https://papers.ssrn.com/sol3/papers.cfm?abstract id=2179944.

80. Fletcher, "Forget the paper trail."

81. I. Allison, "Skuchain: Here's how blockchain will save global trade a billion dollars," *IBtimes*, 2016, http://www.ibtimes.co.uk/skuchain-heres-how-blockchain-will-save-global-trade-trillion-dollars-1540618.

superior performance in terms of speed, their proneness to fraud has been a big concern among banks and other players. Blockchain has clear security benefits in this regard.[82]

Factors are key intermediary players in the global trade finance market. They offer money to exporters. Based on promised future payments, exporters borrow from factors. Exporting firms make an outright sale of accounts receivable to factors in order to maintain liquidity. For instance, a Chinese exporter selling to Walmart can take the invoice for those goods to a factor, which pays the exporter right away. For a US$100 invoice, the factor may pay as little as US$90. The upshot is that buyers such as Walmart pay higher for goods they buy from sellers in the developing world. The global factoring market is estimated at over US$2 trillion annually.[83]

Several companies have created blockchain-based products to address inefficiencies in B2B trade and supply chain financing.[84] The products are expected to eliminate intermediaries and financiers. For example, buyers and sellers agree on the terms of a deal, then blockchain tracks and manages the transaction from start to finish. In March 2017, China's internet financial services company Dianrong and FnConn, the Chinese subsidiary of the Taiwanese electronics manufacturer Foxconn, launched Chained Finance, China's first blockchain platform for supply chain finance.[85] Electronics, auto manufacturing, and clothing companies facing difficulties in accessing supply chain financing are the test markets for Chained Finance.[86] Instead of charging suppliers, Chained Finance charges peer-to-peer (P2P) lenders a fee to access to the system. Using the platform, nonbank lenders can make direct loans in supply chains worldwide.[87] Before launch, the two companies had successfully completed a pilot project and proof of concept to secure US$6.5 mil-

82. Patrick Tan, "Coronavirus hastens trade finance's blockchain moment," *BBN Times*, 2020, https://www.bbntimes.com/technology/coronavirus-hastens-trade-finance-s-blockchain-moment.

83. Tan, "Coronavirus."

84. "Skuchain developing blockchain solutions for $18 trillion trade finance market with funding from Amino," *PRNewswire*, 2016, http://www.prnewswire.com/news-releases/ skuchain-developing-blockchain-solutions-for-18-trillion-trade-finance-market-with-funding-fromamino- dcg-and-fbs-capital-300214205.html.

85. Samburaj Das, "Chinese fintech firms launch blockchain supply chain finance platform," *Cryptocoinsnews*, 2017, https://www.cryptocoinsnews.com/chinese-fintech-firms-launch-blockchain-supply-chain-finance-platform/.

86. Kshetri, "Blockchain's roles in strengthening."

87. K. Peterson, "Foxconn founder: Libra can 'converge' with China's digital currency in Taiwan," *Moon Catcher*, 2019, https://dailynews.bitcoindiamond.org/foxconn-founder-libra-can-converge-with-chinas-digital-currency-in-taiwan/.

lion in funding for Chinese SMEs. Different levels of suppliers are expected to be connected to the system of Chained Finance, and the company aims to expand to other developing economies. As of early 2020, more than 20 electronics suppliers were being paid on Ethereum-based cryptocurrency. The company reported that financing costs reduced from 24% per year to 10% and the time needed to get funding from seven days to one.[88]

Reducing Costs and Increasing Efficiency in International Payment Systems

The transaction costs of remittances, especially small ones, are very high. Immigrants use transfer services such as Western Union, which cost as much as 7% of the transfer amount.[89] To transfer 300 rand from South Africa to neighboring countries, transfer fees varied from 35 to 68.2 rand by bank draft, from 19.2 to 62.5 rand by electronic transfer, and from 25.3 rand by MoneyGram and 6.2 rand by iKobo's services.[90] For small business involved in international trade, the costs of acquiring major international currencies such as dollars and euros are high.

Blockchain fundamentally transforming the international remittance market. As an example, Ripple's On-Demand Liquidity leverages XRP (the payment network's cryptocurrency) to send money faster and at a lower fee.[91] Some users of RippleNet in Global South economies include Vietnam's TPBank, Pakistan's Faysal Bank,[92] the National Bank of Egypt,[93] and Thailand's Siam Commercial Bank.[94] The Siam Commercial Bank has teamed up with the digital money-transfer provider Azimo to use Ripple-

88. "Crypto confidential "bitcoin's guardian angel; the 50 biggest companies in blockchain," *Forbes*, February 23, 2020, https://www.forbes.com/sites/cryptoconfidential/2020/02/23/bitcoins-guardian-angel-the-50-biggest-companies-in-blockchain/#25a0ed083fcf.

89. J. Valenzuela, "Bitcoin remittances to Mexico see huge potential," *Cointelegraph*, 2015, https://cointelegraph.com/news/bitcoin-remittances-to-mexico-see-huge-potential.

90. S. Gupta, C. A. Pattillo, and S. Wagh, "Effect of remittances on poverty and financial development in Sub-Saharan Africa," *World Development* 37 (2009): 104–115.

91. David Rodeck, "What is XRP (Ripple)?," *Forbes*, May 31, 2022, https://www.forbes.com/advisor/investing/cryptocurrency/what-is-ripple-xrp/.

92. "Kiva sets up Sierra Leone blockchain ID system," *Ledger Insights*, 2019, https://www.ledgerinsights.com/kiva-sierra-leone-blockchain-id-system/.

93. Marie Huillet, "Egyptian national bank turns to blockchain to boost remittance business," *Cointelegraph*, 2020, https://cointelegraph.com/news/egyptian-national-bank-turns-to-blockchain-to-boost-remittance-business.

94. Ripple, "Azimo and SCB runs on ripple for instant payments into Thailand," 2020, https://ripple.com/insights/azimo-and-scb-runs-on-ripple-for-instant-payments-into-thailand/.

Net. Using nonblockchain solutions, settling a remittance sent from Europe to Thailand takes more than one business day. With RippleNet, Siam Commercial Bank clears pounds and euros into Thai baht in less than a minute.[95] Using on-demand liquidity, banks can avoid pre-funding;[96] this lets them settle remittances quickly.[97] On-demand liquidity is especially attractive to payment companies and nonbanking institutions that are required to open overseas accounts to fund their transfers. Many such institutions face difficulties opening overseas bank accounts due to concerns related to money laundering.[98]

In November 2019, Thailand's Siam Commercial Bank announced that it launched an AI-based robo-adviser to manage investment portfolios. The robo-adviser makes decisions for investors based on market conditions and their risk preference. Consumers can start using the service with just US$100 in initial investment. By 2019, the bank had invested US$1.3 billion in AI, digital platforms, and other technology as part of its four-year capital spending program.[99] Siam Commercial Bank expected that the new system would result in 100,000 new accounts by 2020.[100]

As another example of a low-cost, fast remittance system based on blockchain, in 2018, Ant Financial introduced a blockchain-based cross-border remittance system. AlipayHK and the Philippines-based mobile money company GCash teamed up to offer real-time money transfer between Hong Kong and the Philippines, with significantly lower fees and higher speed and efficiency than traditional transfer services.[101] Standard Chartered Bank was part of the initiative. Customers make a few clicks with AlipayHK and the money reaches the GCash user's account in seconds. When a user submits a remittance application, all network participants, including AlipayHK,

95. Ripple, "Azimo and SCB."

96. Prefunding involves taking money from the sender and "pushing" the money to a beneficiary partner bank or a beneficiary money transfer operator.

97. Ripple, *On-demand liquidity*, 2020, https://ripple.com/ripplenet/on-demand-liquidity/.

98. "Kiva sets up."

99. Anuchit Nguyen, "Thai bank targets $1 billion spinoff among its fintech units," *Bloomberg*, 2019, www.bloomberg.com/news/articles/2019-11-21/thai-bank-targets-1-billion-spinoff-among-its-fintech-units.

100. Omar Faridi, "People's Bank of China acquires $4.7 million in funding to further develop blockchain-based trade finance platform," *Crowdfund Insider*, 2020, https://www.crowdfundinsider.com/2020/03/158535-peoples-bank-of-china-acquires-4-7-million-in-funding-to-further-develop-blockchain-based-trade-finance-platform/.

101. "Ant Financial," *Fast Company*, https://www.fastcompany.com/company/ant-financial.

GCash, and Standard Chartered Bank, get notification. The sender and receiver can track the money during the entire process.[102]

Crypto-denominated international commerce, which has become increasingly common,[103] has increased efficiency in international payments. Small businesses in developing countries have reported that speed and efficiency can be greatly improved by making payments in cryptocurrencies rather than major international currencies. A Nigerian vendor of phones and accessories, who sources his products from China and the United Arab Emirates, reported that his Chinese suppliers prefer payments in cryptocurrency. He started doing so, which increased his profits because he did not have to buy US dollars or pay expensive fees to money-transfer agencies.

With such practical uses, especially for developing economies, bitcoin's use is growing. For instance, according to U.S. blockchain research firm Chainalysis, monthly cryptocurrency transfers of under US$10,000, which are typically made by individuals and small businesses, to and from Africa increased by more than 55% during June 2019–June 2020, to reach US$316 million.[104]

Chainalysis showed a similar pattern in Latin America. During June 2019–June 2020, Latin America sent US$25 billion worth of cryptocurrency and received US$24 billion.[105] The data showed that East Asia was Latin America's significant counterparty.[106] The blockchain research firm's interviews with Latin America–based cryptocurrency operators indicated that many payments were commercial transactions between East Asian exporters and Latin American importers. A Paraguay-based cryptocurrency exchange explained that businesses in Paraguay import significant goods from China, some of which are then exported to other countries such as Brazil. Many importers make payments using bitcoin because of the speed and ease of settling payments. Due to concerns related to money laundering, banks in Paraguay are reluctant to do businesses with many companies. The banking application process is complex, which requires many supporting

102. "AliPay, GCash launch blockchain cross-border remittance service," *PYMNTS*, June 25, 2018.

103. Mike Orcutt, "Cryptocurrency may be supercharging trade between Latin America and Eastern Asia," *The Block*, 2020, https://www.theblockcrypto.com/post/76839/cryptocurrency-eastern-asia-latin-america-trade-chainalysis.

104. Alexis Akwagyiram and Tom Wilson, "How bitcoin met the real world in Africa," *Reuters*, 2020, https://www.reuters.com/article/us-crypto-currencies-africa-insight/how-bitcoin-met-the-real-world-in-africa-idUSKBN25Z0Q8.

105. Orcutt, "Cryptocurrency."

106. Chainalysis, *How Latin America mitigates economic turbulence with cryptocurrency*, 2020, https://blog.chainalysis.com/reports/latin-america-cryptocurrency-market-2020.

documents and takes a long time. Moreover, even if a business's application to make a payment in international currency is approved, wire transfers are costly. Moreover, by making payments in cryptocurrencies, they can avoid import taxes.[107]

Finance, Banking, and Insurance

In the insurance sector, blockchain may provide risk managers with an effective way to protect individuals and companies from uncertain loss or catastrophe. Insurance and derivatives can be used to control or minimize risk factors associated with unpredictable or uncontrollable events. By supporting decentralized insurance models, blockchain may make derivatives more transparent. A meaningful risk management process can be designed using reputational systems based on individuals' social and economic capital and online behavior.[108] Blockchain-based insurance, for example, is connected to big data, the IoT, and health trackers to ensure better pricing and risk assessment.[109]

The IoT makes it easier for cars, electronic devices, and home appliances to have their own insurance policies. Using blockchain, these devices can be registered and their insurance policies administered by smart contracts. Damage is automatically detected, which triggers the repair process, claims, and payments.[110] Payouts are made against the insurable event, and the policyholder does not have to a make a claim. The insurer does not need to administer claims. The costs of claims processing drops close to zero. Even more important, there is less likelihood of fraud.[111]

Some blockchain-based insurance products have been launched that tar-

107. Chainalysis, *How Latin America.*

108. D. Tapscott, "How will blockchain change banking? how won't it?" *Huffington Post,* 2016, http://www.huffingtonpost. com/don-tapscott/how-will-blockchain-chang_b_9998348.html.

109. M. Ramada-Sarasola, "Want to get an insurer's attention? Just say blockchain," 2016, https://www.willistowerswatson.com/en/insights/2016/06/want-to-get-an-insurers-attention-just-sayblockchain.

110. J. Lorenz, B. Munstermann, M. Higginson, P. B. Olesen, N. Bohlken, and V. Ricciardi, "Blockchain in insurance—Opportunity or threat?," 2016, http://www.mckinsey.com/industries/financial-services/ourinsights/blockchain-in-insurance-opportunity-or-threat.

111. R. Huckstep, "What does the future hold for blockchain and insurance?," 2016, https://dailyfintech.com/2016/01/14/what-does-the-future-hold-for-blockchain-and-insurance.

get the B4B. To take an example, the Mexican mobile payment platform Saldo.mx launched a microinsurance service, Consuelo, which allows users to buy blockchain-powered health and life insurance policies. The target groups are Mexican living in country as well as in the diaspora.[112]

Some innovative blockchain solutions have helped drive financial inclusion. Blockchain-based solutions make peer-to-peer lending possible by directly connecting lenders and borrowers, thereby eliminating the need for intermediaries. Consider Kiva. The company does not make direct loans. While some investors mistakenly think that Kiva offers direct person-to-person connections, it actually works with local microfinance institutions (MFIs). Kiva conducts audits of its field partners to ensure that low-income groups are not exploited. However, due to high overhead costs and other inefficiencies, Kiva field partners charge exorbitantly high interest rates. For instance, according to Female Founder Stories, which publishes interviews and insights from the female alumni of the accelerator YCombinator, a Kiva field partner in Senegal was reported to charge an interest rate of 40% . Such loans could be made more affordable by eliminating intermediaries such as Kiva field partners.

Central bank digital currencies (CBDCs), many of which are based on blockchain, can also promote financial inclusion.[113] The International Monetary Fund argues that CBDCs offer great promise for reaching marginalized groups.[114] A retail CBDC system—in which a central bank issues digital currency directly, without need for traditional bank accounts—could be a game changer in eliminating poverty. This can be achieved through the establishment of an inclusive digital payment ecosystem and creation of financial data identities. For instance, individuals can have a CBDC account on the central bank's ledger. A digital wallet application linked to the account through application program interfaces (APIs) can allow users to access their account and engage in transactions.[115] In China, for instance,

112. Valenzuela, "Bitcoin remittances."

113. S. Allen et al., "Design choices for central bank digital currency: policy and technical considerations," *Brookings*, 2020, https://www.brookings.edu/wp-content/uploads/2020/07/Design-Choices-for-CBDC_Final-for-web.pdf.

114. V. Chatenay, "Facebook-backed Diem has cleared regulatory hurdles to finally launch in Q1," *Business Insider*, 2021, https://www.businessinsider.com/facebook-digital-currency-to-finally-launch-q1-2021-2.

115. N. Raghuveera, "Central bank digital currency can contribute to financial inclusion but cannot solve its root causes," *Atlantic Council*, June 10, 2020, https://www.atlanticcouncil.org/blogs/geotech-cues/central-bank-digital-currency-can-contribute-to-financial-inclusion-but-cannot-solve-its-root-causes/.

since the digital yuan is highly traceable, the country's central bank, People's Bank of China (PBOC), can monitor the flow of money in the economy.[116] This allows the government to deliver targeted programs to improve the well-being of high-risk groups such as SMEs and low-income households.

Blockchain-Based Digital Identity

According to the World Bank's ID4D database, 1 billion people lack any form of identification. An additional 3.4 billion people have some type of identification but lack the ability to use it in the digital world.[117] Identity management is thus a big issue. In financial institutions, the ability to prove someone is who they say they are is very important for increasing the accuracy of risk assessment and reducing fraud.[118] Potential borrowers in many Global South economies cannot prove who they are, which is among the main reasons many low-income people lack access to financial services.

Blockchain might address the lack of formal identity documents and can play a major role in creating secure digital identities.[119] Since ID cards in many countries are paper, which can be easily forged, blockchain solutions have significant potential to reduce fraudulent activities. Blockchain increases the ability to get a secure and authentic identity proof at a low cost. As explained in chapter 1, BanQu's blockchain-based verifiable digital identity for the poverty market is expected to help marginalized groups establish ownership, business assets, and production values and help them engage in economic transactions.

Some LMICs have started working on blockchain-based ID, which is likely to benefit the B4B. In August 2019, Sierra Leone launched the

116. X. Huang, "China's DCEP project launches biggest digital yuan test yet," *Forkast*, 2021, https://forkast.news/china-dcep-digital-yuan-pros-cons/.

117. Olivia White, Anu Madgavkar, James Manyika, Deepa Mahajan, Jacques Bughin, Mike McCarthy, and Owen Sperling, *Digital identification: A key to inclusive growth*, McKinsey, 2019, www.mckinsey.com/business-functions/digital-mckinsey/our-insights/Digital-identification-A-key-to-inclusive-growth?cid=other-eml-alt-mgi-mck&hlkid=ecd8822bafc44de78b1f2670c2979652&hctky=2259579&hdpid=0945f28e-3aa8-4f73-a111-6ba86e377b51.

118. "5 ways that insurance will be disrupted by the blockchain," *BoostVC*, 2016, https://medium.com/ boost-vc/5-ways-that-insurance-will-be-disrupted-by-the-blockchain-8ffc33674713#.4ng8nof0w.

119. Nir Kshetri, "Blockchain-based financial technologies and cryptocurrencies for low-income people: Technical potential versus practical reality," *IEEE Computer* 53, no. 1 (2020): 18–29.

blockchain-based National Digital Identity Platform developed by Kiva.[120] Kiva worked with the UN Capital Development Fund and the UN Development Program to develop the platform. Kiva's blockchain protocol aims to address two major barriers that hinders low-income people's access to financial services: formal identification and verifiable credit history.[121]

Blockchain Diffusion

Activities of Foreign Multinationals

Blockchain companies from the industrialized world are making inroads into developing economies. In this section, two examples are given to illustrate how such activities have facilitated the use of blockchain in these economies.

In 2017, the British-Dutch multinational consumer goods company Unilever announced that it had teamed up with British supermarket chain Sainsbury, packaging company Sappi and three global financial services companies—BNP Paribas, Barclays, and Standard Chartered—to develop a blockchain platform to track sustainability practices in its supply chains. The initial year-long project started with US$700,000 in funding from the U.K.'s Department for International Development and private sources to track two categories of products: tea supplies used by Unilever and Sainsbury's and wood fibers in certain Sappi packaging solutions. The plan was to start with a system to track and verify contracts for farmers in Malawi who supplied tea to Unilever and Sainsbury. Financial incentives were offered to the tea farmers for feeding social or ecological data into the blockchain system.[122]

Tea is Malawi's second-biggest agricultural export, after tobacco. The initiative is expected to reach up to 10,000 farmers. The group announced that preferential pricing would be applied to farms that engage in sustainable

120. Cooper Inveen, "San Francisco crowdfunder Kiva sets up Sierra Leone credit database," *Reuters*, 2019, www.reuters.com/article/us-leone-kiva/san-francisco-crowdfunder-kiva-sets-up-sierra-leone-credit-database-idUSKCN1VB262.

121. Catherine Cheney, "In Sierra Leone, new Kiva Protocol uses blockchain to benefit unbanked," 2019, *Devex*, www.devex.com/news/in-sierra-leone-new-kiva-protocol-uses-blockchain-to-benefit-unbanked-95490.

122. "Sappi teams up with major global brands to explore the potential of innovative blockchain technology in enhancing the sustainability of global supply chains," Sappi, 2019, https://www.sappi.com/sappi-teams-up-with-major-global-brands-to-explore-the-potential-of-innovative-blockchain-technology.

farming methods, which increase harvests without using more land. The banks would finance farms that utilize sustainable farming methods.[123]

Provenance, the blockchain-enabled fintech startup Halotrade, and the real estate development company Meridia developed the "Trado model," which facilitates data-sharing among producers, consumers, and supply chain players.[124] The Trado model was first piloted in the tea supply chain in collaboration with the Lujeri Tea Estate in Malawi. It made farming data from smallholder farmers directly accessible to Unilever. Data related to tea leaf production, social impact, and sustainability credentials were recorded in the Provenance platform. Using data related to the availability of goods, Unilever supported the release of payment earlier. The 18-month pilot found that Unilever and Sainsbury's could increase supply chain transparency with the data.[125] It reduced the period of expensive financing for upstream partners. The transactions take place using a bank's regular supply chain finance process, so results in minimal disturbance to the banks' businesses.[126]

The resulted savings were invested to fund projects such as a field school for local farmers to provide training in sustainable practices. Local NGOs confirmed payment distributions to the school and recorded impact progress on the blockchain.[127]

As a second example, consider Easy Trading Connect (ETC), the trade finance solution launched in February 2017 by the multinational bank ING and the French multinational investment bank and financial services company Société Générale. The trial took place on a transaction between the two banks and Mercuria, a Switzerland-based global commodities trading corporation, involving an oil cargo shipment containing African crude sold to the Chinese petrochemical company ChemChina. According to the two banks, the solution performed well in terms of assessment criteria, such as elimination of document fraud, lower costs, and improved efficiency. The trade involved the banks, traders, an agent, and an inspector.[128] The prototype allowed all parties

123. Heather Clancy, "Unilever teams with big banks on blockchain for supply chain," *Greenbiz*, December 13, 2017, https://www.greenbiz.com/article/unilever-teams-big-banks-blockchain-supply-chain.

124. "Leveraging new technologies to fund fair, sustainable smallholder farming," Provenance, https://www.provenance.org/case-studies/unilever.

125. Sarah George, "Blockchain-enabled supply chain sustainability scheme hailed 'successful' by business giants," *edie*, September 17, 2019, https://www.edie.net/news/8/Blockchain-enabled-supply-chain-sustainability-scheme-hailed--successful--by-business-giants/.

126. "Banks back pilot bidding to unlock finance for sustainability in supply chains," *finextra.com*, September 17, 2019, https://www.finextra.com/newsarticle/34401/banks-back-pilot-bidding-to-unlock-finance-for-sustainability-in-supply-chains.

127. "Leveraging new technologies."

128. BI Intelligence, "ING and SocGen partner on blockchain solution," *Business*

to execute their roles directly on the platform and reduced the time for the bank to execute its role in the transaction to 25 minutes from an average of three hours.[129] While this test was for an oil transaction, the steps involved in trades of other commodities are the same, which would allow the platform to easily scale.[130] It also reportedly led to cost savings up to 30%.[131]

ING and Société Générale were negotiating with traders in the liquefied natural gas industry to test the solution.[132] Other economies in Africa and other parts of the world, as well as other commodities sectors, are thus likely to benefit from solutions like this.

An enhanced version of the ETC was used to conduct a live agricultural commodity trading transaction in January 2018 that involved 60,000 tons of soybeans sold by the global merchant Louis Dreyfus Company to China's Shandong Bohi Industry. All relevant documents, such as sales contract, LC, and certificates, were digitized. The U.S. Department of Agriculture provided data on phyto-sanitary certificates. Russell Marine Group and Blue Water Shipping issued other required certificates. ING, Société Générale, and ABN Amro had issued and confirmed the LC.[133] It led to a shorter cash cycle.

The three banks—ING, Société Générale and ABN AMRO—which started ETC were among the 15 shareholders of the independent venture komgo, started in August 2018 to digitize and streamline commodity trade finance.[134] Among major clients, Sberbank Switzerland AG, a subsidiary of Russia's largest bank, Sberbank, signed an agreement with komgo to apply its blockchain-powered trade financing service.[135]

Insider, February 24, 2017, http://www.businessinsider.com/ing-and-socgen-partner-on-blockchain-solution-2017-2.

129. BI Intelligence, "ING and SocGen."

130. Noelle Acheson, "Overtaking banking in the race to blockchain," *CoinDesk*, February 27, 2017, http://www.coindesk.com/overtaking-banking-race-blockchain/.

131. K. Schaps, C. Steitz, and V. Eckert, "Exclusive: ING, SocGen to test LNG trading with blockchain in months," *Reuters*, 2017, http://www.reuters.com/article/us-energy-companies-blockchain-idUSKBN16G1UU.

132. Schaps et al., "Exclusive: ING, SocGen to test."

133. Sanne Wass, "Major banks and traders test blockchain platform for commodity trade," *Fintech*, January 24, 2018, https://www.gtreview.com/news/fintech/major-banks-and-traders-test-blockchain-platform-for-commodity-trade/.

134. Sanne Wass, "Komgo releases two new blockchain-based products, reveals future plans," *Fintech*, June 25, 2019, https://www.gtreview.com/news/fintech/komgo-releases-two-new-blockchain-based-products-reveals-future-plans/.

135. Helen Partz, "Russia's largest bank joins blockchain trade finance platform," *Cointelegraph*, September 15, 2020, https://cointelegraph.com/news/russia-s-largest-bank-joins-blockchain-trade-finance-platform.

Local Entrepreneurial Activities

In some LMICs, entrepreneurial activities in blockchain are rapidly rising. For instance, China has become a global epicenter of blockchain-related activities. As early as in 2018, the country had more new blockchain companies than the United States.[136] According to the blockchain and crypto data platform LongHash, as of August 2020, there were more than 84,410 registered blockchain companies in China, and 29,340 of them were in operation. About 10,000 of them were registered in the first seven months of 2020.[137] Likewise, as of 2021, Africa had 60 active cryptocurrency exchanges platforms that focused on diverse activities such as P2P transactions and trade financing.[138] To take an example, the Kenyan fintech firm Pezesha, which specializes in micro-, small-, and medium-sized enterprise (MSME) credit scoring and loan origination, developed a crypto-based solution to allow global lenders to invest in Africa. Foreign lenders can send stablecoin in U.S. dollars for conversion to Kenyan shillings. As of March 2022, Pezesha had facilitated 3,751 loans in Kenya and 344 in Ghana.[139] Since MSMEs bring benefits to the poor,[140] startups such as Pezesha play a key role in poverty alleviation.

Innovative Activities in Blockchain

Some LMICs have invested in intangible assets such as patents and research and development. For instance, China has consistently ranked as the world leader in blockchain patent filing. While some argue that patents are irrelevant in blockchain, others argue that patents reflect technological or innovative activities and output, which are key in private and hybrid block-

136. Marko Vidrih, "The blockchain international standardization organization: China will dominate," *Good Audience*, May 22, 2018, https://blog.goodaudience.com/the-blockchain-international-standardization-organization-china-will-dominate-b7d423904bfb.

137. Ting Peng, "More than 10,000 new blockchain companies established in China in 2020," *Cointelegraph*, August 9, 2020, https://cointelegraph.com/news/more-than-10-000-new-blockchain-companies-established-in-china-in-2020.

138. Baobab Network, *60 cryptocurrency-enabled fintech's—Africa market map*, February 7, 2021, https://insights.thebaobabnetwork.com/2021-02-cryptocurrency-africa-market-map/.

139. Bitange Ndemo, "The role of cryptocurrencies in sub-Saharan Africa," Brookings, March 16, 2022, https://www.brookings.edu/blog/africa-in-focus/2022/03/16/the-role-of-cryptocurrencies-in-sub-saharan-africa/.

140. O. U. Asikhia, "SMEs and poverty alleviation in Nigeria: Marketing resources and capabilities implications," *New England Journal of Entrepreneurship* 13, no. 2 (2010): 57–70, https://doi.org/10.1108/NEJE-13-02-2010-B005.

chains.[141] China's global dominance in blockchain patents has implications for blockchain-based solutions in the global market. For instance, according to the intellectual property media outlet IPRDaily and the patent database incoPat, in 2019, Chinese companies took the top three spots (Alibaba, Tencent, and Ping An) in the number of blockchain patents filed. Chinese companies also accounted for 7 of the top 10 and 19 of the top 30 firms.[142]

Favorable Policies for Developing Blockchain Ecosystems

Some LMICs have formulated and implemented favorable policies for developing blockchain ecosystems. For instance, China's 13th Five-Year Plan on Informatisation from 2016 to 2020 has focused on advanced technologies, such as big data, AI, and blockchain. The plan has listed blockchain technologies among the main development directions.[143] In the second half of 2016, the Chinese government also published a white paper that laid out key milestones for blockchain's development and emphasized the importance of "quick response and reasonable planning" to influence the international standardization of blockchain. The report set a deadline of April 2017 for pilots using blockchain standards developed by the group.[144] Likewise, the influential policy document—No. 1 Central Document,[145] released on February 5, 2017, has given due recognition to agricultural innovation as a priority area, especially quality supervision and standards. Policy analysts have argued that this is an indication of a very high level of official support for blockchain deployment in the food industry.[146]

---

141. Will Heasman, "China dominates global blockchain patent applications," *Decrypt*, April 24, 2020, https://decrypt.co/26589/china-dominates-global-blockchain-patent-applications.

142. "World's top 10 patent holders by blockchain inventions," *China Daily*, September 18, 2020, https://www.chinadaily.com.cn/a/202009/18/WS5f63e5c0a31024ad0ba7a3fa_1.html.

143. Lester Coleman, "China to support blockchain development under new five-year plan," *Cryptocoinsnews*, December 29, 2016, https://www.cryptocoinsnews.com/china-support-blockchain-development-new-five-year-plan/.

144. Joon Ian Wong and Zheping Huang, "China's tech giants will shape international blockchain standards, with Beijing's backing," *Quartz*, October 21, 2016, https://qz.com/813248/chinas-tech-giants-will-shape-international-blockchain-standards-with-beijings-backing/.

145. The "No. 1 Central Document" is the first policy statement released by the central government in a given year. This document is seen as an indicator of the government's policy priorities.

146. Noelle Acheson, "Counting chickens: can blockchain restore trust in China's food supply?," *Coindesk*, June 26, 2017, http://www.coindesk.com/counting-chickens-can-blockchain-restore-trust-in-chinas-food-supply/.

Likewise, the government of Mauritius has devoted resources to develop the blockchain ecosystem. Mauritius has collaborated with the private sector in country and internationally.[147] The government and the blockchain software technology firm ConsenSys were exploring potential collaboration to establish the foundational elements of a blockchain ecosystem, including "know your customer," or KYC, rules, digital identity, and title registries. In the subsequent phase, it plans to help the country to build a talent pool of developers, entrepreneurs, executives, and regulators to further enrich the ecosystem.[148] The Mauritius government and ConsenSys were reported to be exploring the possibility of opening a ConsenSys Academy in Mauritius.[149] ConsenSys Academy Dubai's first class of Ethereum blockchain developers graduated in 2017.

Conclusion

Although it can be argued that it is the interest of providers of blockchain-related services to exaggerate the potential benefits of this technology, the analysis here suggests that blockchain, in combination with other technologies, such as the IoT and cloud computing, can drive economic, social, and political transformations in developing economies. Among the most attractive features of blockchain is that once a record is created, it is almost impossible to be tampered with or forged. Blockchain will thus make data secure. Transactions can also be conducted to achieve any degree of privacy or openness. Some application areas include land registration and donation and payment tracking. Cryptocurrencies as an interoperable system can more easily convert various currencies and facilitate cross-border trade.

Programs such as Kiva's blockchain-based IDs are a first step to improve access to finance for low-income populations. True decentralization will

---

147. Government of Mauritius, *Blockchain technology and its impact on digital transformation of Mauritius*, February 7, 2019, http://www.govmu.org/English/News/Pages/Blockchain-Technology-and-its-impact-on-Digital-Transformation-of-Mauritius.aspx.

148. Aaron Stanley, "Mauritius: The tropical paradise looking to become a blockchain hub," CoinDesk, May 7, 2017, https://www.coindesk.com/mauritius-the-tropical-paradise-looking-to-become-a-blockchain-hub.

149. "Mauritius plans to create an ethereum island," *NEWSBTC*, 2017, https://www.newsbtc.com/2017/07/16/mauritius-plans-create-ethereum-island/.

be complete when impact investors and philanthropic funders can directly reach low-income groups with cryptocurrencies.

Blockchain helps prevent corrupt officials from engaging in fraudulent activities. Organizations can make sure that their business partners play by rules. Services providers can make sure that people are who they say they are when they enroll and participate in various services. These features are of special interest in developing economies, which can lack effective and trustworthy institutions.

LMICs need to deal with various challenges and bottlenecks in successful deployment of blockchain. Powerful actors that are against transparency and openness may oppose blockchain. In the land ownership example, blockchain can increase the transparency of land ownership and records and make it difficult or impossible for corrupt officials to alter land registries after the records are on the blockchain. Nonetheless, blockchain cannot address corruption in decisions about how land is registered in the ledger.

The widespread adoption of blockchain may enhance a country's image. For instance, the Republic of Georgia has been promoting itself as a corruption free-country with modern and transparent governance.

Blockchain is particularly suitable for detecting widespread fraud in industries such as insurance and banking. Blockchain makes it possible for donors to directly make payments to the causes they are passionate about, and without depending on other organizations to act as intermediaries. In this way blockchain helps prevent the misuse and abuse of donor money.

Blockchain is especially likely to make contract enforcements more efficient and effective. For instance, a blockchain-based life or health insurance contract can be a powerful tool: it is possible to automatically activate a policy based on diagnosis. For instance, if a diagnosis indicates the existence of a triggering condition for the policy that is written in the smart contract, the information is fed to the blockchain. The smart contract automatically authorizes payments based on the policy. Smart contracts can also act as a warranty for a down payment to the medical service provider, and there is no need to have a previous contractual relationship between the medical service provider and the insurance company. In this way, smart contracts drastically reduce administrative costs.

Blockchain applications are in a nascent stage of development. Rather than viewing them as a self-contained phenomenon, they must be seen against the backdrop of economic and institutional realities facing developing economies. There are many possible uses of the blockchain, and several channels and mechanisms through which developing economies may benefit. In practice, however, a number of challenges stand in the way of implementation and practical results. Blockchain-based innovations and business models are as yet far from inclusive with regard to SMEs in developing economies. If the technology is properly developed, utilized, and implemented, some of the institutional bottlenecks can be alleviated. Overall, blockchain can unlock entrepreneurship opportunities.

CHAPTER 4

Remote Sensing and Satellite Imagery

During the COVID-19 lockdown in Togo in March 2020, the country's president wanted to send cash payments of around US$20 per person to about 525,000 Togolese households that were poor or vulnerable to poverty. However, a big challenge was that Togo lacks accurate data related to the economic situations of specific households.[1] Togo conducted its fourth General Population and Housing Census in 2010, its first census carried out in 29 years. The Census planned in 1991 and others after that were not held due to social unrest and other national political situations. Major international donors had suspended operations in the country.[2] And collecting required information during a pandemic is extremely challenging.

The Peruvian government's efforts to help vulnerable groups with emergency funds during the COVID-19 pandemic faced similar challenges. About 10 million Peruvians depend on agriculture for their livelihood. They lacked channels to sell their produce due to high costs or unavailability of transportation during the lockdown. On March 16, 2020, the president announced a plan to send 380 soles (US$110) to each family in poverty or extreme poverty, and to independent workers who were unable to work.[3] Subsequently, the Ministry of Labor and Employment Promotion provided

1. Joshua Blumenstock, "Machine learning can help get COVID-19 aid to those who need it most," *Nature*, May 14, 2020, https://www.nature.com/articles/d41586-020-01393-7.

2. UN Population Fund, *1st census in 29 years has been crucial for 2013 elections in Togo*, August 8, 2013, https://wcaro.unfpa.org/en/news/1st-census-29-years-has-been-crucial-2013-elections-togo.

3. "Martín Vizcarra anuncia bono de 380 soles a familias que dependan de trabajos diarios," *La República*, March 16, 2020, https://larepublica.pe/economia/2020/03/16/martin-vizcarra-anuncia-bono-de-380-soles-a-personas-que-cumplan-trabajos-diarios/.

additional 380 soles, increasing the total amount to 760 soles.[4] The two direct payments did not reach vulnerable populations, such as indigenous Andean farmers.[5] A major difficulty in this was that the government lacked information about the families targeted by the emergency funds and where they lived.

The governments of Togo and Peru turned to satellite imagery and remote-sensing data to locate the vulnerable groups. Man-made satellites are machines launched into space that orbit Earth.[6] Remote sensing involves the satellite (or an aircraft) detecting and monitoring an area's physical characteristics by measuring and interpreting the radiation reflected from it. Typically, cameras installed in satellites or aircraft are used for this purpose.[7] Remote sensing is another major 4R technology that can overcome the limitations associated with the unavailability of important data.

Currently, about 2,200 satellites orbit Earth.[8] They are of many types, such as Earth observation, communications, navigation, and weather. In mid-2019, there were 768 satellites that produced images.[9] Telecommunications companies are launching dozens of new satellites with a goal of providing high-speed broadband internet access to the entire world population. By 2025, an additional 1,100 satellites are expected to be added annually, compared to 365 in 2018.[10] Estimates suggest that 50,000 additional satellites will be launched during 2020–2030.[11]

---

4. El Estado Peruano, *Consultar si te corresponde el Bono Independiente*, May 15, 2020, https://www.gob.pe/8883-consulta-si-te-corresponde-el-bono-independiente-de-s-380-soles.

5. Traveling and Living in Peru, "Lack of government support forces vulnerable populations to walk home," April 30, 2020, https://www.livinginperu.com/lack-of-government-support-forces-vulnerable-populations-to-walk-home/.

6. NASA, "What is a satellite?," February 12, 2014, https://www.nasa.gov/audience/forstudents/5-8/features/nasa-knows/what-is-a-satellite-58.html.

7. US Geological Survey, *What is remote sensing and what is it used for?*, 2019, https://www.usgs.gov/faqs/what-remote-sensing-and-what-it-used?qt-news_science_products=0#qt-news_science_products.

8. "A crowd in space: Tens of thousands of satellites planned for orbit," *Pittsburgh Post-Gazette*, May 20, 2020, https://www.post-gazette.com/opinion/editorials/2020/05/20/space-satellites-crowded-junk-exploration-impact/stories/202002190057.

9. Christopher Beam, "Soon satellites will be able to watch you everywhere all the time," *MIT Technology Review*, June 26, 2019, https://www.technologyreview.com/2019/06/26/102931/satellites-threaten-privacy/.

10. Tate Ryan-Mosley, Erin Winick, and Konstantin Kakaes, "The number of satellites orbiting Earth could quintuple in the next decade: The coming explosion of constellations," *MIT Technology Review*, June 26, 2019, https://www.technologyreview.com/2019/06/26/755/satellite-constellations-orbiting-earth-quintuple/.

11. "A crowd in space."

Every day, Earth observation satellites gather huge amounts of data about Earth's physical, chemical, and biological characteristics. The data are used for civil, military, and commercial activities. These satellites provide the same richness of data for remote areas of the least developed countries as for urban areas of developed countries. Thus, the data divide does not exist for data produced by Earth observation satellites.

Satellites can generate rich data that can be used to address, social, humanitarian, economic, and environmental problems facing the B4B. For instance, using high-resolution commercial satellite data, it is possible to identity small objects, such as individual houses and cars.[12] One company providing such solutions is the Colorado-based space technology company Maxar, which sells the highest-resolution imagery to its clients, which include governments, corporations, and nonprofits. In one case, its WorldView-2 satellite produced images with a resolution of 46 centimeters.[13]

Remote sensing holds special significance for developing countries, especially in the face of two data-related challenges: outdated or low-quality census data and the lack of essential details in standard maps. First, in many developing countries, there is a lack of accurate and current data about the population. For instance, Nigeria's population is estimated in the range of 180 million to 200 million, but not even the president knows the country's exact population size. The country's last census took place in 2006. The plan for the next census has not been implemented due to lack of funding. It has also been argued that, due to survey malpractices, the 2006 census lacks validity and cannot be trusted.[14]

Second, the quality of standard maps suffers because these countries lack resources, technology, and skilled staff. While maps are constructed to represent physical space, the reality is that objects represented by a map are different from the truth on the ground. What exists on a map is a function of map producers' goals, incentives, motivations, and economic and strategic environments.[15] Also important are the costs of mapmaking, the nature of the demand for maps, and innovations in mapmaking technology.[16]

12. *Satellite data: What spatial resolution is enough?*, April 12, 2019, https://eos.com/blog/satellite-data-what-spatial-resolution-is-enough-for-you/.

13. Geoff Brumfiel, "Trump tweets sensitive surveillance image of Iran," *NPR*, August 30, 2019, https://www.npr.org/2019/08/30/755994591/president-trump-tweets-sensitive-surveillance-image-of-iran.

14. Yomi Kazeem, "African governments need to fix their problematic relationship with data on their own countries," *Quartz*, October 16, 2019, https://qz.com/africa/1725537/africa-has-a-data-problem-says-mo-ibrahim-governance-report/.

15. Abhishek Nagaraj and Scott Stern, "The economics of maps," *Journal of Economic Perspectives* 34, no. 1 (2020): 196–221.

16. Nagaraj and Stern, "The economics."

Developing and developed countries clearly demonstrate differences in terms of these factors. The U.S. Census Bureau combines advanced satellite technology, other geospatial mapping tools, and visits by real people in its survey. Each year, the U.S. spends more than US$250 million on the American Community Survey (ACS), a door-to-door study to collect demographic factors such as race, gender, education, occupation, and unemployment rates.[17]

With these efforts, even the smallest settlement in the most remote area and all homes in the U.S. can be mapped.[18] These levels of detail are not shown in standard maps of most developing countries. There has been a lack of efforts and resources to develop useful maps in these countries. For instance, despite some efforts in the past to extract health service providers by OpenStreetMap and the Global Healthsites Mapping Project (https://www.healthsites.io), Sub-Saharan Africa has no georeferenced, comprehensive public health facility inventory.[19] Overall, 4 billion people in the world live in places that have no street names or numbers.[20]

This discussion suggests that current standard maps and survey data do not provide answers to simple questions about how many people live in a location and who they are. Moreover, the added value of satellite data stems from the fact that remote communities and vulnerable groups are less likely to generate contents such as social media posts which help evaluate, respond to, and monitor the effects and recovery from pandemics. For instance, in communities that are connected to the internet, online media reports, social media activities, and clinical reports can help understand the patterns and predict the spread of infections. They can also be used to plan and track the impact of different policies. On the contrary, a large proportion of popula-

17. Timnit Gebru, Jonathan Krause, Yilun Wang, Duyun Chen, Jia Deng, Erez Lieberman Aiden, and Li Fei-Fei, "Using deep learning and Google Street View to estimate the demographic makeup of neighborhoods across the United States," *PNAS*, December 12, 2017, 13108–1311.

18. *Mapping the most remote areas and out-of-the-way homes*, August 12, 2019, https://www.census.gov/library/stories/2019/08/from-horseback-to-satellite-how-us-census-bureau-checks-addresses.html.

19. Joseph Maina, Paul O. Ouma, Peter M. Macharia, Victor A. Alegana, Benard Mitto, Ibrahima Socé Fall, Abdisalan M. Noor, Robert W. Snow, and Emelda A. Okiro, "A spatial database of health facilities managed by the public health sector in sub-Saharan Africa," *Nature*, July 25, 2019, https://www.nature.com/articles/s41597-019-0142-2.

20. "How 3 words could put 4 billion people without addresses on the map," *CBC Radio*, June 21, 2016, https://www.cbc.ca/radio/thecurrent/the-current-for-june-21-2016-1.3644974/how-3-words-could-put-4-billion-people-without-addresses-on-the-map-1.3645088.

tions living in remote areas do not generate any digital footprints. According to the International Telecommunication Union (ITU), as of 2019, only 57% of the world's population was using the internet. The proportions of populations with internet access were 47% for developing countries and 12% for LDCs.

Remote-sensing technologies hold great promise in this regard, representing an important source of information critical for health, survival, and well-being for vulnerable groups living in remote communities. Predicting pandemics and disasters, and helping remote communities affected by such risks, is among the most compelling role of these technologies.

From the perspective of remote communities, remote-sensing technologies have many attractive features. Remote-sensing data are available at a global scale. Satellites cover Earth's entire surface, including hard-to-reach areas. Satellite data are collected with significantly higher frequency than household survey data. A satellite may also have observation data on the same location at around the same time every day. It is thus possible to collect data systematically over time.[21] Such data offer more granular geographic insights into site-specific settings such as hospitals, schools, and homes.[22]

In the past, most developing countries could not utilize satellite imagery data due to technological and financial barriers. One reason is that the satellite image files often are of large size. Only well-equipped labs could process and analyze the satellite data. In this regard, recent scientific and technological developments have been highly encouraging, and this has increased the attractiveness of satellite imagery data to address diverse challenges facing the developing world. For instance, cloud computing platforms such as Google Earth Engine (GEE) have changed the unfavorable economics of satellite images for remote communities. A desktop computer with an internet connection is sufficient to access and analyze satellite data. GEE provides large and rich datasets that include data from satellites such as NASA and the U.S. Geological Survey's Landsat and MODIS technology, and the intergovernmental European Space Agency's (ESA) Sentinel 1 and 2 satellites.[23] It also has nonsatellite data related to climate and topography.

---

21. OECD, "Chapter 7: Bringing space to earth with data-driven activities," *The space economy in figures*, https://www.oecd-ilibrary.org/sites/07d50927-en/index.html.?itemId=/content/component/07d50927-en.

22. N. A. Wardrop, W. C. Jochem, T. J. Bird, et al., "Spatially disaggregated population estimates in the absence of national population and housing census data," *Proceedings of the National Academy of Sciences of the United States of America*, 115, no. 14 (2018): 3529–3537.

23. Sentinel-1 consists of two satellite constellations that monitor land and ocean

Moreover, GEE is free for noncommercial uses.[24] Despite these encouraging developments, scientists, policymakers, and businesses have not even scratched the surface of satellite imagery data for the economic and social benefits of remote communities. However, the COVID-19 pandemic has highlighted the importance of remote sensing and satellite imagery.

How Remote Sensing Works

Figure 4.1 shows the key steps and processes involved in remote sensing. First, an energy source illuminates electromagnetic energy to the target of interest or provides such energy.[25] The energy then travels from its source to the target by interacting with and passing through the atmosphere. The energy then interacts with the target. The energy that has been scattered by or emitted from the target, which again interacts with and passes through the atmosphere, is then recorded by a remote sensor.

The energy recorded by the remote sensor is transmitted to a receiving and processing station, often in electronic form. The data are processed into an image. The next step involves interpreting and analyzing the processed image to extract information about the target. The final step is to apply the information extracted from the imagery about the target. The information can help better understand the image and may reveal information that was not known before. The information can also help in solving a particular problem.

Amplified Benefits of Combining Technologies

Remote-sensing technologies have amplified benefits beyond their direct effects when combined with other technologies. Specifically, the usefulness

---

(https://earth.esa.int/web/guest/missions/esa-operational-eo-missions/sentinel-1). Sentinel-2's two twin satellites, Sentinel-2A and Sentinel-2B, support a broad range of services and applications: agricultural monitoring, emergency management, land cover classification, and water quality. These satellites acquire high spatial resolution imagery.

24. Giriraj Amarnath, Surajit Ghosh, and Carolyn Fry, "Satellite maps can help nations make critical food production decisions amid coronavirus," *Geospatial World*, May 13, 2020, https://www.geospatialworld.net/blogs/satellite-maps-can-help-nations-make-critical-food-production-decisions-amid-coronavirus/.

25. Natural Resources Canada, *Fundamentals of remote sensing—Introduction*, November 19, 2015, https://www.nrcan.gc.ca/maps-tools-and-publications/satellite-imagery-and-air-photos/tutorial-fundamentals-remote-sensing/introduction/9363.

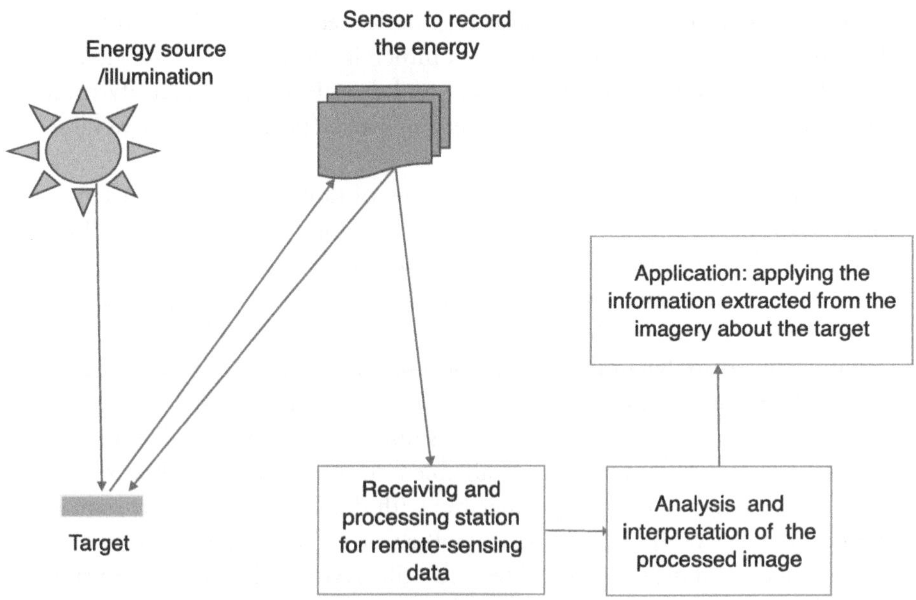

Figure 4.1. Key steps in remote sensing

of remote-sensing data is improved by recent developments in ML and AI.[26] For instance, in the case of crisis, such as the COVID-19 pandemic, the use of ML and AI to analyze satellite images could revolutionize the way vulnerable groups can be tracked and assistance provided.

Image data are inherently high-dimensional because each image contains thousands of pixels. Each pixel represents different information or traits[27] High levels of accuracy have been achieved using satellite imagery data combining with ML. The global security and aerospace company Lockheed Martin has claimed that its system has over 90% accuracy in identifying characteristics of an object or target. This accuracy is comparable to the commonly used deep-learning algorithm in the medical field—known as convolution neural network—that assists in disease diagnosis and is reported to yield accuracy of over 90% in diagnosing and providing treatment sugges-

26. N. Jean, M. Burke, M. Xie, W. M. Davis, D. B. Lobell, and S. Ermon, "Combining satellite imagery and machine learning to predict poverty," *Science* 353, no. 6301 (2016): 790–794.

27. F. Jiang, Y. Jiang, H. Zhi, Y. Dong, H. Li, S. Ma, Y. Wang, Q. Dong, H. Shen, and Y. Wang, "Artificial intelligence in healthcare: Past, present and future," *Stroke and Vascular Neurology*, 2017, https://svn.bmj.com/content/2/4/230.full.

tions.[28] Lockheed Martin's self-learning model recognizes ships, airplanes, buildings, seaports, and many other commercial categories.[29] Likewise, in a study that combined satellite imagery and deep learning to identify rural and remote communities, the researchers reported a positive predictive value of 86.47%.[30]

Blockchain can further increase the value of satellite data. For instance, blockchain-based smart contracts make it possible for owners of satellites and those that need their services to autonomously negotiate. Transactions can be completed based on predetermined criteria, such as the price a customer pays to access and download an image. It is also possible for users and satellite owners to dynamically create new services, with the revenue used to pay for launching them, insurance premiums, and other costs.[31]

Some companies are also developing solutions involving satellite images, blockchain, and AI that reward sustainable farming practices (In Focus 4.1). One such example is Oracle's partnership with the World Bee Project to help farmers manage bee populations and pollinator habitats. The plan is to take farm images with drones or satellites and utilize AI-based image recognition to evaluate whether the way a farmland is managed sustainably supports bee colonies and other pollinators.[32] Research has indicated that farms that allocate a certain proportion of their land to plant flowering crops can increase crop yields by up to 79% due to pollination by bees. An eco-label certificate can be issued to farmers, too. The certification can be stored in a blockchain so that all supply chain partners can see its journey to the retailer.

---

28. Jiang et al., "Artificial intelligence."

29. Ishveena Singh, "Lockheed Martin develops AI model for satellite imagery analysis," *Geo Awesome*, June 18, 2019, https://geoawesomeness.com/lockheed-martin-artificial-intelligence-model-satellite-imagery-analysis/.

30. Emilie Bruzelius, Matthew Le, Avi Kenny, Jordan Downey, Matteo Danieletto, Aaron Baum, Patrick Doupe, Bruno Silva, Philip J. Landrigan, and Prabhjot Singh, "Satellite images and machine learning can identify remote communities to facilitate access to health services," *Journal of the American Medical Informatics Association* 26, nos. 8–9 (2019): 806–812.

31. World Economic Forum, "The future may owe itself to blockchain technology: Here's why," *Futurism*, April 4, 2017, https://futurism.com/the-future-may-owe-itself-to-blockchain-technology-heres-why/.

32. Justin Charness, "How Oracle and the World Bee Project are using AI to save bees," *Oracle AI and Data Science Blog*, January 14, 2019, https://blogs.oracle.com/datascience/how-oracle-and-the-world-bee-project-are-using-ai-to-save-bees-v2.

In Focus 4.1: Bext360's Tracking of Coffee Supply Chains through Satellite Images, Blockchain, and AI

The Denver, Colorado-based startup Bext360 has kiosks in Uganda that evaluate coffee beans using "Bextmachines." A Bextmachine is a Coinstar-like device that employs smart image recognition, machine vision, AI, IoT, and blockchain to grade and track coffee beans. It takes a three-dimensional scan of each bean's outer fruit.[33] Bextmachines analyze farmers' coffee cherries and coffee parchment deposited at collection stations and sort them to assess quality. Farmers that supply bigger and riper cherries are paid more. Bext360's systems store data related to transaction time, date, and location, as well as amount of payment, indicators related to sustainable sourcing, and satellite imagery to show if producers are polluting water.[34]

The Bextmachines link the output to cryptotokens that represent the coffee's value. New tokens are automatically created when the product passes through the supply chain. The values of tokens increase at each successive stage of the supply chain.[35]

Using a mobile app, parties can negotiate a fair price.[36] Farmers get paid immediately via the app, which also verifies the identity of the person selling the products. Using Bext360's API, intermediaries such as wholesalers and retailers embed the technology into their websites, marketing, and point of sale. This level of transparency may not be possible without blockchain. In April 2018, the world's first blockchain-traced coffee tracked by Bext360's solution was sold.

The company started its pilot program in November 2017. In the same month, it teamed up with the Dutch startup Moyee Coffee and the FairChain Foundation to produce the blockchain-traced coffee Token. By June 2018, 60,000 kilograms of coffee from Ethiopia were exported to Amsterdam. The system provides proof

33. Z. Cadwalader, "Trace your coffee using blockchain," *Sprudge*, 2018, https://sprudge.com/132380-132380.html.

34. C. Zhong, "Innovator BanQu builds blockchain and bridges for traceability, small farmers' livelihoods," *Greenbiz*, 2019, https://www.greenbiz.com/article/innovator-banqu-builds-blockchain-and-bridges-traceability-small-farmers-livelihoods.

35. Moyee Coffee, *World's first blockchain coffee project*, 2018, https://moyeecoffee.ie/blogs/moyee/world-s-first-blockchain-coffee-project.

36. Bitcoin Magazine, "Innovation percolates when coffee meets the blockchain," *Nasdaq*, April 17, 2017, http://www.nasdaq.com/article/innovation-percolates-when-coffee-meets-the-blockchain-cm774790.

that the farmer was paid a living wage.[37] Moyee's plan is to launch Token in coffee-producing countries such as Kenya, Colombia, and Rwanda.[38]

The program also includes indicators related to sustainable sourcing and satellite images to show if producers are polluting water.[39] The combination of different technologies can help reveal the truth about sustainability practices.

Disasters Like the COVID-19 Pandemic

In Table 4.1, I summarize the usefulness of satellite imagery to remote communities to predict, evaluate, respond to, and monitor the effects and recovery from infectious diseases and pandemics. As the table makes clear, the value of satellite data was especially evident during the COVID-19 pandemic. As noted, this technology has attracted the attention of governments in some developing countries and NGOs. Data are being utilized to address health, social, humanitarian, economic, and environmental problems facing the world during the pandemic.

Predicting and Preventing Pandemics and Infectious Diseases

Infectious diseases are estimated to kill over 15 million people worldwide every year.[40] A significant proportion of those deaths can be prevented with the appropriate use of remote-sensing technologies. One example is onchocerciasis, or "river blindness," a tropical disease that infects over 20 million people each year and blinds about 1 million. Black flies that breed in fast-flowing rivers and streams in remote communities transmit it. Real-time satellite imagery of river flows can help assess the risk of parasitic transmission and send health workers to high-risk areas. The Earth imaging company

37. Bext360, "Bext360 and Coda Coffee release the world's first blockchain-traced coffee from bean to cup," *GlobeNewswire*, April 16, 2018, https://globenewswire.com/news-release/2018/04/16/1472230/0/en/bext360-and-Coda-Coffee-Release-The-World-s-First-Blockchain-traced-Coffee-from-Bean-to-Cup.html.

38. Howard Bryman, "Fully blockchained coffee brand token invites consumers into the matrix," *Daily Coffee News*, July 11, 2018, https://dailycoffeenews.com/2018/07/11/fully-blockchained-coffee-brand-token-invites-consumers-into-the-matrix/.

39. Zhong, "Innovator."

40. Maxar, *Infectious disease response planning with geospatial data*, April 9, 2020, https://explore.maxar.com/Infectious-Disease-Response-Planning-with-Geospatial-Data-Webinar-APAC.html.

Table 4.1. Usefulness to remote communities of satellite imagery to predict, evaluate, respond to, and monitor the effects and recovery from pandemics

Use	Explanation	Example
Predicting and preventing pandemics and infectious diseases before they start	Identifying areas that have high probabilities of humans coming in contact with wild animals Looking at "proxy indicators" to model and monitor the growth of viruses and pathogens that lead to pandemics	Indicators associated with phytoplankton blooms used as a proxy measure for the expansion of V. cholerae-associated copepods in the Bay of Bengal.
Stopping the spread of pandemics	Satellite technologies can be used to people's compliance with quarantine, track patients' position and guarantee large-scale disinfection missions	China: BeiDou used to dispatch over 10,000 disinfection drones to perform aerial disinfectant spraying South Korea: "self-quarantine safety protection" app used satellite navigation
Sending emergency relief assistance such as cash and essential items	Using deep learning algorithms to identify an area with extreme poverty and houses in the area and means to reach there.	Using satellite imagery HOT found buildings and roads in rural areas in Peru that cannot be seen on standard maps.
Implementing early intervention such as treatment	Finding locations of hospitals and pharmacies	During COVID-19, HOT volunteers' tasks included finding hospitals in countries such as Iran, the Philippines and Turkey and business and practice details.
Monitoring the effects and recovery from pandemics	Comparing satellite images of economic and other activities before, during and after a pandemic.	WeBank used neural network to analyze images from Sentinel-2 and other satellites to observe steel manufacturing activities inside a plant.
Ensuring that basic human values are respected during a pandemic	Satellite imageries can be used to monitor if nefarious actors are taking advantage of public health emergencies to engage in human rights abuses and environmental destruction	Amnesty International: investigated accusation against the Ethiopian government regarding the demolition ed homes of casual laborers Satellite imagery found visual evidence of significant reduction in rainforest in Brazil's *Rondônia state*

Planet's advanced analytical feeds automatically identify roads and buildings in its satellite imagery. These tools play a key role in identifying populations that face risk of exposure to pandemic such as COVID-19. Satellite imagery can also locate and track the construction of testing and health facilities, mobilization centers, new hospitals, and other public health infrastructure.[41]

Satellite data can play a key role in stopping pandemics before they start. Most viruses causing pandemics spread from animals to humans either directly or indirectly (e.g., through livestock). Satellite imagery can help identify areas that have high probabilities of humans coming in contact with wild animals. For instance, it has been found that deforestation in Uganda has led to an increase in the emergence of animal-to-human diseases.[42] Deforestation destroys the natural habitat of wild animals and increase the chances that they come in contact with humans and transmit infectious diseases.

Viruses and pathogens that transmit diseases are not visible to the satellites. However, scientists have identified various proxy indicators that satellites can see that play a key role in modeling, and monitoring the growth of viruses and pathogens that lead to pandemics.[43] To take an example, cholera is transmitted by *V. cholerae*, a bacterium that is attached to small organisms known as copepods that drift in salt water and fresh water.[44] Epidemic risk from cholera increases if people drink untreated water contaminated with *V. cholerae*-associated copepods. Copepods feed upon microscopic marine algae known as phytoplankton, which grow and reproduce more rapidly during seasons with high sea surface temperature (SST).[45] Researchers used data from the NASA Sea-Viewing Wide Field of View Sensor (SeaWiFS) and temperature data from the Advanced Very-High-Resolution Radiometer (AVHRR) of the National Oceanic and Atmospheric Administration.

---

41. Andrew Zolli, "How satellite data can help with COVID-19 and beyond," *Planet*, April 14, 2020, https://www.planet.com/pulse/how-satellite-data-can-help-with-covid-19-and-beyond/.

42. Marie Quinney, "COVID-19 and nature are linked: So should be the recovery," *World Economic Forum*, April 14, 2020, https://www.weforum.org/agenda/2020/04/covid-19-nature-deforestation-recovery/.

43. Zolli, "How satellite data."

44. James M. Wilson, "Use of satellite imagery for epidemic surveillance and response," *19th Conference on IIPS, Session 11, Satellite IIPS and Applications*, 2003, https://ams.confex.com/ams/annual2003/techprogram/session_14254.htm.

45. B. Lobitz, I. Beck, A. Huq, B. Wood, G. Fuchs, A. S. Faruque, and R. Colwell, "Climate and infectious disease: Use of remote sensing for detection of Vibrio cholerae by indirect measurement," *Proceedings of the National Academy of Sciences of the United States of America* 97, no. 4 (2000): 1438–1443.

The authors demonstrated that phytoplankton blooms can be used as a proxy measure for the expansion of *V. cholerae*-associated copepods in the Bay of Bengal.

Stopping the Spread of Pandemics

In China, satellite technologies and space-based systems have been used to track patients' position and monitor cargo transports. Drones guided by Global Navigation Satellite System are also used for large-scale disinfection missions. The Chinese satellite navigation system BeiDou especially played a critical role. BeiDou's navigation and positioning capabilities were used to dispatch over 10,000 disinfection drones to perform aerial spraying nationwide with accuracy at the centimeter level. Before the pandemic, the drones were used for spraying in farms.

To mitigate risks for commuters, the HaiGe Smart Epidemic Prevention Management Platform was developed, based on BeiDou. The system tracked the location history of patients and connected the information to a regional database. The platform displayed locations of infection in real time, which helped users avoid areas that were potentially contagious.

In South Korea, the "self-quarantine safety protection" app was developed by the Ministry of the Interior and Safety using satellite navigation to make sure people comply with quarantine.[46]

Law enforcement agencies and drivers of medical supplies and daily necessities relied on GNSS-based vehicle network guides to adapt routes in response to transport restrictions. China's transport management department required all vehicles used in Wuhan with BeiDou mobile satellite positioning devices. Since satellite connections also work in remote locations. real-time communication with isolated drivers was possible.[47]

Sending Emergency Relief Assistance

Remote-sensing tools have helped emergency responders to make decisions and take actions quickly. The Argentine startup Dymaxion Labs uses advanced ML, AI-powered geospatial analytics, and computer vision to do surveying faster and at a lower cost.[48] Cameras installed in satellites detect

---

46. "What we can learn from the Coronavirus crisis with satellite data," *Eurisy*, March 25, 2020, https://www.eurisy.org/article-what-we-can-learn-from-the-corona-crisis-with-satellite-data_46.

47. "What we."

48. Mapbox, "Detecting informal settlements in South America: How I built it,"

and monitor the physical characteristics of cities and settlements, so human surveyors do not need to visit the sites. Moreover, machines observe features that human eyes cannot and process information faster. Dymaxion's AP LatAm tool uses satellite imagery algorithms and ML techniques to monitor informal settlements in Latin America. It provides information about communities' locations and changes in movements on a real-time basis, including informal settlements in Argentina, Guatemala City, and Tegucigalpa in Honduras, and Asunción in Paraguay.[49] The goal is to improve decision-making and provide rapid disaster response.[50]

Satellite imagery data can be used to develop reliable methods to send urgent emergency relief assistance such as cash and essential items like medicines and foodstuffs. To send emergency cash assistance to vulnerable groups during COVID-19, the government of Togo explored how satellite imagery, big data, and machine learning could help find people who desperately needed relief payments. Deep-learning algorithms can be trained by using different datasets to look at commercial aerial photography and high-resolution satellite imagery covering the entire country to identify an area with extreme poverty.[51]

About half of Togo's 8 million people live on less than US$1.90 a day. Most are employed in the informal sector, and COVID-19 restrictions cut off their livelihood. An aid system called Novissi ("solidarity" in the Ewe language) was launched in March 2020. Men received 10,500 CFA francs per month (about US$20) and women received 12,250 CFA francs (about US$20). As of December 2020, about 600,000 people had received $22 million through Novissi. To create a fine-grained map of Togo from satellite images, ML algorithms were trained to analyze images. A 2018 household survey, which had reached part of the country, was used to calibrate the algorithms, which utilized indicators related to wealth and poverty, such as roofing materials and road surfaces. Phone call data provided by Togo's two primary cell networks were used to estimate the wealth of subscribers, with variables such as calling patterns and credit top-ups. Data from a phone survey conducted in September 2020 with about 10,000 people in the poorest regions, flagged by an analysis of satellite images, was also used to train ML

Medium, February 28, 2018, https://blog.mapbox.com/detecting-informal-settlements-in-south-america-how-i-built-it-cb139a870816.

49. Mapbox, "Detecting informal."

50. Benjamin Kumpf, "Who is writing the future? Designing infrastructure for ethical AI," *Medium*, May 31, 2018, https://medium.com/@UNDP/who-is-writing-the-future-designing-infrastructure-for-ethical-ai-4999620db295.

51. Blumenstock, "Machine learning."

algorithms. The Relative Wealth Index discussed in chapter 2 was also used to identify low-income rural populations.[52] In addition, a team gathered information on communities in need.[53]

In Peru, the Humanitarian OpenStreetMap Team (HOT) worked with the government to find such groups in the Cusco region. HOT mobilizes volunteers at the request of a government or an NGO to map places for which geospatial information is unavailable. The volunteers are organized to respond to disasters such as floods, earthquakes, hurricanes, and disease outbreaks. HOT's crowdsourced mapping efforts add to the OpenStreetMap, a free and editable representation of the world.[54]

HOT volunteers can work from anywhere in the world to improve the maps using details from satellite images. For instance, they draw the buildings and roads on top of existing maps using a simple online tool. If it is feasible, team members can go into the field to collect map data. As of June 7, 2020, 1,486 HOT volunteers had contributed to COVID-19-related mapping in Peru. These community members made 267,277 edits, tagged 236,995 buildings, and traced 3,962 kilometers of road). The Peruvian projects used imagery donated from Maxar, which helps humanitarian efforts like HOT. As more details are added to the maps, it is easier to know where and how people live and how to reach them.[55]

Implementing Interventions

People affected by and responding to disasters need to know the locations of hospitals, pharmacies, and stores. It is also important to know where there are transport infrastructures such as roads and railways to make food and medicine available to the needy population. They must be able to count the houses so that aid workers can estimate the population and know how many

52. Laura McGorman, Guanghua Chi, and Han Fang, "How AI-powered poverty maps can increase equity in the COVID-19 response," *Brookings*, May 7, 2021, https://www.brookings.edu/blog/future-development/2021/05/07/how-ai-powered-poverty-maps-can-increase-equity-in-the-covid-19-response/.

53. Tom Simonite, "A clever strategy to distribute covid aid—with satellite data," *Wired*, December 17, 2020, https://www.wired.com/story/clever-strategy-distribute-covid-aid-satellite-data/.

54. Sarah Scoles, "Satellite data reveals the pandemic's effects from above," *Wired*, April 9, 2020, https://www.wired.com/story/satellite-data-reveals-the-pandemics-effects-from-above/.

55. Scoles, "Satellite"; "Our work in Peru," https://www.hotosm.org/where-we-work/peru/.

vaccines to bring.[56] During COVID-19, HOT volunteers' tasks included finding hospitals in countries such as Iran, the Philippines, and Turkey; they included relevant information about business and practice details such as addresses, opening hours, services provided, outlines of hospital building, and helipads.[57]

As mentioned, the level of details of geographic information available on standard maps constructed from conventional survey data of a place depend on a number of factors, including the demand for such information.[58] With a lack of demand, resources, skills, and technologies, maps of remote areas often are lacking key details. High-resolution maps that are produced by ML using satellite imagery provide more detailed information.[59] For instance, HOT found that buildings and roads are not visible on maps of rural Peru, but they are clear in satellite imagery.

Monitoring Recovery from the Pandemic

One might expect that once a pandemic is under control, economic activities start recovering. However, different economic activities may recover at different paces, which can be captured by remote-sensing tools. By analyzing satellite images of activities such as animal movements and agricultural land use, it is possible to know whether economic activities are recovering. The resulting insights can be used to better target policy measures to address the most serious problems.

The Chinese online banking company WeBank used neural networks to analyze images from various satellites, including the Sentinel-2 satellite. Using high resolution images, it was possible for the system to observe steel manufacturing inside a plant. Blast furnaces and electric furnaces used in steel mills need to be operated at very high temperatures, which increase the temperatures of their surroundings. In images taken by certain satellite, these furnaces appear as clusters of bright red dots. These allow researchers to monitor operations and outputs of the mills.

A deep-learning framework known as SolarNet, originally designed to detect solar farms using large-scale satellite imagery data, was used for this purpose.[60] The analysis indicated that in the early days of the COVID-19

56. Scoles, "Satellite."
57. HOT, 2020, Volunteer mappers.
58. Nagaraj and Stern, "The economics."
59. Blumenstock, "Machine learning."
60. Xin Hou, Biao Wang, Wanqi Hu, Lei Yin, and Haishan Wu, "SolarNet: A deep

outbreak, steel manufacturing capacity had dropped by 29%. By February 9, it recovered to 76%.[61] To build a complete picture of the state of manufacturing and commercial activity, the system combined satellite imagery with GPS data from mobile phones and also social media posts. Using the data gathered, the system predicted that most Chinese workers outside Wuhan would return to work by the end of March and that China's economic growth in the first quarter of 2020 would fall by 36%.[62]

WeBank researchers also looked at other types of manufacturing and commercial activities. By counting cars in large corporate parking lots in Shanghai, it found that Tesla's Shanghai car production had fully recovered by February 10, but Shanghai Disneyland and other tourist attractions were still shut down then.[63] WeBank analyzed visible, infrared, near-infrared, and short-wave infrared images from Sentinel-2 and other satellites. These are mainly in the resolution range of 10 to 30 meters, considered medium resolution.

Respecting Human Rights during a Pandemic

Satellite imagery can help ensure that basic human rights and environmental protections are being respected, especially during the pandemic. Human rights NGOs such as Amnesty International have used such data to investigate atrocities by state forces. An accusation was made against the Ethiopian government, for example, that it demolished the homes of many informal laborers near the Addis Ababa Bole International Airport. The victims worked in construction sites that were closed due to the COVID-19 shutdowns. To investigate the accusation, Amnesty International compared satellite images of the area before the pandemic with those from the end of April 2020. The analysis indicated that 40 recently built structures had been

---

learning framework to map solar power plants in China from satellite imagery," *arxiv*, December 10, 2019, https://arxiv.org/abs/1912.03685.

61. Tekla S. Perry, "Satellites and AI monitor Chinese economy's reaction to coronavirus," *IEEE Spectrum*, March 10, 2020, https://spectrum.ieee.org/view-from-the-valley/artificial-intelligence/machine-learning/satellites-and-ai-monitor-chinese-economys-reaction-to-coronavirus.

62. Ashley Johnson, *How artificial intelligence is aiding the fight against coronavirus*, Center for Data Innovation, March 13, 2020, https://www.datainnovation.org/2020/03/how-artificial-intelligence-is-aiding-the-fight-against-coronavirus/.

63. Perry, "Satellites and"; EOS Data Analytics, "Satellite data: What spatial resolution is enough?," April 12, 2019, https://eos.com/blog/satellite-data-what-spatial-resolution-is-enough-for-you.

destroyed in the last three weeks of April 2020. Satellite images before the pandemic clearly showed permanent homes; images from after consisted of tents and other temporary structures.[64]

Satellite imagery can also be used to track and respond to criminal and illegal activities. For instance, illegal logging in the Brazilian Amazon became even more widespread during the COVID-19 lockdown. According to Brazil's National Institute for Space Research, deforestation in the area increased by 55% in the first four months of 2020 over the same period in 2019. Destruction in April 2020 was 64% higher than in April 2019.[65] It has been a challenge to police the powerful bad actors, and it is reported that some tribal villagers are also involved in the illegal activity. Government agencies have shown little interest or capacity in addressing the issue, and there is a fear that criminals may engage in retaliation.[66]

The extent of the problem was assessed through satellite imagery. The Verification Unit of NBC News found visual evidence of significantly reduced rain forest in the Northwest state of Rondônia after the pandemic started in 2020.[67] A comparison of images from January 21 and April 25 of that year showed felled rain forest about the size of the state's capital city, Porto Velho, equivalent to 448 football fields.[68]

The space technology company Maxar has a two-decade-old library of images that covers about every square mile of Earth. Its analysts can compare today's images with ones from any time in the last 20 years to see what has changed. The information can be used by governments and other interested parties to enforce quarantines or decide the locations to build, for example, new hospitals.[69]

---

64. Amnesty International, *Ethiopia: Forced evictions in Addis Ababa render jobless workers homeless amid COVID-19*, April 29, 2020, https://www.amnesty.org/en/latest/news/2020/04/ethiopia-forced-evictions-in-addis-ababa-render-jobless-workers-homeless-amid-covid19/.

65. Jamie Roberton and Lorand Bodo, "Deforestation of the Amazon has soared under cover of the coronavirus," *NBC News*, May 11, 2020, https://www.nbcnews.com/science/environment/deforestation-amazon-has-soared-under-cover-coronavirus-n1204451.

66. Evan Simon and Aicha El Hammar Castano, "Deforestation of Amazon rainforest accelerates amid COVID-19 pandemic," *ABC News*, May 6, 2020, https://abcnews.go.com/International/deforestation-amazon-rainforest-accelerates-amid-covid-19-pandemic/story?id=70526188.

67. Roberton and Bodo, "Deforestation of the."

68. Roberton and Bodo, "Deforestation of the."

69. Michael Hardy, "Satellites and AI monitor Chinese economy's reaction to coronavirus you can see the coronavirus from space," *Wired*, March 10, 2020, https://www.wired.com/story/coronavirus-satellite-imagery/.

Space Agencies and the Importance of Earth Observation Data

Space agencies have called for ideas and project proposals to use the data they collect to respond to COVID-19.[70] The ESA teamed up with the European Commission to launch a special edition of the Sentinel Hub Custom Script Contest, which is a remotely run hackathon that involves data and remote sensing scientists, students, and the general public. The goal is to produce new algorithms to process Earth Observation data, managed by Euro Data Cube group. The contest had a simple goal: find new algorithms and ideas for using satellite data to monitor and mitigate the situation during and after the crisis. The group invited remote-sensing experts, ML scientists, and the general public(. The contest invites participants to analyze patterns and changes in three categories: industrial activities, such as factories, mining and oil extraction, supermarkets, transport networks, and commercial ports; distribution of human activities, such as cars parked in urban areas and social distancing; and agriculture activities, such as unattended fields and crops and supply chain disruptions.[71]

NASA, the ESA, the Japan Aerospace Exploration Agency (JAXA), the Canadian Space Agency (CSA), and the National Centre for Space Studies (CNES) also organized a two-day virtual hackathon, the Space Apps Challenge, focused on the COVID-19 crisis in May 2020. In the competition, participants worked in virtual teams and use Earth observation data to develop solutions to respond to COVID-19. Thematic areas included in the challenges were local response to the pandemic, changes, and solutions; impacts of the pandemic on the Earth system (i.e., physical, chemical, and biological processes) and its responses; and economic opportunity, impact, and recovery during and after COVID-19.[72]

---

70. James Dacey, "Space agencies redirect resources to support COVID-19 studies," *Physics World*, May 8, 2020, https://physicsworld.com/a/space-agencies-redirect-resources-to-support-covid-19-studies/.

71. European Space Agency, *Euro Data Cube Custom Script Contest—COVID-19 edition*, April 6, 2020, https://eo4society.esa.int/2020/04/06/euro-data-cube-custom-script-contest-covid-19-edition/; "Sentinel hub contest," https://www.sentinel-hub.com/contest/.

72. *Space apps COVID-19 challenge*, Space Apps, May 30, 2020, https://covid19.spaceappschallenge.org/.

Usefulness during Noncrisis Times

While satellite technologies are highly useful especially in times of crisis, they can also promote the social and economic well-being of marginalized populations in normal times. In this section, several important uses of remote-sensing technology during noncrisis times are considered. Some examples of future uses are also given.

Farming and Animal Husbandry

Remote-sensing tools have helped improve farming and animal husbandry. For instance, in northeastern China's Heilongjiang Province, a modern big-data center has transformed the Qixing farm, the largest paddy farm in the province, covering 81,300 hectares. The center uses data collected from high-resolution Gaofen-1 satellites, meteorological machines, underground water-level monitoring equipment, and other sources.[73] Farming decisions are made based on real-time data about air temperature, humidity, wind direction, soil temperature, and humidity. Farmers need to install a mobile app, Modern Agriculture Platform (MAP), developed by the state-owned enterprise Sinochem Group, to access the real-time data.[74]

As a second example, take Indonesia's CI-Agriculture, whose precision farming technique uses big data, the IoT, and other technologies. Data on soil conditions from aerial photographs, sensors, and drones, and weather data obtained from satellite imagery are analyzed. Data are used to calculate a field's production potential with higher accuracy and to make more efficient use of fertilizer and pesticides. The technology is scalable, so it is possible to use sensors for a large area. The solution is expected to reduce loan costs for smallholder farmers. CI-Agriculture provides insurance to Indonesian farmers based on its calculations and weather data for up to ten years.[75]

As discussed in chapter 2, AI tools such as PlantVillage are available to help farmers predict near-term crop productivity. The UN's portal Water Productivity through Open Access of Remotely Sensed Derived Data is the main data source for PlantVillage Nuru,[76] and its database uses NASA's satel-

73. "Big data reshaping harvest for Chinese farmers," *Xinhua Net*, 2018, http://www.xinhuanet.com/english/2018-11/29/c_137640065_2.htm.

74. Mi Xue, "Feeding 1.4 billion: Smart farming in China's big grain warehouse," *CGTN*, December 13, 2019, https://news.cgtn.com/news/2019-12-13/Feeding-1-4-Billion-Smart-farming-in-China-s-big-grain-warehouse-MohBFcaajK/index.html.

75. "About CI Agriculture," *E27*, 2020, https://e27.co/startups/ci-agriculture/.

76. FAO water productivity open-access portal (WaPOR), *Relief web*, June 1, 2018, https://reliefweb.int/report/world/fao-water-productivity-open-access-portal-wapor.

lite data to compute relevant metrics for crop productivity. By integrating diverse data, the AI assistant can provide information about crops' drought tolerance and suitability of crops in different areas.[77]

These technologies have also helped farmers take care of their domesticated animals. For instance, about 40,000 shepherds in Burkina Faso have benefited from the Geodata for Agriculture and Water (G4AW) Mobile Data for Moving Herd Management (MODHEM) project, supplied by the Dutch company Satelligence. In Mali, the system is called the G4AW project Sustainable Technology Adaptation for Mali's Pastoralists (STAMP), in use by 60,000 shepherds. Another Dutch company, Hoefsloot Spatial Solutions (HSS), provides the satellite technology. The information is supplied by the European satellite program Copernicus. The shepherds use cellphones to connect with a call center. Various sensors are deployed to measure indicators related to grass and water resources. For surfaces of 10 meters by 10 meters, the resolution provided by HSS's imaging systems provide daily information about the quality of plant biomass, such as grass and availability of surface water.[78]

Counting People for Community Health Activities

Community health workers (CHWs) mostly provide home-based, rather than facility-based, healthcare. A key challenge that CHWs face in remote areas is their limited ability to efficiently and accurately count and find the population.[79] This is because in remote areas, census data are often missing or outdated. It is highly challenging to identify and map communities, as this requires a lot of time and resources.

Manual techniques often lead to underidentification. That is, such methods fail to identify some people in the communities. The gap in data availability can be narrowed through the use of publicly available satellite data.[80] Satellite data can thus be extremely useful in providing a range of services to rural communities. Computer analysis of satellite photos can help vac-

77. "AI-based tool can predict climate change stress for farmers," *The Quint*, September 23, 2019, https://www.thequint.com/news/hot-news/ai-based-tool-can-predict-climate-change-stress-for-farmers.

78. Remco Takken, "Satellite services help African shepherds through Sahel," *Geo Spatial World*, February 14, 2020, https://www.geospatialworld.net/blogs/satellite-services-help-african-shepherds-through-sahel/.

79. Bruzelius et al., "Satellite images."

80. Luis Gonzalez Morales, Yu-Chieh Hsu, Jennifer Poole, Benjamin Rae, and Ian Rutherford, *A world that counts: Mobilising the data revolution for sustainable development* (New York: UN Data Revolution, 2014), https://www.undatarevolution.org/wp-content/uploads/2014/11/A-World-That-Counts.pdf.

cination workers and mosquito insecticide sprayers increase coverage. Such techniques allow them to reach remote communities and people living in rapidly growing informal dwellings not found on traditional maps.[81]

Fighting Injustice and Human Rights Abuses

Satellite data can also be used to fight human rights abuses such as slavery (In Focus 4.2). Data from satellites and improvements in AI algorithms can boost initiatives to end modern slavery. The International Labour Organization estimated that the slavery industry has victimized about 40.3 million people. One estimate suggested that one-third of all activities associated with slavery are visible from space, including brick kilns, illegal mines, and fish-processing camps.[82]

In Focus 4.2: Satellite Technology's Potential to Fight Forced Labor in the Marine Fishing Industry

In February 2021, the food-focused news outlet *The Counter* published a story of the 37-year-old Indonesian fisherman Darmaji, who was on a Taiwanese tuna-fishing vessel for two years. Darmaji described his experience as "a prison at sea." His employer paid much less than what Darmaji was promised in his contract. Darmaji was also forced to pay a US$1,200 security deposit before he received his monthly salary.

The crew of 22, which largely consisted of Indonesians, worked 18 hours a day even when there were seven-meter waves that flooded the interior of the boat. They were allowed to sleep for only three hours a day. Darmaji and other workers were verbally abused daily. They mostly ate sticky rice and boiled chicken or fish, and sometime bait. The crew mostly drank distilled salt water. They needed to pay for any other food items they consumed.[83] Thousands of workers in such boats also suffer physical abuse. One worker who told his story to the NGO Environmental Justice Foundation

81. Andrew Jack, "AI set to transform healthcare in world's poorer regions," *Financial Times*, May 14, 2020, https://www.ft.com/content/cdc166d4-6845-11ea-a6ac-9122541af204.

82. Sarah Scoles, "Researchers spy signs of slavery from space," *Science*, February 19, 2019, https://www.sciencemag.org/news/2019/02/researchers-spy-signs-slavery-space.

83. Virginia Gewin, "How new technology is helping to identify human rights abuses in the seafood industry," The Counter, February 25, 2021, https://thecounter.org/new-technology-helping-identify-human-rights-abuses-seafood-industry-forced-labor/.

described being locked in a freezer and then electrocuted with a tool for killing tuna.[84]

Such stories are more common than most people imagine. Recent advances in satellite technologies and related areas provide some hope for victims of forced labor in the fishing industry. A model tested on the satellite vessel-monitoring data of the non-profit public technology platform Global Fishing Watch correctly predicted forced labor in over 90% of reported cases of high-risk activity. An analysis of 16,000 industrial fishing vessels during 2012–2018 by a research team from GFW, emLab at University of California (Santa Barbara), and the charitable organization Liberty Shared found that up to 26% of analyzed vessels showed risk indicators for forced labor, victimizing about 100,000 individuals. The team compiled behavioral data for 27 vessels and characteristics associated with forced labor that could be observed with Global Fishing Watch's monitoring. Among key indicators of high-risk vessels were greater distance from ports, higher engine power, longer fishing hours per day, longer time fishing on the high seas, and fewer fishing trips per year. In 2018, high-risk vessels visited ports in 79 countries, mainly in Africa, Asia, and South America, including 39 countries that are signatories to the Food and Agriculture Organization's Agreement on Port State Measures, which aims to control illegal, unreported, and unregulated fishing by strengthening state control of ports.[85]

Satellite technologies, however, can only help identify unjust acts. The fate of victims such as Darmaji ultimately depends on the strength of national regulative and enforcement measures to punish the perpetrators of injustice.

Chapter Summary and Conclusion

Pandemics can trigger severe distress among vulnerable individuals, as COVID-19 made clear. One of the lessons the COVID-19 crisis has made very clear is that satellite imagery must be used to address future challenges,

84. Gewin, "How new."

85. Freedom Collaborative, *How satellite data is being used to root out forced labour on fishing vessels around the world,* January 12, 2021, https://freedomcollaborative.org/newsletter-archive/how-satellite-data-is-being-used-to-root-out-forced-labour-on-fishing-vessels-around-the-world.

especially those facing vulnerable communities in remote areas. In public health emergencies, accurate information that is updated frequently is critical for the delivery of healthcare and emergency relief to isolated and remote communities. The ramifications of such distresses can be serious. With traditional tools and methods such as household surveys and censuses, it is difficult to accurately map and identify where rural communities live. Standard maps are not much help either. For mapmaking organizations and governments, remote areas do not have the same level of economic and strategic importance as major cities in developed countries, so they have less detailed information collected about them. Remote sensing is the most effective way to predict, prevent, detect, and respond to pandemics among remote populations in developing countries.

Satellite-based analyses are the most effective tools for rapidly and accurately identifying and enumerating the populations in remote areas and planning the delivery of public services. Once populated areas are identified, their density and activity can be mapped, the distance from an active pandemic site can be calculated, and impacts can be assessed based on the current activity. The information can also help in determining the best method of providing supports such as medical assistance and supplies.

While the real-world use of remote sensing in the developing world currently is significantly below its potential, the situation is likely to improve in the future. Remote-sensing tools have a number of unexploited future opportunities that are expected to further improve health, environment, security, and economic well-being of people living in poverty. More important for future uses are new approaches that combine remote-sensing technologies with AI and ML. New methods, such as deep learning, could play a critical role in reaching remote communities in a timely manner and providing assistance to people in a life-and-death situation.

In a globalized world, pandemics have negative spillover effects on countries other than where they begin. Rich countries should recognize such effects and help developing countries make better use of remote-sensing tools. Doing so is not only a moral imperative but also a prerequisite for protecting rich countries' own national security and economic interests.

CHAPTER 5

Internet of Things

Internet of things (IoT) systems involve collecting data and sending it to centralized platforms via sensors or gateways. The platforms aggregate, process, store, analyze, and visualize the data to create valuable insights, which can be used to improve operational efficiencies of firms and processes.

The IoT is a major 4R technology. In the UK market research company Euromonitor International's "Voice of the Industry: Digital Survey" conducted in 2020, participants ranked the IoT second among the technologies most impacting businesses, just behind cloud computing.[1] The survey covered 30 countries and 60 use cases that involved various technologies, market potential (e.g., scaling up, expanding into new services and markets), and innovativeness (e.g., potential to offer a solutions and be adopted by more players). For the B4B populations, these IoT features can provide access to necessary resources to engage in economic activity, facilitate economic exchange and consumption, and increase productivity.

The IoT's key features and insights gained from their use make possible taking action and making decisions to enhance productivity and quality of life in poor communities. For instance, IoT can help agronomists and farm owners monitor livestock and farm sites and measure key performance indicators. The IoT has helped consumers in Kenya and Tanzania use mobile payments to access cooking gas; a smart meter on a gas cylinder allows people to pay when they need gas. This level of accessibility and tailoring of products to local needs has made a big impact on the welfare of the poor.[2]

1. Euromonitor International, *Top 10 digital innovations of 2020*, March 2021, London.

2. International Telecommunication Union (ITU), *Digital trends in Africa*, 2021, https://www.itu.int/dms_pub/itu-d/opb/ind/D-IND-DIG_TRENDS_AFR.01-2021-PDF-E.pdf.

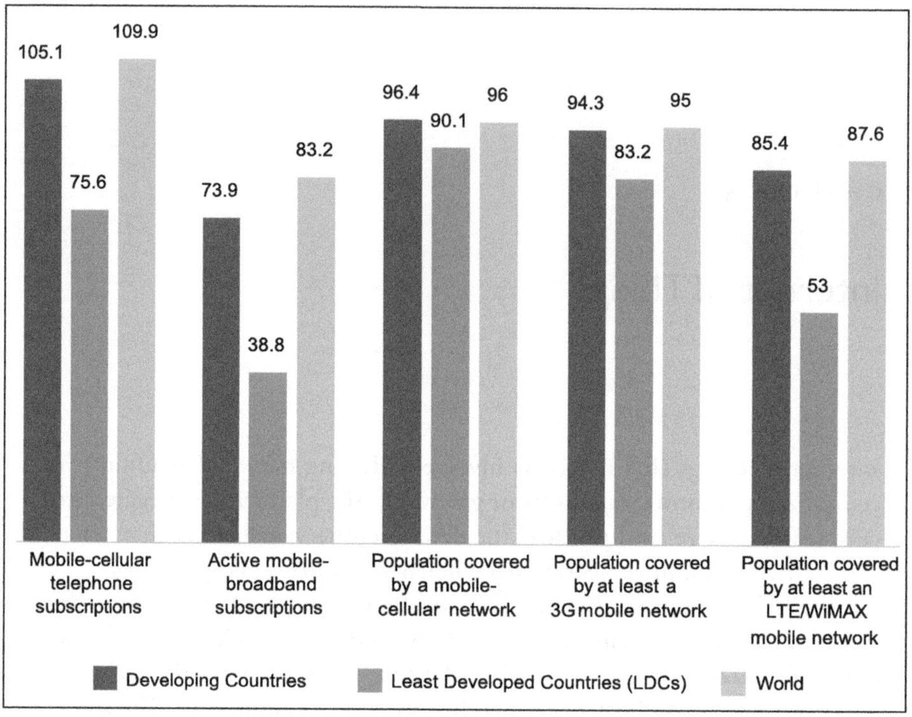

Figure 5.1. Percentage of the world's population subscribing to or covered by cellphones and cellular mobile networks (2021)

An encouraging trend is that prices of IoT sensors and hard drives are falling (see Figure 1.1), which has made it increasingly affordable to develop and deploy IoT solutions to address economic and social challenges facing the B4B population. Consequently, countries with high poverty rates are rapidly adopting IoT. The rapid adoption of IoT is also facilitated by the availability of cell phones and cellular mobile networks (Figure 5.1).

Major investments are also being made to develop IoT solutions targeted to the B4B. In 2019, IoT was the second most popular investment sector after social media for AI investment in the Middle East and Africa, and the IoT sector accounted for 28% of investment transactions.[3]

As noted in chapter 1, Asia-Pacific (particularly South Asia), Sub-Saharan Africa, and Latin America have high concentrations of poor people. The IoT industries in these economies are of substantial sizes (Table 5.1). In 2019,

3. Microsoft and E&Y, *Artificial intelligence in Middle East and Africa*, 2018, https://info.microsoft.com/rs/157-GQE-382/images/report-SRGCM1065.pdf.

Table 5.1. The IoT industry in regions with high concentrations of poor people

Region	Number of IoT connections (2025)	Remarks
Asia Pacific	11.5 billion	5.2 billion in 2019
Latin America	1.3 billion	510 million in 2018
SSA	300 million	Cellular IoT connections (2019, millions): 16.7 million

Data sources: Asia Pacific (GSM Association, The mobile economy Asia Pacific 2020, https://www.gsma.com/mobileeconomy/wp-content/uploads/2020/06/GSMA_MobileEconomy_2020_AsiaPacific.pdf; Latin America (Statista, New IoT connections in Latin America by sector 2025, January 20, 2022, https://www.statista.com/statistics/1190496/internet-things-connections-sector-latin-america/); SSA: IoT connections (GSMA Intelligence, The Internet of things by 2025, 2018, https://www.gsma.com/iot/wp-content/uploads/2018/08/GSMA-IoT-Infographic-2019.pdf). For cellular IoT (GSM Association, The mobile economy: Sub-Saharan Africa 2020, https://www.gsma.com/mobileeconomy/wp-content/uploads/2020/09/GSMA_MobileEconomy2020_SSA_Eng.pdf).

India was estimated to have approximately 250 million IoT-connected devices, and this was expected to reach over 2 billion by 2021.[4]

Development institutions such as the World Bank have also realized the importance of the IoT in addressing key socioeconomic challenges facing LMICs. In 2018, the World Bank Group announced a partnership with GSMA and mobile network operators worldwide to use big data from IoT devices to help end extreme poverty and boost economic growth. Anonymized data collected by mobile network operators through IoT devices and from smartphones is likely to provide new insights. Operators and governments are also expected to benefit from increased use of big data for development. Such data can help provide better services and create new indicators and statistics. A mobile-enabled IoT and big-data project in India has been cited as among the successful examples: bangle-shaped sensors are wearable IoT data-collection devices that allow users to monitor emissions from their stoves. The data they generate help understand the level of harmful emissions and drive a shift to cleaner cookstoves.[5]

This chapter highlights the economic, health, and environmental effects

4. Shangliao Sun, *Number of Internet of Things (IoT) connected devices in India in 2019, with forecast for 2021(in million)*, July 2, 2021, https://www.statista.com/statistics/1184091/india-number-of-iot-connected-devices/.

5. World Bank, *World Bank Group and GSMA announce partnership to leverage IoT big data for development*, February 26, 2018, https://www.worldbank.org/en/news/press-release/2018/02/26/world-bank-group-and-gsma-announce-partnership-to-leverage-iot-big-data-for-development.

of the IoT. A key focus is on the IoT's roles in measuring, enforcing, and designing contracts to create value. It gives special consideration to the IoT's impacts on entrepreneurial activities and the creation of new markets.

How IoT Works

There are three main components of an IoT service: the edge, the platform, and the user (Figure 5.2). The IoT edge device is the location where data originates or is aggregated. An IoT edge device consists of a sensor, which is a transducer or an electronic device that converts energy from one form to another.[6] A sensor measures the parameters of interest (e.g., air quality, soil moisture). The device consists of electronic components that support its functions. Data collection may also be reduced to the essential. In some cases, the devices may analyze data.

IoT edge devices are connected to a gateway using a communication protocol to a server. The wide availability of cellular networks makes frequent and near-real-time data transmission possible. The data then goes to the platform, which is typically in the cloud. Analytics are often performed in the cloud using algorithms. Data can be queried, cleaned, and analyzed. Relevant insights help inform decisions and take actions.[7] A real-time data stream helps users decide if some actions need to be taken right away or if the data needs to be stored for future use. The user then engages in a business action based on the data.

There are three possible ways the analyzed data can move from the IoT platform to a user to implement the IoT solutions: (1) the user deploys an application program interface (API) to call or query the data, which specifies how software components of the user and platform should interact; (2) if the IoT finds a predetermined set of events, it can announce or signal to the business user; (3) some combination of 1 and 2 (Laskowski, 2016).[8]

---

6. Eugene Y. Song and Kang Lee, "Understanding IEEE 1451-Networked smart transducer interface standard—What is a smart transducer?," *IEEE Instrumentation & Measurement Magazine* 11, no. 2 (2008): 11–17, https://ieeexplore.ieee.org/abstract/document/4483728?casa_token=566L0P8H0n8AAAAA:2wB3tQat2kgsOKIXbHkcAILWLK7-U58QFBVaI14frrNxaQJuNeA2sld9VfgWVI1v2HVxqi6WQb-zMA.

7. Anish P. Antony, Kendra Leith, Craig Jolley, Jennifer Lu, and Daniel J. Sweeney, "A review of practice and implementation of the internet of things (IoT) for smallholder agriculture," *Sustainability*, March 1, 2020, https://doi.org/10.3390/su12093750.

8. N. Laskowski, "Delving into an enterprise IoT initiative? Read this first" *TechTarget*, September 15, 2015, https://www.techtarget.com/iotagenda/feature/Delving-into-an-enterprise-IoT-initiative-Read-this-first.

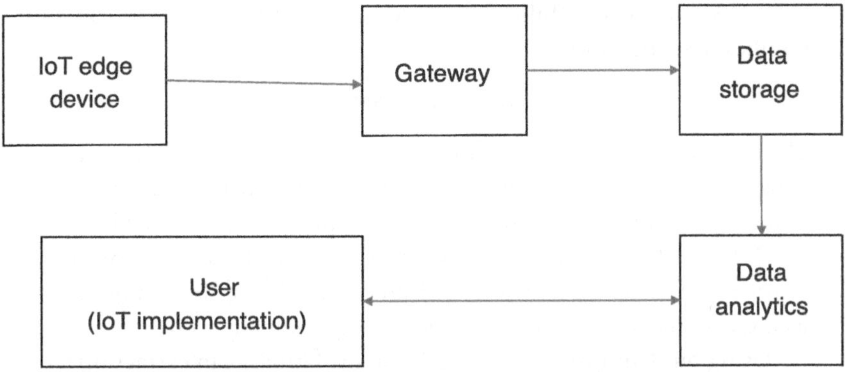

Figure 5.2. How IoT works

Economic, Health, and Environmental Effects

By introducing new technologies such as the IoT to the production process, productivity can be increased. The IoT can facilitate various mechanisms that may contribute to increased productivity. IoT solutions have provided vulnerable groups such as smallholder farmers access to necessary resources to engage in economic activity and increase productivity. For instance, smallholder productivity is limited by a variety of constraints, including lack of access to water for irrigation.[9] IoT-based irrigation and solar pumping solutions are helping improve the availability of water for farmers (In Focus 5.1).

In Focus 5.1: Togo's Government Teams Up with SunCulture and Bboxx to Offer IoT-Based Irrigation and Solar Pumping Solutions to Farmers

According to the International Finance Corporation (IFC), more than 43 million small farmers in SSA are not connected to the power grid[10] Many of these farmers live above near-surface aquifers but lack the means to tap the water. They are vulnerable to crop failures even though water might be just a few meters away.

In Africa, 80% of families depend on agriculture for their liveli-

9. S. Yalew, M. Mul, A. van Griensven, E. Teferi, J. Priess, C. Schweitzer, and P. van Der Zaag, "Land-use change modelling in the Upper Blue Nile Basin," *Environments* 3, no. 4 (2016): 21.

10. International Finance Corporation, *Off-grid solar market trends report 2020*, 2020, https://www.lightingglobal.org/wp-content/uploads/2020/05/VIVID_OCA_2020_Off_Grid_Solar_Market_Trends_Report_Full_High-compressed.pdf.

hood. Due to the instability of the electricity grid, only 4% of those families use irrigation.

SunCulture's Solutions

The solar irrigation company SunCulture, based in Nairobi, Kenya, aims to address this problem by providing smallholder farmers irrigation and solar pumping solutions.[11] The company was founded in 2013. SunCulture ranked No. 2 in *Fast Company*'s 10 most innovative European, Middle Eastern, and African companies of 2021. Among its high-profile investors is the French utility company EDF.[12]

Using SunCulture's off-grid technology, farmers can extract up to 3,000 liters of water per hour from wells up to 70 meters deep.[13] After four major design iterations, SunCulture's robust system cost about US$950 in 2021.[14] Solar and battery prices have further reduced hardware costs. New digital financing tools make it possible for farmers to buy these solutions.

The system's built-in algorithms study the weather to optimize performance. Based on predicted weather patterns, the smart machines send phone and text alerts to farmers to advise them about the timing of irrigation.[15]

If farmers pump too much water out of the ground, pressure in the aquifer gradually decreases and will dry out the water supply. Other IoT solutions can be used to evaluate the optimal amount of water. For instance, farmers can use sensors to detect soil moisture, then pump water only when needed. Innovative farming techniques thus minimize water consumption, which would increase the long-term potential of the business.[16]

11. EEP Africa, *SunCulture wins project of the year*, December 19, 2019, https://eepafrica.org/sunculture-wins-project-of-the-year/#:~:text=SunCulture%20is%20based%20in%20Nairobi.

12. ImpactAlpha, *SunCulture secures backing from EDF Group*, July 30, 2018, https://impactalpha.com/sunculture-secures-backing-from-edf-group/.

13. "The 10 most innovative European, Middle Eastern, and African companies of 2021," *Fast Company*, March 9, 2021, https://www.fastcompany.com/90600369/europe-middle-east-africa-most-innovative-companies-2021.

14. Peter Fairley, "Off-grid solar's killer app," *IEEE Spectrum*, May 25, 2021, https://spectrum.ieee.org/offgrid-solars-killer-app.

15. Edward Mungai, "AI is a game changer for small companies," *Africa Sustainability Matters*, February 18, 2021, https://africasustainabilitymatters.com/ai-is-a-game-changer-for-small-companies/.

16. Fairley, "Off-grid."

Starting from the Home Country

SunCulture's home market Kenya is attractive for its irrigation and solar pumping solutions. The market size for solar pumps for small farmers is estimated at US$1 billion in the country.

In the initial phase, SunCulture sold its integrated energy generation and water-pumping systems, at about US$5,000, to Kenya's middle-income farmers. These farmers, also referred to as "telephone farmers," have jobs in big cities such as Nairobi and own land in rural areas. The cost of a system then fell to the US$500-US$1,000 range. The company also offered a "pay-as-you-grow" model, making the system accessible smallholder farmers.[17]

Internationalization

In 2020, SunCulture started a partnership with the Togolese government and the UK-based pay-as-you-go solar provider Bboxx to bring its solution to thousands of farmers.[18] Bboxx describes itself as a next-generation utility startup. It manufactures and distributes decentralized solar-powered systems in developing countries and provides financing for consumers to purchase them. As of July 2020, Bboxx had served more than 1 million people in over 35 markets.[19]

In Togo, SunCulture water pumps are integrated with Bboxx Pulse, which utilizes the IoT for remote management and monitoring. The Bboxx Pulse IoT Device Management System is also used to manage Bboxx's Solar Home Systems (SHSs). The services are provided on the pay-as-you-go (PAYG) model.[20] BBOXX uses IoT provider Aeris's global subscriber identity module (SIM) to provide reliable connectivity.[21] Aeris supports major cellular

17. Jonathan Shieber, "SunCulture wants to turn Africa into the world's next bread basket, one solar water pump at a time," *Tech Crunch*, December 4, 2020, https://techcrunch.com/2020/12/04/sunculture-wants-to-turn-africa-into-the-worlds-next-bread-basket-one-solar-water-pump-at-a-time/.

18. "The 10 most."

19. Febrice Iranzi, "TV access forging ahead through BBOXX SHS and CANAL+ partnership," *RegionWeek*, July 27, 2020, https://regionweek.com/tv-access-forging-ahead-through-bboxx-shs-and-canal-partnership/.

20. N. P. Zyl, "Solar-powered farming in Togo sees Bboxx, EDF and SunCulture team up," *ESI-Africa.com*, 2020, https://www.esi-africa.com/industry-sectors/renewable-energy/partnership-cultivated-to-deliver-solar-powered-farming-in-togo/.

21. Aeris, *BBOXX to install Aeris's IoT solutions to provide electricity in remote areas*, April 27, 2018, https://www.aeris.com/news/post/bboxx-to-install-aeris-iot-solutions-to-provide-electricity-in-remote-areas/.

technology, such as GSM, CDMA, and LTE. Aeris's single global access point name (APN) makes it possible for the BBOXX system to be deployed anywhere in the world.[22] The APN defines the network path for cellular-data connectivity. Cellular carriers may require a user to enter APN settings to set up cellular services.[23] Aeris's global APN removes step required to configure local network settings, which means that the BBOXX system can be used with a simple plug and play.

It would be unreasonable to expect that most African farmers can or will commit financial resources to pay for expensive assets such as SunCulture's US$950 irrigation system. However, as part of the partnership, Togo's government provided a 50% subsidy to buy irrigation systems for 5,000 farmers. The government also offered tax exemptions on import duties and VAT on the water pumps.[24]

SunCulture's plan is to distribute its solar water pumps in other African markets, such as Ethiopia, Uganda, Zambia, Senegal, and Ivory Coast.[25]

In agriculture, the IoT can help maintain appropriate soil moisture conditions and nutrient availability to maximize the effect of fertilizers. Kenya's IoT-based smart irrigation system Illuminum Greenhouses, which runs on solar power, offers a case in point. It was built at low cost using local materials.[26] The company's greenhouses are powered by solar panels and sensors, which work together to create an optimal environment for growing crops. Farmers can control temperature, humidity, and soil moisture using mobile phones. If the sensors detect that the soil is dry and the crops need water,

---

22. MarketScreener, *BBOXX, Aeris collaborate to deliver clean energy to millions of people*, April 27, 2017, https://www.marketscreener.com/news/latest/BBOXX-Aeris-Collaborate-to-Deliver-Clean-Energy-to-Millions-of-People--24276435/.

23. Apple Support, *View and edit your Access Point Name (APN) on your iPhone and iPad*, October 27, 2021, https://support.apple.com/en-us/HT201699.

24. EDF France, *Bboxx, EDF, and SunCulture team up with Togo Government to accelerate access to sustainable solar-powered farming*, December 18, 2020, https://www.edf.fr/en/the-edf-group/dedicated-sections/journalists/all-press-releases/bboxx-edf-and-sunculture-team-up-with-togo-government-to-accelerate-access-to-sustainable-solar-powered-farming.

25. Anita Anyango, "SunCulture to distribute solar-powered water pumps across Africa," *Pumps Africa*, December 9, 2020, https://pumps-africa.com/sunculture-to-distribute-solar-powered-water-pumps-across-africa/.

26. Murithi Mutiga, "Kenya's smart greenhouse texts when your tomatoes need watering," *The Guardian*, January 5, 2016, https://www.theguardian.com/global-development/2016/jan/05/kenya-smart-greenhouse-tomatoes-watering-farming.

an automated watering system supplies a precise amount of water. The solutions are cost-efficient, so accessible for farmers with small land areas.[27]

IoTs can also help reduce corruption such as pilferage and theft, among the key challenges facing LMICs. For instance, Nigeria's downstream oil industry suffers from losses, risks, and theft.[28] The IoT is already being used to fight problematic workplace behaviors (In Focus 5.2).

In Focus 5.2: Usangu Logistics Uses IoT-Enabled Solution to Fight Oil Pilfering

Tanzania's heavy transport company Usangu Logistics owns a fleet of more than 100 trucks and tankers to deliver oil, lubricants, and bulky products to its clients. A challenge the company faced was that, after a tanker is loaded with product for transportation, the drivers would often pilfer the oil and sell the stolen oil on the black market.

A non-IoT solution used by the company was a lock system to prevent the unauthorized opening of the tank's hatch, which was not effective. The company had no way to know when, where, and how much oil has been stolen.

The company thus decided to deploy an IoT application involving radio-frequency identification (RFID). The seals are fitted with RFID-enabled tags and fastened to the hatch. An IoT-enabled gateway device, which acts as a "gate" between two networks (one in the truck and one in the main office), is attached to the cabin area. The tags transmit a signal to the main office gateway device every eight seconds. Data stored in the IoT platform include the seal status and the trucks' location to allow for real-time monitoring of the truck location and seal information. Any unauthorized attempt to open the hatch is recorded by the IoT system, which makes it possible to know the perpetrator immediately. The IoT-enabled solution led to a significant drop in oil pilfering.[29]

---

27. Julie Stewart, "Challenges surrounding IoT deployment in Africa," *Compare the Cloud*, 2019, https://www.comparethecloud.net/articles/challenges-surrounding-iot-deployment-in-africa/.

28. Adeyemi Adepetun, "IoT technology adoption in Nigeria rises as global spending hits $128b," *The Guardian Nigeria*, August 19, 2021, https://guardian.ng/business-services/iot-technology-adoption-in-nigeria-rises-as-global-spending-hits-128b.

29. Vizocom, *6 IoT applications that improved people's lives in Africa—a story of 6 countries*, 2019, https://www.vizocom.com/ict/6-iot-applications-that-improved-peoples-lives-in-africa-a-story-of-6-countries/.

Addressing Environmental Challenges

Various environmental risks can be reduced and managed using IoT solutions. Attempts have been made to help consumers switch from dirty and dangerous fuels such as open fires, traditional stoves, and kerosene to clean-burning stoves and cleaner-burning liquefied petroleum gas (LPG).

LPG is more expensive than traditional fuels such as wood, coal, charcoal, animal waste, and kerosene. For instance, the first-time user needs to acquire equipment, such as an LPG cylinder, stove, regulator, and hose. The lack of local credit facilities such as microfinance constitutes a significant entry barrier for the poor. In many countries, there is a widely held perception that LPG is unsafe, which has been difficult to change.[30] A non-functioning safety check system led to dangerous LPG cylinders being in circulation,[31] especially in SSA.[32]

Unregulated street sellers that bring LPG to end users' homes are a part of the problem. Such sellers account for about 30%–40% of LPG distribution in Nigeria. A poor distribution infrastructure and network has also been a challenge for distributing LPG cylinders.[33]

The IoT can address many of these barriers by facilitating the switch to LPG. This technology has already made LPG more affordable and increased people's access to cooking gas in rural Africa. Companies such as PayGo Energy (In Focus 5.3) and M-Gas in Kenya and KopaGas in Tanzania provide gas-as-a-service solutions, which allow users to pay daily through mobile credits to use LPG through a smart meter attached to a gas cylinder.[34]

---

30. World LPG Association (WLPGA), *Accelerating the LPG transition global lessons from innovative business and distribution models*, https://www.wlpga.org/wp-content/uploads/2018/10/Accelerating-the-LPG-Transition.pdf.

31. Caesar Keluro, "Reimagining LPG delivery in Africa: Retailing through IoT," *business am live*, May 17, 2021, https://www.businessamlive.com/reimagining-lpg-delivery-in-africa-retailing-through-iot/.

32. Ideas to impact, *Design of a low-cost absorption refrigerator energy access*, 2017, https://docslib.org/doc/2475268/design-of-a-low-cost-absorption-refrigerator-energy-access-lpg-cylinder-prize-winning-solution-3-7.

33. Keluro, "Reimagining."

34. Vincent Matinde, "Africa's IoT network: Are we building it the wrong way?," *IDG Connect*, September 14, 2020, https://www.idgconnect.com/article/3583362/africa-s-iot-network-are-we-building-it-the-wrong-way.html.

In Focus 5.3: PayGo Energy's IoT Solution for LPG Cylinders

In Kenya, about 80% of households cook with dirty fuels.[35] In Democratic Republic of Congo, only 4% of the population has access to clean cooking facilities.[36] To facilitate the switch to clean-burning fuels, PayGo Energy uses the IoT to allow users to pay daily using mobile credits. Other companies providing similar solutions include M-Gas in Kenya and KopaGas in Tanzania.

In developing countries, the cost of a cylinder is a significant obstacle. PayGo reached 400 customers through a pilot in Kenya in 2018 and supplied 2,000 meters in 2020.

The Solution

PayGo has built hardware and software solutions for downstream LPG distributors. The focus is on developing markets. The solutions are for two related problems: making clean cooking affordable and accessible for low-income households and also making the supply chain more efficient and expansive.

PayGo has two core products. The Cylinder Smart Meter (CSM) is an IoT device that attaches to most LPG cylinders. It measures gas flow as customers use it for cooking. It enables households to monitor and manage consumption and purchase gas in any amount, using mobile credits. It automatically shuts off when a customer's credit reaches zero. The CSM allows consumers to pay for LPG by the gram.[37] Customers can top up as little as one shilling at a time.[38]

Its Tag & Trace software platform allows suppliers to better manage cylinder distribution, track distribution metrics, and extend reach. It digitizes LPG distribution and retail data. The platform is customizable, with a range of features such as cylinder tracking, inventory management, customer relationship management, and payments.[39] Distributors can monitor consumption and replace cyl-

35. Argus Media, *Q&A: PayGo's plans to unlock energy poverty areas*, 2020, https://www.argusmedia.com/en/news/2169225-qa-paygos-plans-to-unlock-energy-poverty-areas.

36. https://cleancookingalliance.org/news/05-19-2021-paygo-energy-striking-strategic-partnerships-to-scale-up-clean-cooking/.

37. Matinde, "Africa's IoT network."

38. Hilary Kimuyu, "PayGo Energy launches prepaid cooking gas service," *Nairobi News*, February 29, 2020, https://nairobinews.nation.co.ke/paygo-energy-launches-prepaid-cooking-gas-service/.

39. Clean Cooking Alliance, "PayGo energy."

inders before they become empty. Tracking the cylinder movement to the market, warehouse, retailers, and customers provides powerful data that is used to lower distribution costs.[40]

In October 2019, the Kenyan Bureau of Standards launched the world's first technology standard for LPG metering. PayGo's meter was the first fully compliant in the market. The startup's device also conforms with ATEX directives—a European Union safety certification required of explosive equipment.[41]

Partnership with Safari Supa Gas

PayGo partnered with the cooking solutions provider Safari Supa Gas to better serve the Kenyan market. These two companies' solutions complement each other. Through the PayGo technology, Safari knows the meter's geolocation and the rate at which gas is being consumed. Once the level is low, Safari customer care calls the homeowner to arrange for a replacement cylinder, which is dropped off free of charge. The gas automatically shuts off when the customer runs out of credit. When additional credit is added, the account is instantly credited and the gas service resumes.[42]

An IoT-based LPG distribution platform also has the potential to transform the distribution of LPG by improving the procedure for replacing LPG cylinders for consumers. In this way it is likely to create a new market. The platform can monitor LPG weight and automatically trigger a new cylinder booking if the gas cylinder is about to be empty. Consumers can do so by logging onto a mobile app or the IoT LPG service provider dashboard using a GSM/Bluetooth system.[43] IoT solutions can also handle cylinder revalidation and recertification by regulators.

Another important way the IoT can contribute to environmental conservation and protection is by facilitating the monitoring of environment and resources such as endangered animals in remote areas. Noncellular IoT technologies have been found to be especially appropriate for such purpose.

---

40. Kimuyu, "PayGo energy."
41. *TechCabal,* "Kenya-based PayGo Energy is exporting its gas meter tech to Asia," *TechCabal,* June 16, 2020, https://techcabal.com/2020/06/16/paygo-energy-cylinder-gas-meter-saisan-japan/.
42. Kimuyu, "PayGo energy."
43. Keluro, "Reimagining."

To take an example, in 2017, the UK-based social enterprise Smart Parks started to attach sensors based on long-range (LoRa) technology to the horns of black rhinos in Mkomazi, Tanzania. By 2020, Smart Parks had operations in Malawi, Tanzania, Rwanda, Kenya, Congo, India, Zambia, Namibia, and the Netherlands. The enterprise works with conservation organizations such as African Parks, Peace Parks, and the World Wildlife Fund.

The LoRa transmitter attached to the animal feeds data into the wide area network (WAN) to provide near real-time tracking.[44] The trackers can be configured remotely. The accelerometer provides detailed status updates on temperature, battery usage, and movement. Settings can be optimized for particular species' behavior patterns.[45]

In Rwanda's Akagera National Park, where Smart Parks has deployed the solution, 100 solar-powered LoRaWan sensors have been installed. With gateways at high altitude around the site, the sensors regularly relay signals to a control room. Connection to the LoRa network inside the park is reliable.[46] Such data allow the park to track animals, staff, and tourist vehicles and check the state of electric fencing and other security equipment.

While a LoRaWan system is less secure than cellular solutions such as NB-IoT and LTE-M, it is more secure than traditional radio systems. It was reported that the radio signals used to track endangered species were intercepted by poachers.[47] Also, the cost of LoRa solutions is significantly less than traditional satellite GPS collars. For instance, commercial GPS tracking collars are estimated to cost between US$1,500 and US$5,000, and they are often unreliable and perform poorly in terms of data storage and battery life.[48]

---

44. Charles McLellan, "GPS collars for lions and cheetahs: How IoT and open source are protecting rare animals," *ZDNet*, June 15, 2020, https://www.zdnet.com/article/gps-collars-for-lions-and-cheetahs-how-iot-and-open-source-are-protecting-rare-animals/.

45. McLellan, *GPS collars*.

46. "In developing countries, the LoRa network is the bearer of progress—Hello Future Orange," *Hello Future*, November 19, 2018, https://hellofuture.orange.com/en/developing-countries-lora-network-bearer-progress/.

47. "In developing countries."

48. Pat Clark, David Spencer, Douglas Johnson, and Fred Pierson, *Cutting-edge technologies: GPS/Satellite communications-based tracking*, Agricultural Research Service, US Department of Agriculture, February 8, 2009, https://www.ars.usda.gov/research/publications/publication/?seqNo115=232384.

Using Contracts to Create Value

Measurement and Enforcement

The IoT has also stimulated economic exchange and transactions. Among the key factors that influence transaction costs are costs of measuring and costs of enforcement.[49] Measuring the dimensions and attributes of goods and services being exchanged or the performance of agents is not an easy task.[50] The IoT, in combination with other technologies, can provide for accurate measurements to describe precisely what the parties engaged in a transaction are exchanging and what performance characteristics can be expected.

For enforcement, in a society characterized by a perfect contract enforcement, a neutral third party impartially evaluates disputes and awards compensation to the party affected by a violation. In such a situation, opportunism, shirking, and cheating are not attractive options for the transacting parties. However, the real world is far from ideal. The high costs of measurement often make it difficult to determine whether a contract has been violated and by whom. In countries with a strong rule of law, well-developed court systems, and coercive power of the state to enforce judgments, complex contracting is an important mode of formal governance. Many developing economies such as those in SSA lack such mechanisms.

Enforcement can be implemented at three levels: first, second, and third party. First, it is important to emphasize that third-party enforcement mechanisms, which are often formal coercive enforcement measures by the state, have been relatively ineffective in many LMICs. This increases the relative importance of the first two types of enforcement.

IoT can improve the mechanisms of self-enforcement to make more productive use of factors of production and foster entrepreneurial activity. For instance, in Nigeria, most banks' smoke-alarm systems cannot transmit information effectively in case of a fire outbreak. Many banks lack visibility of their ATM systems, and batteries could have low power supply.[51] IoT devices can be used to measure levels of remaining ATM batteries to ensure uninterrupted operations or to detect smoke to enable a rapid and effective response to fire.

IoT devices have helped companies identify problems before they occur, such as reducing machine downtime and cutting service costs. South African

49. North, "Dealing with a nonergodic."
50. North, "Dealing with a nonergodic."
51. Adeyemi Adepetun, "IoT technology."

Breweries (SAB), a subsidiary of AB InBev, manages more than 100,000 beverage coolers in remote and urban areas across South Africa. The company relies on the IoT to track multiple metrics across beverage coolers and maintain them.[52] In addition to a location sensor, the company's IoT solutions include temperature sensors inside the cooler and on the cooler's motor. Relevant data are sent to Microsoft's Azure IoT Central to provide SAB with valuable insights for improving business outcomes. Azure can analyze the patterns in the temperature data inside the cooler and on the motor. ML algorithms can help AB InBev predict when a technician needs to visit a location to repair a cooler. More accurate measurements lead to a more effective enforcement.[53]

In second-party enforcement, one party retaliates against the other. The IoT provides a low-cost mechanism for this type of enforcement. For instance, Kenya's M-KOPA Solar, which offers solar home systems to poor households on credit (chapter 1), can switch off a home's device remotely using the IoT when the prepaid amount is used up.[54]

Contract, Value Creation, and Value Capture

An exchange can simply and clearly help create value. In the sale of M-KOPA Solar's IoT-based solutions, the buyer values the solutions more than the seller. However, the created value depends on how the contract between buyer and seller is created. A properly designed contract may create additional value.

A *contract* is a "legally binding promise to act in the future."[55] In the M-KOPA Solar example, the company promises to deliver the IoT-based solution that functions as specified in the contract in exchange for its customers' promise to make a specified daily payment. M-KOPA Solar can switch off devices remotely when the prepaid amount runs out, which incentivizes users to make timely payments. Simpa Networks and Fenix International have similar capabilities.[56]

---

52. Microsoft Customers Stories, *AB InBev South Africa improves cooler management with Azure IoT Central*, https://customers.microsoft.com/es-es/story/776768-sab-ab-inbev-consumer-goods-azure.

53. North, "Institutions."

54. Alejandro Moreno and Asta Bareisaite, *Scaling up access to electricity: Pay-as-you-go plans in off-grid energy services*, World Bank, 2015, https://documents1.worldbank.org/curated/en/687851468320946678/pdf/93786-REVISED-LW34-fin-logo-OKR.pdf.

55. Robert E. Scott and George G. Triantis, "Incomplete contracts and the theory of contract design," *Case Western Reserve Law Review* 56, no. 1 (2005): 187–202.

56. World Bank, "Scaling up."

Table 5.2 presents the IoT's role in including features in a contract to create and capture value. Various mechanisms have been proposed to explain how a contract can create and capture additional value. One such mechanism is increased investment by one or both parties in anticipation of the exchange. Such an investment can increase the exchange value by lowering the cost of performance and/or raising the benefit from performance.[57] For instance, M-KOPA's proprietary technology platform called M-Kopanet is used to process payments, monitor the system's functionality, and address problems. An M-KOPA customers in Machakos County near Nairobi was reported to save about KES 20 per day on power, about one-third of what they were spending. The system also frees users from the hassle of buying batteries and kerosene and of cleaning kerosene residue from the walls and ceiling.

Similarly, Kenya's Azuri Technologies combines the IoT with AI and ML to provide smart solar power to low-income households. The solution consists of a yellow box called HomeSmart. Each box is the size of a landline phone. HomeSmart uses AI to learn a home's energy needs. Power output is automatically adjusted to ensure that sufficient battery power is available to meet light and other energy needs for the entire day. For instance, the system automatically dims lights and TV screens, lowers speaker volume, and slows a fan's motor if the energy supply is low.[58] In such solutions, the payment of the last instalment includes a special code that unlocks the system and the electricity generated after that will be free for the customer.[59]

A contract also creates value for the parties because they can benefit from shifting risks between each other.[60] In South Africa, where 48,306 vehicles were stolen in 2019 and only one in five stolen vehicles is recovered, IoT is helping to deal with this situation. A slow search process allows thieves to dismantle stolen cars or ship them to far-off locations. Even when a stolen car is found, insurers can refuse to compensate the victim in the absence of physical evidence of a break-in. Discovery Insure uses IoT to address this problem. Using IoT sensors, it is possible to locate stolen cars even if hidden in enclosed or underground locations.[61]

---

57. Scott and Triantis, "Incomplete contracts."

58. Adam Popescu, "AI helps Africa bypass the grid," *Bloomberg*, June 11, 2018, https://www.bloomberg.com/news/articles/2018-06-11/ai-helps-africa-bypass-the-grid.

59. Secretariat of the UN Framework Convention on Climate Change, *Mobisol Smart Solar Homes | Rwanda and Tanzania*, https://unfccc.int/climate-action/un-global-climate-action-awards/information-and-communications-technology-solutions/mobisol-smart-solar-homes-rwanda-and-tanzania.

60. Scott and Triantis, "Incomplete contracts."

61. Patrick Cason, "Insurance and IoT: The perfect match," *Insurance Thought Leader-*

Table 5.2. The IoT's roles in including features in a contract to create and capture value

	Features of a contract	Value creation/capture
Consumers	Consumer-friendly contracts can be formulated	Features can be adjusted to enhance value delivery
Companies (sellers of goods and services)	Can enforce contracts more efficiently at a low cost.	Production costs can be reduced and revenues can be increased by more efficient maintenance, such as reduced downtime and monitoring costs

The central focus of economic contract theory is on managing the conflicts between the shared and private incentives of the contracting parties. A challenge for parties designing contracts is to have an idea about possible outcomes and courses of action beforehand. Or they must be at least able to handle any future renegotiation in a way that maximizes ex ante and ex post efficiency.[62]

In the SAB example, ML and IoT help the company predict when a technician needs to visit a location to repair a cooler. This means that coolers in bars, taverns, and restaurants do not go unrepaired for a long time. Thus, production costs can be reduced and revenues can be increased by more efficient maintenance, such as reduced downtime and reduced monitoring costs. All these lead to an increase in ex post efficiency.

In economic contract theory, the cost of enforcement is an important exogenous factor in contract design.[63] The speed and ease with which contracts can be enforced have important implications for the creation of new markets. The IoT can help enforce contracts more efficiently at a low cost.

Entrepreneurial Activities and the Creation of New Markets

Serving low-income consumers often requires creating a new market, generating new business opportunity, and discovering and adopting new business

ship, June 10, 2021, https://www.insurancethoughtleadership.com/insurance-and-iot-the-perfect-match/.

62. Scott and Triantis, "Incomplete contracts."

63. Scott and Triantis, "Incomplete contracts."

models.[64] New markets, however, do not emerge or appear.[65] Technological, political, or regulatory changes can open up new markets.[66]

Entrepreneurial firms' activities can play a key role in creating new markets by transforming the existing realities into new possibilities.[67] To do so, firms need to correctly sense latent needs and offer solutions to meet that need. An obvious way to create a new market is to offer a novel product or service that addresses unmet needs. In some cases, such needs may not even be sensed.[68] New market creation requires the implementation of various activities that integrate technological knowledge and market knowledge.[69]

To successfully commercialize innovations and technological know-how, it is important to combine them with other capabilities.[70] From an innovator's perspective, an innovation's commercial success hinges on the innovator's ability to combine distinctive technological capabilities with specialized complementary assets.[71] Such assets are often controlled by large incumbent firms, which makes it important for startups to form commercial alliances.[72] Supports of "a new network of stakeholders" are thus critical for the exploitation of new possibilities.[73] More to the point, to make a new market a reality, a firm needs to convince other economic actors (e.g., firms providing complementary solutions) and stakeholders to shift resources to that firm's solution.[74]

---

64. Venkatesh Bala and Sanjev Goyal, "The birth of a new market," *Economic Journal* 104 (1994): 282–290; J. G. March, "Exploration and exploitation in organizational learning," *Organization Science* 2, no. 1 (1991): 71–87.

65. Erin Anderson and Hubert Gatignon, "Firms and the creation of new markets," *Handbook of New Institutional Economics* (New York: Springer, 2008), 401–431.

66. Bala and Goyal, "The birth."

67. March, "Exploration and exploitation."

68. Anderson and Gatignon, "Firms and."

69. Stefania Migliori, Daniel Pittino, Augusta Consorti, and Lorenzo Lucianetti, "The relationship between entrepreneurial orientation, market orientation and performance in university spin-offs," *International Entrepreneurship and Management Journal* 15 (2017): 793–814.

70. David J. Teece, "Profiting from technological innovation: implications for integration, collaboration, licensing, and public policy," *Research Policy* 15 (1986): 285–305.

71. Teece, "Profiting from."

72. Massimo G. Colombo, Luca Grilli, and Evila Piva, "In search of complementary assets: The determinants of alliance formation of high-tech start-ups," *Research Policy* 35, no. 8 (2006): 1166–1199.

73. S. D. Sarasvathy and N. Dew, "New market creation through transformation," *Journal of Evolutionary Economics* 15, no. 5 (2005): 533–565.

74. E. Anderson and H. Gatignon, "Firms and the creation of new markets," *Handbook of New Institutional Economics*, 2008, 401–431.

The off-grid solar market in Africa is relevant for analyzing how IoT firms are creating a new market. The potential market size of off-grid solar is estimated at 20 million households in East Africa alone.[75] IoT firms in the region have transformed this reality into new possibilities by forming alliances with other companies that have complementary assets.

From the perspective of an IoT startup, economies vary in terms of opportunities for establishing successful alliances. The business models of companies such as Kenya's M-KOPA and KOKO Networks (In Focus 5.4) and Tanzanian scalable off-grid electric company Off.Grid:Electric, which run micro-asset-based finance, would not be feasible without the existence of a digital retail payment system such as M-PESA. Without such systems, the cost of collecting micropayments from customers is extremely high.[76]

A well-developed technological ecosystem has facilitated the availability of such opportunities in Kenya. In the Kenyan IoT Industry and market there is active support by and collaboration with appropriate networks of stakeholders. The country's IoT solutions providers have formed networks with mobile financial services such as M-PESA. For instance, M-KOPA's business model relies on Safaricom's mobile money system M-PESA to collect customer payments. M-KOPA's founders view the dependency as an essential and effective option for its offerings because M-PESA is reliable and has been a key component of the Kenyan economy. While an interorganizational network of stakeholders facilitated the growth of M-KOPA in Kenya, such networks are not fully developed or functional in other markets such as Tanzania and Uganda, where M-KOPA operates. M-KOPA faces challenges in collecting in these markets, and consumers lack understanding of the payment systems for their M-KOPA system.[77]

To take another example, Kenya's largest telecommunications provider, Safaricom, launched an NB-IoT network in 2017. The telecommunications company has partnered with the IoT solutions provider Upepo Technology and real-time mapping platform provider Esri Eastern Africa to enable remote monitoring of water supplies for Kenyan water utility Embu Water and Sanitation Company. As of October 2019, the solutions were deployed in 20 households in Kenya's Embu region. Water flow is measured, and real-time readings are transmitted over Safaricom's NB-IoT network. The information helps estimate Embu's water balance and usage and know the

75. John Aglionby, "Lightbulb moment for M-Kopa," *Financial Times*, March 17, 2016, https://www.ft.com/content/ccfaa1ba-d0f1-11e5-831d-09f7778e7377.

76. A. Costa and T. Ehrbeck, "A market-building approach to financial inclusion," *Innovations: Technology, Governance, Globalization* 10 (2015): 53–59.

77. Aglionby, "Lightbulb moment."

proportion of water collected from rivers, dams, and boreholes that reach to customers' meters. The solutions helped Embu to identify optimal water flow and match supply and demand.[78] Estimates suggest that due to physical leaks and other losses, 52% of water collected by Kenya's water utility companies is not accounted for.[79] Embu hopes to reduce water leakage by monitoring the usage. The smart water meters also send SMS alerts if they are tampered with.[80]

Two types of business models have been identified to serve the low-income market: isolated and interactive. Isolated business models are characterized by an exploitation strategy, which entails leveraging a firm's own resources and capabilities to enhance efficiency of production factors. The goal is to reduce costs and fix the price below the consumer's willingness to pay. In most of the examples discussed above, interactive business models have been utilized, which entail an exploration strategy. Firms utilizing such business models rely on external resources. Learning and innovation processes are key to such models. Firms need to reconfigure their business models and establish comanaged value chains through a network of alliances.[81] Technology businesses have attempted to focus on payment terms rather than price as a more reasonable approach.

The government is also a key stakeholder for this purpose. For instance, the Government of Togo provided subsidies to buy irrigation systems and offered tax exemptions on import duties and VAT on the water pumps.

In Focus 5.4: IoT-Enabled and Cloud-Connected Koko Points to Deliver Ethanol in Kenya's Capital

The venture-capital-backed technology company Kenya-based KOKO Networks has built dense networks of cloud-connected KOKOpoints in the stores of its network of KOKO Agent shopkeepers to deliver consumers fuels using IoT and cloud solutions.[82]

---

78. Joseph Purnell, "Kenyan waterco taps Safaricom for IoT data stream," *TelcoTitans*, June 19, 2020, https://www.telcotitans.com/vodafonewatch/kenyan-waterco-taps-safaricom-for-iot-data-stream/1842.article.

79. "Smart meters are helping to track water on the internet: Here is how," *Safaricom Newsroom*, June 3, 2020, https://newsroom.safaricom.co.ke/smart-meters-are-helping-to-track-water-on-the-internet-here-is-how/.

80. Anfernee Onamu, "Safaricom's new technology helps keep track of water spent at home," *Gadgets Africa*, June 4, 2020, https://gadgets-africa.com/2020/06/04/safaricom-smart-meter-keeps-track-of-water-spent-at-home/.

81. Joan E. Ricart and Pablo Sanchez, "Business model innovation and sources of value creation in low-income markets," *European Management Review* 7, no. 3 (2010): 138–154.

82. Tom Jackson, "Meet the African companies making pioneering strides in the carbon

KOKOpoints are ethanol refueling stations to power home stoves. Ethanol is cleaner to burn and safer to store and handle than solid or gaseous fuels. Customers purchase ethanol via M-PESA. Just like a soft drink dispenser, a customer presses a button on a high-tech screen and fills up a smart cannister. They then take the canister home and dock it in their KOKO Cooker.[83]

The Market

KOKO Networks was launched in September 2019. By September 2020, the company had over 700 outlets, with more than 600 agents in downtown Nairobi dispensing cooking fuel. By that time, 50,000 households were using KOKO's IoT solution for ethanol.[84] This is arguably Africa's largest deployment of IoT for consumer fuels that enables cloud-based tracking of fuel inventory across the supply chain. Each KOKOpoint is connected to the KOKO Cloud software platform.[85]

The company's target customers are in the US$150–300 per month household income range. They have electricity in their homes, TV, and smartphones but cook with charcoal or kerosene. The cost of a KOKO Cooker with a canister costs about US$65. The company makes the network hardware in Kenya and the stoves and canisters in India. The cost to fill one canister with 2.3 liters of bioethanol is about US$2. For an average family of four, a canister would last about a week as cooking fuel. The cost of bioethanol was about US$0.95 per liter in 2018. Cooking with charcoal was estimated to cost about 40% more.[86]

Customers can buy fuel in small quantities for daily cooking.[87]

credits space,*" Disrupt Africa*, June 29, 2021, https://disrupt-africa.com/2021/06/29/meet-the-african-companies-making-pioneering-strides-in-the-carbon-credits-space/.

83. Expogroup, *KOKO introduces first network of 700 "KOKOpoints" in Nairobi*, October 2, 2021, https://expogr.com/detail_news.php?newsid=5759&pageid=2.

84. Advanced BioFuels, *50,000th Nairobi household switched on by IoT & KOKO fuel*, September 8, 2020, https://advancedbiofuelsusa.info/50000th-nairobi-household-switched-on-by-iot-koko-fuel/.

85. "Kenyan cooking start-up uses tech to cut costs and saves lives," *Media Africa*, November 6, 2019, https://www.media4africa.com/kenyan-cooking-start-up-uses-tech-to-cut-costs-and-saves-lives/.

86. Bryony Collins, "Africa's $40B market for cooking fuel is being cleaned up: Q&A," *Bloomberg Finance*, September 9, 2019, https://about.bnef.com/blog/africas-40b-market-cooking-fuel-cleaned-qa/.

87. Appsafrica, "*Koko launches new smart e-commerce platform for Africa*," April 24, 2017, https://www.appsafrica.com/koko-launches-new-smart-e-commerce-platform-for-africa/.

They can purchase a next-generation KOKO Cooker, a two-burner ethanol stove, with a KOKO Agent or via the myKOKO smartphone app. After placing the order, the KOKO Cooker can be picked up in a shop within 24 hours.[88] They can top up their KOKO Fuel account via M-PESA and then access bioethanol at a KOKOpoint Fuel ATM.[89]

The Solution

Most of the bioethanol comes from Kenyan molasses-based ethanol plants. Molasses is a viscous substance and waste product of sugar refining. It produces only water vapor and no emissions when burned. Some bioethanol is also imported from Uganda and Tanzania.[90]

While bioethanol is a sustainable by-product of the local sugar industry, it has been difficult to package, transport, and make available to customers in a safe and cost-efficient way. KOKO's model removes the need for single-use plastic bottles and reduces distribution costs.[91]

KOKO's IoT hardware tracks fuel all the way from fuel delivery trucks to the last-mile distribution to the KOKOpoint Fuel ATMs.[92] Each ATM unit has backup power and is installed securely to the wall.[93]

KOKOpoints are fitted with a wide range of internal electronic sensors to send real-time updates to a network operations center. They measure their "technical health," fuel inventory levels, video advertising inventory, and other indicators.[94] The KOKOpoints are used to gather customer data for targeted advertising when they refill their canisters.[95]

88. A. Amt, "E-commerce at Kenya's kiosks," June 15, 2018, https://germanyinafrica.diplo.de/zadz-en/-/2106898.

89. Advanced BioFuels, "KOKO's technology production facility inaugurated in Nairobi," March 14, 2019, https://advancedbiofuelsusa.info/kokos-technology-production-facility-inaugurated-in-nairobi/.

90. BloombergNEF, *Africa's $40B*.

91. Envirotech Online, *Flow meter technology aids the dispensing of ethanol cooking fuel in Kenya*, November 26, 2020, https://www.envirotech-online.com/news/business-news/44/titan-enterprises-ltd/flow-meter-technology-aids-the-dispensing-of-ethanol-cooking-fuel-in-kenya/53805.

92. Arne Siegmund, "KOKO Networks has built technology to transform the market for urban cooking fuel," KfW, November 30, 2020, https://www.kfw.de/stories/economy/companies/koko-networks/.

93. Appsafrica, "Koko."

94. Appsafrica, "Koko."

95. "Op-Ed: Africa's high-tech path to energy inclusion," ESI-Africa.com, October 31,

Partners

KOKO's fuel distribution partner Vivo Energy Kenya sources bio-ethanol from local and regional suppliers and delivers it to service stations in Nairobi that have dedicated underground storage tanks for cooking fuel. KOKO's smart microtankers transport fuel from petrol stations to KOKOpoint Fuel ATMs.[96]

As of September 2019, ten Shell petrol stations were used for storing bioethanol and ten smart microtankers for transport. A KOKOpoint fuel ATM is within five minutes of every house in Nairobi.[97]

Technological Readiness of LMICs for the IoT

Overall, LMICs' IoT readiness is gradually increasing. The essential conditions for developing and implementing IoT solutions are improving. For instance, IoT connectivity is widely available in most parts of the world given growing coverage of cellular networks (Figure 5.1). Noncellular IoT connectivity is also expanding. For instance, as of 2020, Africa's data, voice, and IP provider Liquid Telecom's low-power wide-area IoT network through Sigfox OG covered up to 85% of Kenya's population.[98]

LMICs are becoming more heterogeneous over time in terms of the type of IoT solutions they use and the features they look for.[99] In the case of IoT, this translates to individuals' and organizations' demands of IoT solutions with different features and performance levels in terms of key metrics such as latency or lag time, data throughput, quality of service (QoS), and cybersecurity. A variety of IoT solutions have been implemented to tailor according to the need and requirements of specific use cases of the heterogeneous market. For instance, in the utility industry in rural locations, GSM is the

---

2018, https://www.esi-africa.com/top-stories/op-ed-africas-high-tech-path-to-energy-inclusion/.

96. Tech in Africa, *Nairobi startup introduces home cooking fuel powered by IoT technology*, 2019, https://www.techinafrica.com/nairobi-startup-introduces-home-cooking-fuel-powered-iot-technology/.

97. BloombergNEF, "Africa's $40B."

98. Kenn Abuya, "Liquid Telecom and Twiga Food's partnership intros precision agriculture to Kenya," *Techweez*, June 16, 2020, https://techweez.com/2020/06/16/liquid-telecom-and-twiga-foods-partnership-intros-precision-agriculture-to-kenya/.

99. Sergio Tezanos Vázquez and Andy Sumner, "Revisiting the meaning of development: A multidimensional taxonomy of developing countries," *Journal of Development Studies* 49, no. 12 (2013): 1728–1745.

most widely used machine-to-machine technology to transfer data over long distances. A drawback of GSM is that it consumes a lot of power and network coverage is unreliable in rural areas. Utility companies are thus switching to narrowband (NB) IoT for its lower cost and low power consumption. NB IoT is ideal for utility applications, which mostly require occasional connectivity and minimal throughput is sufficient.[100] The NB IoT standard is emerging as a popular standard.

Cellular low-power wide-area network (LPWAN) solutions such as NB IoT and Long Term Evolution for Machines (LTE-M), which were developed by the 3rd Generation Partnership Project (3GPP), offer low latency or lag time, high data throughput, higher QoS, and strong cybersecurity.[101] Some applications that require these features include remote healthcare, water and gas metering, and smart home systems.

The features offered by cellular IoT connections often translate into higher deployment costs, though. Moreover, many rural areas lack coverage for mobile broadband technologies such as LTE and Worldwide Interoperability for Microwave Access (WiMAX) (Figure 5.1). Noncellular IoT technologies are appropriate in such cases, especially in rural settings, for which cost efficiency is important but data throughput, speed, and cybersecurity are less critical. For many of these applications, it is sufficient to exchange data in batches. For instance, cost efficiency is important in smart farming and livestock management. Smart farming systems also require long battery life.[102]

Noncellular IoT connections can be deployed in areas that lack cellular coverage. A drawback of such connections, however, is that they are unlicensed and operate on a free frequency band. Unlike the GSM network and NB IoT, these networks do not have their own frequencies. This means that other networks in the same area may theoretically interfere with operations of noncellular IoT connections such as SIGFOX and LoRa networks.[103]

---

100. Ihuoma Atanga. *The Internet of everything water*, United Nations Africa Renewal, https://www.un.org/africarenewal/magazine/may-july-2017/internet-everything-water.

101. Thibault Werlé, Rachid ElAmeri, Rüdiger Schicht, and Rodolphe Frugès, "Unlocking telcos IOT potential how telecom operators should strategically navigate the complex landscape of iot lpwan technologies," Boston Consulting Group, February 28, 2021, https://web-assets.bcg.com/07/74/21620e1145dba9aae7b2f9f4dcfb/me-unlocking-telcos-iot-potential.pdf.

102. Werlé et al., "Unlocking telcos."

103. Lionel Anciaux, "From domotics to autonomous sensors based on SigFox, LoRaWAN or NB-IOT: Realistic today?," *IOT Factory*, September 23, 2018, http://iotfactory.eu/from-domotics-to-autonomous-sensors-based-on-sigfox-lorawan-or-nb-iot-realistic-today/.

LMICs are also developing local technological capabilities. Off-the-shelf and ready-made solutions are often ineffective or too costly for addressing specific local needs. Consider IoT technology used in the agricultural sector. An off-the-shelf soil moisture sensor in Kenya costs about US$500, beyond the reach of most farmers. To address this, local IoT companies are developing their own solutions tailored to their customers' requirements. In most cities, analog circuit components (e.g., wires, resistors, capacitors, inductors, transistors) can be found locally. However, digital components such as integrated circuits (ICs) are not available. Hardware developers rely on e-commerce sites to source components.[104]

The IoT's contribution to the socioeconomic development of the B4B faces several challenges. Despite LDCs' increasing connectivity, mobile or cellular coverage is lower than the world average. The gap in coverage between LDCs and other economies is especially large for broadband wireless systems such as LTE (Figure 5.1).

The lack of economies of scale has hindered the development of the IoT market. In 2019, only about 9% of homes in South Asia had IoT or connected devices, compared to the worldwide average of 40%.[105]

The economies-of-scale barrier has also been found in upper-middle-income countries such as South Africa. For instance, South Africa's licensed SIGFOX operator SqwidNet, which was launched in November 2016, reported starting voluntary restructuring or winding down in mid-2021. SqwidNet was the country's largest operator to establish an IoT national connectivity network. The company reported that it was not possible to fund its operating costs due to the early stage of IoT development and an underdeveloped IoT ecosystem.[106] COVID-19 was cited as a main trigger for the restructuring of SqwidNet.

104. Anish Paul Antony, Daniel Sweeney, and Jennifer Lu, "Seeds of silicon: How the Internet of Things can improve agricultural outcomes," *Agrilinks*, January 16, 2020, https://www.agrilinks.org/post/seeds-silicon-how-internet-things-can-improve-agricultur al-outcomes.

105. Neelam Dimri, "10% of the South Asian homes incorporate IoT device," *IoT Avenue*, June 27, 2019, https://www.iotavenue.com/10-of-the-south-asian-homes-incorp orate-iot-device.

106. Admire Moyo, "Lack of IOT growth in SA was SqwidNet's biggest undoing," ITWeb, June 2, 2021, https://www.itweb.co.za/content/JBwErvnBdAav6Db2.

Chapter Summary and Conclusion

IoT devices are diffusing rapidly in LMICs and being used as a tool for the socioeconomic development of the B4B population. Data from IoTs can create valuable insights for improving operational efficiencies of firms, processes, and key economic activities.

This technology helps increase productivity through various mechanisms, including reducing waste of key resources (e.g., minimizing losses due to physical leaks, meter tampering, and other factors at Embu), reducing downtime of machines (e.g., AB InBev's use of IoT and ML to predict cooler repairs), and ensuring optimum conditions (e.g., soil moisture and nutrient availability to increase agricultural productivity). By enabling accurate measurement of product and service attributes and economic agents' actions, the IoT can also facilitate exchange and transactions. The examples in this chapter show that the IoT can facilitate first- and second-party enforcement. This aspect is especially relevant to economies in Sub-Saharan Africa due to weak mechanisms for third-party enforcement of contracts.

This technology can increase the welfare of vulnerable groups such as smallholder farmers and help the population in LDCs participate in the market economy. For instance, IoT technologies can be used to provide real-time data and analysis to help improve farmers' decision making, which can reduce the risk of crop failure, decrease costs, increase farm yields, and provide market access to farm products.

IoTs can help LMICs overcome barriers related to deficient infrastructures and markets. Many examples discussed here (e.g., StoveTrace, M-KOPA, KOKO Networks) show how economic and technical considerations affect the adoption and diffusion of IoT solutions. For instance, M-KOPA was able to create a market for its products by leapfrogging East Africa's poor infrastructure. To do so, it combined mobile payment technology with the latest solar systems.

IoT solutions can help dynamically adjust a product to consumers' requirements. In this way, the IoT can influence a party's willingness to enter into a contract. It is evident from examples such as Azuri Technologies' HomeSmart and SAB that the IoT's impact can be further increased by combining with advanced technologies such as AI and ML. Only with IoT is it possible to know the device status such as "in use" or "not in use" (e.g., StoveTrace). By combining IoT with AI, proactive maintenance can be planned and scheduled, and corrective actions can be initiated before the equipment experiences a catastrophic failure (e.g., SAB's solutions).

Several innovative solutions are being developed by local firms, such as

Kenya's Illuminum Greenhouses. Such initiatives also promote the local economy by using locally available materials. The case of this startup also illustrates how affordable and simple IoT systems have the potential to improve living conditions for the B4B.[107] The case of Illuminum Greenhouses indicates that locally made intermediate inputs such as parts and components play a key role in developing IoT solutions. They also develop software locally, so there is a need for building local skills and capacity to develop the IoT industry.

The deployment of technologies tends to diffuse from large and more resourceful organizations to small or less resourceful organizations—known as the rank effect.[108] Programs such as the Togolese government's subsidy indicate that the government can accelerate diffusion by making sustainable agriculture affordable. In this way, regulators can take measures to ease new technologies' path to market.

The IoT can make first-party enforcement or even self-enforcement attractive and help individuals and enterprises live up to contracts and promises, thus reducing transaction costs. The SAB example indicates that the IoT can help increase reliability of customer service by minimizing problems such as equipment failure. Thanks to IoT solutions, firms do not need to rely on expensive and inefficient enforcement measures. In the M-KOPA example, if a customer fails to make a payment, the system can be shut off remotely and switched on after the customer pays.

107. Stewart, "Challenges."

108. Georg Götz, "Monopolistic competition and the diffusion of new technology, *Rand Journal of Economics* 30, no. 4 (1999): 679–693.

PART 3

The 4R in Economic and Social Developments

CHAPTER 6

Healthcare and Pandemic Preparedness

A healthcare system has three primary goals: to keep people healthy, to treat people who are sick, and to provide healthcare services that are affordable, efficient, and cost-effective.[1] The 4R has tremendous potential to bring about positive development in the healthcare systems of LMICs. For instance, the deployment of 5G-enabled healthcare in hospitals and clinics can help provide digital healthcare technologies in rural areas of LMICs.[2] Likewise, digital tools such as AI-driven health solutions can help LMICs leapfrog high-income countries.[3] The first of the direct benefits of AI to developing economies stems from the democratization and decentralization of medical knowledge and excellence. For instance, AI algorithms can analyze patients' symptoms and vital signs, then compare that information with the history for the patient, the patient's family, and hundreds of millions of other patients. In this way, causes of illness can be correctly and efficiently identified.[4]

Deep learning for visual tasks is advancing at a breathtaking pace. In

1. Prasad Godbole and Matthew Kurian, "Models of healthcare in developed and developing countries," *Hospital Transformation*, Springer, May 14, 2019, https://link.springer.com/chapter/10.1007%2F978-3-030-15448-6_3.

2. Joyce Mwangama, Bessie Malila, Tania Douglas, and Molebogeng Rangaka, "What can 5G do for healthcare in Africa?," *Nature Electronics*, 2020, https://doi.org/10.1038/s41928-019-0362-7.

3. Ann Aerts and Doreen Bogdan-Martin, "Opinion: Digital tool to help countries leapfrog via AI-driven health solutions," *Devex*, June 21, 2021, https://www.devex.com/news/sponsored/opinion-digital-tool-to-help-countries-leapfrog-via-ai-driven-health-solutions-100151.

4. Ben Dickson, "What is the difference between artificial and augmented intelligence?," *TechTalks*, December 4, 2017, https://bdtechtalks.com/2017/12/04/what-is-the-difference-between-ai-and-augmented-intelligence.

2017, an article in *Nature* described a system that had the same performance as dermatologists for classifying cancer.[5] Similar systems are being developed to assess other diseases such as diabetic retinopathy (a cause of blindness), stroke, bone fracture, and Alzheimer's. Such programs can be installed on smartphones to provide low-cost universal access to diagnose diseases.

This chapter focuses on the roles of 4R technologies in helping LMICs achieve key goals of a healthcare system. It also looks at the roles in these technologies in fighting the COVID-19 pandemic.

Keeping People Healthy

Many 4R technologies provide mechanisms to keep people healthy through disease prevention and health promotion programs, such as environmental protection, public health, immunization, and nutrition and diet. Nutritional deficiency is a major concern for the B4B. About 243 million people in Africa are estimated to be malnourished. Many meals are based on a high-starch diet, which has been reported to lead to the deaths of at least 15,000 children a day.[6]

For instance, consider anemia, a health condition characterized by low hemoglobin (HGB) counts.[7] Anemia is most commonly caused by nutritional deficiency.[8] In Africa, 35% of women of reproductive age are affected by anemia,[9] and even more in low-income countries such as Ethiopia[10]

Micronutrient deficiency leads to cognitive development issues, which

5. Apurv Mishra, "Deep-learning networks rival human vision," *Scientific American*, June 26, 2017, https://www.scientificamerican.com/article/deep-learning-networks-rival-human-vision1/.

6. Carlos Escueta, "Rolex honours people changing lives with projects that fight climate change and malnutrition in its 2021 Awards for Enterprise," *South China Morning Post*, August 26, 2021, https://www.scmp.com/magazines/post-magazine/fashion/article/3146415/post-edit-rolex-honours-people-changing-lives.

7. Teshager Weldegiorgis Abate, Biruk Getahun, Mekuriaw Mesfin Birhan, Getasew Mulatu Aknaw, Sefealem Assefa Belay, Dessalegn Demeke, Dagninet Derebe Abie, Adela Memberu Alemu, and Yirga Mengiste, "The urban–rural differential in the association between household wealth index and anemia among women in reproductive age in Ethiopia, 2016," *BMC Women's Health* 21, no. 1 (2021): https://doi.org/10.1186/s12905-021-01461-8.

8. Nynke van den Broek, "Anaemia and micronutrient deficiencies," *British Medical Bulletin* 67, no. 1 (2003): 149–160, https://doi.org/10.1093/bmb/ldg0049.

9. World Health Organization, *The global prevalence of anemia in 2011*. 2015, https://apps.who.int/iris/bitstream/handle/10665/177094/9789241564960_eng.pdf.

10. Aklilu Alemayehu, Lealem Gedefaw, Tilahun Yemane, and Yaregal Asres, "Prevalence, severity, and determinant factors of anemia among pregnant women in South Sudanese Refugees, Pugnido, Western Ethiopia," *Anemia* (2016): 1–11.

has been estimated to result in a loss of as much as 5% of GDP in LMICs. The direct healthcare costs related to such deficiencies are estimated at US$20 billion–$30 billion annually. IoT solutions have been deployed to address this issue (In Focus 6.1).

In Focus 6.1: Sanku's Project Healthy Children Uses IoT Solutions for Small-Scale Fortification

Sanku, a nonprofit social enterprise, runs Project Healthy Children, which aims to address micronutrient deficiency. Sanku's dosifier enables small mills in rural Africa to fortify flour in a sustainable and cost-effective manner. The mills can add precise amounts of nutrients during the milling process. By 2017, over 100 mills were operating across Tanzania. At that time, the only way to ensure that the mills were using dosifier machines correctly was to drive for hours to visit each one.[11]

Sanku can now remotely monitor equipment and production. The program can precisely measure and mechanize the process of adding iron, folic acid, vitamin B12, and zinc to cereal grains when milled[12] Each mill's production data is also stored and sent to Sanku dashboards every five minutes, then linked to Sanku's inventory management system, which allows the company to automate deliveries of flour bags and premixed nutrients. The data also helps monitor miller compliance and fortification accuracy. Moreover, the mills can alert Sanku of dosifier maintenance issues.

Partnership with Vodafone

Sanku teamed up with Vodafone to develop IoT solutions to enable the mills' fortification processes remotely. Sanku dosifiers are connected with Vodafone's IoT SIM. Vodafone's IoT solutions involve installing a "cellular module" in each dosifier. The millers pay Sanku using Vodafone's M-PESA mobile payment service[13]

Vodafone's in-country roaming services are available even in the most remote areas, which makes it possible to access real-time

11. Vodafone, *How Sanku uses IoT to fight malnutrition, helping children have a brighter future*, 2020, https://www.vodafone.com/business/news-and-insights/blog/gigabit-thinking/how-sanku-is-using-iot-to-fight-malnutrition-helping-the-children-of-tomorrow-have-a-brighter-future.

12. Vodafone, *Flour power: How IoT is connecting mills and saving lives*, October 16, 2019, https://www.vodafone.com/news/technology/flour-power-iot-connecting-mills-saving-lives.

13. "Sanku Dosifier: The 100 Best Inventions of 2019," *Time*, 2019, https://time.com/collection/best-inventions-2019/5733123/sanku-dosifier/.

information related to maintenance, power supply, and machine tracking. Sanku can accurately monitor fortification levels to ensure that consumers receive the benefits of fortified flour.[14]

Key Milestones

Sanku's dosifiers for small-scale fortification were introduced in 2013 and made the list of "Time's Best Inventions" in 2019.[15] As of 2018, Sanku was working with 150 flour mills and serving 1 million consumers. As of November 2020, Sanku's Project Healthy Children reached 2 million people with its fortified flour in five countries across East and South Africa.[16] By August 2021, Sanku had outfitted 300 flour mills in East Africa. The goal is to reach 15,000 more by 2025.[17] Sanku estimates that if it can replace 15,000 traditional mills with Sanku dosifiers, 100 million consumers' lives will be positively impacted.[18]

According to the WHO, polluted air contributes to one in eight deaths worldwide via lung damage, heart disease, stroke, and cancer. The WHO also estimates that indoor air pollution in homes in Africa contributed to about 600,000 deaths in 2012. As discussed in chapter 5, attempts have been made to help consumers switch from using dirty fuels such as open fires, traditional stoves, and kerosene to clean-burning stoves (In Focus 6.2) and cleaner-burning LPG.

In LMICs, exposure to air pollution is often extreme, and levels of indoor and outdoor air pollution are significantly higher than in high-income economies[19] Indoor and outdoor air pollution ranked as the 10th and 13th leading causes of mortality in the 2001 WHO Global Burden of Disease (GBD) Report.[20]

---

14. Juan Pedro Tomás, "Vodafone helps connect flour mills in Africa with IoT technology," *Enterprise IoT Insights*, June 8, 2018, https://enterpriseiotinsights.com/20180608/internet-of-things/vodafone-helps-connect-flour-mills-africa-iot-technology-tag23.

15. "Sanku Dosifier."

16. "Project healthy children introduces nutrition technology," *Borgen*, November 10, 2020, https://www.borgenmagazine.com/project-healthy-children/.

17. Escueta, "Rolex Awards."

18. Vodafone, *Sanku fights against malnutrition*, 2019, https://www.vodafone.com/business/news-and-insights/case-study/sanku-fights-against-malnutrition.

19. Sanjay Rajagopalan and Robert D. Brook, "Indoor-outdoor air pollution continuum and CVD burden: An opportunity for improving global health," *Global Heart* 7, no. 3 (2012): 207, https://www.ncbi.nlm.nih.gov/pmc/articles/PMC3501678.

20. Alan D. Lopez, Colin D. Mathers, Majid Ezzati, Dean T. Jamison, and Christopher

Indoor air pollution is a major public health hazard for large numbers of the world's poorest and most vulnerable people. It may be responsible for a similar proportion of the global burden of disease as risk factors such as tobacco use and unsafe sex.

Appropriate interventions can significantly decrease exposure to indoor air pollution, including by improved stoves, cleaner fuels, and behavioral changes. To be effective, such interventions should meet domestic energy and cultural needs, and also improve safety, fuel efficiency, and environmental protections.[21] They should also be affordable and sustainable. Solutions that are associated with income generation and credit arrangements are preferable. Cleaner fuels such as liquefied petroleum gas are among the best long-term options (In Focus 5.3). Nonetheless, most poor communities that are currently using biomass are likely to make the transition to LPG eventually.[22] For such communities, IoT-based clean-burning stoves can reduce indoor air pollution (In Focus 6.2).

4R technologies are also being deployed to fight outdoor air pollution. Air-quality sensors have been deployed in several cities to track levels and changes in pollutants. One such project is Fresh Air in Benin, which utilizes a network of air-quality sensors to capture and send data every 20 minutes via GSM connectivity.[23]

In Focus 6.2: StoveTrace's IoT-Based Clean-Burning Stoves

As of 2018, about 3 billion people cooked food on open fires or traditional stoves, which consist of a few mud bricks on the floor. These emit black carbon, which is noxious and adds to the global greenhouse effect. According to the World Health Organization, about 4 million people die every year from illnesses attributed to household air pollution. Black carbon is arguably second only to carbon dioxide (CO_2) in its contributions to global warming.

J. L. Murray, "Global and regional burden of disease and risk factors, 2001: Systematic analysis of population health data," *Lancet*, 367, no. 9524 (2006): 1747–1757, https://doi.org/10.1016/S0140-6736(06)68770-9.

21. Nigel Bruce, Rogelio Perez-Padilla, and Rachel Albalak, "Indoor air pollution in developing countries: A major environmental and public health challenge," *Bulletin of the World Health Organization*, 2021, https://www.scielosp.org/article/bwho/2000.v78n9/1078-1092/en.

22. Bruce et al., "Indoor air pollution."

23. *Harnessing IoT global development*, 2016, International Telecommunication Union, https://www.itu.int/en/action/broadband/Documents/Harnessing-IoT-Global-Development.pdf.

The Los Angeles–based nonprofit Nexleaf Analytics' StoveTrace project has introduced clean-burning stoves to address these challenges in India's Odisha state. The stoves are various types of "forced draft" appliances that look like metal cylinders and make a fire burn hotter and cleaner. They are also more efficient and use less wood.

A local "energy entrepreneur" provides loans to village women to buy the stoves, which cost about US$80. Each stove has a thermal sensor wired to a wall-mounted communications module that records data when the stove is in use. In the first iteration of the project, the modules used the cellular network to send information to the cloud in real time. The subsequent version used a cheaper, simpler gadget to store data, which are not sent in real time. In this version, the local entrepreneur visits the house and connects to the module via Bluetooth to download the data.

The Global Alliance for Clean Cookstoves, which has been backed by the United Nations Foundation and dozens of governments, corporations, and foundations, distributed over 80 million stoves during 2010–2016. But it is not possible to know the proportion of them that are working and being used regularly. StoveTrace's sensors make it possible for local energy entrepreneurs to know if the stoves are being used. If any are not being used, they can visit households for troubleshooting or repairs.

The sensors have an additional function. Based on each household's stove usage, small monthly payments are made to households via a mobile banking app. For instance, a household that cooks regularly on the new stove receives about US$4 per month, considered a "climate credit." The money can be used to make payments on the loan to buy the stove.[24] StoveTrace formed a partnership with Vodaphone M-PESA in India to make direct payments to women's mobile phones. Each woman receives notification of the payment. Customers without a phone receive direct payments from the cash-out agent.[25]

4R technologies can also help keep people healthy by preventing illness. Life sciences technologies such as biotechnology and genomics are expected to help eradicate diseases such as malaria and Zika virus. For instance, the

24. Elisa Strickland, "How sensors and mobile payments are getting Indian women to use cleaner cookstoves," *IEEE Spectrum*, June 4, 2018, https://spectrum.ieee.org/how-sensors-and-mobile-payments-are-getting-indian-women-to-use-cleaner-cookstoves.

25. Nexleaf Analytics, *IoT for development: How data can steer clean cooking*, September 26, 2020, https://nexleaf.org/impact/iot-for-development-stovetrace.

gene drive process can be used to force certain genes to pass from one generation to the next and ultimately throughout the population of a species. By spreading edited genomes through an entire population, it is possible to prevent species of mosquitos that are vectors for such diseases from reproducing. The same process can help engineer endangered species so that they are less affected by adverse climate changes.[26] With rapid advances in genomic sciences, such possibilities are increasingly feasible. Scientists can test very rapidly and cheaply how specific genetic variations generate particular traits and diseases.[27] The first sequencing of the whole human genome in 2003 cost about US$2.7 billion and took 13 years among thousands of scientists. In 2012, it took several months to sequence a person's genome.[28] In 2018, it became possible to map someone's genome 30 times in 40 hours, which made it possible to eliminate scanning errors. It cost less than US$1,000.[29] Illumina, a developer of life science tools, says that "one day," this can be done for less than US$100.[30]

While clinical datasets such as genomes are biased against people of African ancestry, some groundbreaking genome-editing technologies focusing on this population have been developed thanks to growing African genome data. Genome editing involves making changes to the DNA of a cell or an organism. By doing so, scientists can test how specific genetic variations generate particular traits and diseases. Looking at the patterns in huge datasets of genetic information and medical records, scientists can examine mutations and linkages in diseases. They can then personalize treatment and protocols for a patient. This type of approach is referred to as precision medicine; it relies on genetic information of a person or group for the diagnosis or treatment for that person or group. In chronic disease, this may play a critical

26. Scott Moore, "China's biotech boom could transform lives—or destroy them," *Foreign Policy*, November 8, 2019, https://foreignpolicy.com/2019/11/08/cloning-crispr-he-jiankui-china-biotech-boom-could-transform-lives-destroy-them/.

27. Klaus Schwa, *The Fourth Industrial Revolution*, World Economic Forum, 2016, https://luminariaz.files.wordpress.com/2017/11/the-fourth-industrial-revolution-2016-21.pdf.

28. Bonnie Rochman, "TIME explains: Genome sequencing Time's Bonnie Rochman explains how doctors are mapping out all the genes in a person's body," *Time*, October 22, 2012, http://healthland.time.com/2012/10/22/time-explains-genome-sequencing/.

29. Sandoz, *Genomics and big data—unlocking the code to new therapies*, 2018, https://www.sandoz.com/stories/access-medical-information/genomics-and-big-data-unlocking-code-new-therapies.

30. Sarah Buhr, "Illumina wants to sequence your whole genome for $100," *Tech Crunch*, January 10, 2017, https://techcrunch.com/2017/01/10/illumina-wants-to-sequence-your-whole-genome-for-100/.

role in increasing treatment efficacy. In December 2020, scientists successfully used gene editing to treat sickle cell disease, an inherited red-blood-cell disorder common among people of African ancestry. To treat persons with this disease, scientists needed to identify and cut out a gene that instructs the production of the defective sickle-shaped cells in those persons. Doing so requires sequencing of the entire genome that makes up the DNA of people of African ancestry.[31] Ten participants had received the treatment, which did not require hospitalization or blood transfusion. They were also free from pain and side effects.[32]

These technologies have also helped administer vaccines in a timely and targeted manner to prevent COVID-19 infection and hepatitis C. The WHO has been monitoring vaccination records, treatments, and prevention methods in Mongolia with Rymedi's blockchain system.[33] Rymedi's focus has been on deploying hepatitis C vaccines in Mongolia.[34] About 9% of the population in Mongolia has hepatitis C.[35] The process includes screening the population for hepatitis C, rapid field diagnostics, viral load testing, e-prescription, treatment delivery, and monitoring patient outcomes. With Kadena's blockchain technology, Rymedi can capture, track, and share data related to these processes faster and more securely. The data are linked to electronic medical records. These processes have been reported to strengthen the mechanisms for quality assurance in healthcare and improve supply chain coordination.[36] Medicinal products being tracked by Rymedi and Kadena include antibiotics, hepatitis C cures, cannabis products, kidney dialysis, and treatments for cancer, HIV, and neonatal pulmonary issues. The companies have deployed their solutions in Kenya, Mongolia, and the

31. Uwagbale Edward-Ekpu, "Nigeria boosts genetics medicine for Africa with new labs," *Quartz*, 2020, https://qz.com/africa/1945960/nigerian-labs-acegid-54gene-boost-genetics-medicine-in-africa/.

32. H. Frangoul et al., "CRISPR-Cas9 gene editing for sickle cell disease and β-Thalassemia," *New England Journal of Medicine* 384, no. 3 (2021): 252–260.

33. Luke Fitzpatrick, "Fantom's blockchain tech is being trialed in Afghanistan to solve 'a surprising' World Health Organization (WHO) problem," *Forbes*, July 6, 2020, https://www.forbes.com/sites/lukefitzpatrick/2020/07/06/fantoms-blockchain-tech-is-being-trialed-in-afghanistan-to-solve-a-surprising-world-health-organization-who-problem/#6bd9eec23844.

34. Eli Richman, "JPMorgan spinoff Kadena puts blockchain on AWS—and on the table for small and medium-sized businesses," *Fierce Healthcare*, January 23, 2019, https://www.fiercehealthcare.com/tech/jp-morgan-spinoff-kadena-puts-blockchain-development-platform-aws.

35. Charles Brett, "Kadena and Rymedi partner to validate medicinal products quality," *Enterprise Times*, December 9, 2019, https://www.enterprisetimes.co.uk/2019/12/09/kadena-and-rymedi-partner-to-validate-medicinal-products-quality/.

36. Brett, "Kadena and Rymedi."

United States. As of 2019, they were planning to expand services to the Caribbean and Central America.[37]

Treating Sick People

Four billion people in the world lack access to basic health services.[38] The global shortfall in health workers is expected to exceed 12.9 million by 2035.[39] 4R technologies have provided a solution for an overburdened healthcare system. AI tools have become increasingly powerful for increasing access to healthcare services in LMICs (In Focus 6.3). One such solution is Germany-based Ada Health's chatbot symptom checker app, launched in 2016. As of May 2021, the app had attracted more than 11 million users worldwide, including about 2 million in Brazil and India and more than 3 million in LMICS.[40] In a study of the eight most popular online symptom assessment apps, the most comprehensive app was Ada (in terms of possible conditions and user types), which provided a condition suggestion in 99% of cases.[41]

The AI tool as an "agnostic" platform, which can be tailored and integrated into existing health systems.[42] The app is free to download. Users input symptoms and preexisting medical conditions. AI-based questioning then pinpoints possible diagnoses and recommends next steps, such as resting or seeking professional help.[43]

Ada Health has been working in Tanzania since 2017, partnering with

37. Brian Njuguna, "Authentication of medicinal products made possible by blockchain technology," *Blockchain News*, December 10, 2019, https://blockchain.news/news/authentication-of-medicinal-products-made-possible-by-blockchain-technology.

38. "Global shortage of health workers expected to keep growing, UN agency warns," *UN News*, November 11, 2013, https://news.un.org/en/story/2013/11/455122-global-shortage-health-workers-expected-keep-growing-un-agency-warns.

39. "Global shortage."

40. Sam Shead, "Samsung and Bayer invest in A.I. doctor app Ada Health," *CNBC*, May 27, 2021, https://www.cnbc.com/2021/05/27/samsung-and-bayer-invest-in-ai-doctor-app-ada-health.html; Ben Turner, "Tanzania's digital doctor learns to speak Swahili," *Financial Times*, May 16, 2020, https://www.ft.com/content/7ed03336-6a0a-11ea-a6ac-9122541af204.

41. Chloe Kent, "Significant disparities found in symptom assessment apps," *Medicaldevice-network.com*, December 16, 2020, https://www.medicaldevice-network.com/news/symptom-assessment-apps/?utm_source=Army%20Technology&utm_medium=website&utm_campaign=Must%20Read&utm_content=Image.

42. Adeola Eribake and Suzi Ring, "Covid-19 telehealth boom picks up pace in East Africa," *Fin24*, August 24, 2021, https://www.news24.com/fin24/companies/covid-19-telehealth-boom-picks-up-pace-in-east-africa-20210820.

43. Turner, "Tanzania's digital."

the Swiss philanthropic organization Fondation Botnar. Tanzania has one doctor per 25,000 people. Ada Health partnered with Botnar to launch a localized version of its app in Swahili in 2019.[44] The Ada app is reported to optimize 160 disease models to ensure that the app correctly factors in conditions and symptoms more common in Tanzania and East Africa than in other parts of the world.[45]

In Focus 6.3: Babylon Health Offers Diagnosis and Video Appointments in LMICs

The London-based health startup Babylon Health offers a digital healthcare app for diagnosis and video appointments. As of June 2021, it covered 24 million people in four continents.[46] Babylon Health partnered with Samsung to incorporate its Ask an Expert video doctor appointments to Samsung's Health app.[47] Babylon also operates in LMICs such as Cambodia, India, Indonesia, Laos, Malaysia, the Philippines, Thailand, Rwanda, and Vietnam.[48]

Patients who need care are directed to an in-person consultation or digital and phone service, which helps relieve some in-person burden. It can also speed up processes such as validating insurance. Babyl plans to launch an AI-powered tool for health centers.[49]

Babylon launched in Rwanda under the rebranded name Babyl. In early 2018, Babyl announced "the first ever fully digital healthcare service in east Africa using artificial intelligence." The service included a chatbot "to take the power of a doctor's brain and put it on a mobile phone for medical advice and triage." By early 2021, it had 2 million users in Rwanda and handled 3,500 daily consultations.[50]

44. Eribake and Ring, "Covid-19."

45. Ada, *World's first AI health guidance app in Swahili*, 2019, https://ada.com/press/191119-worlds-first-ai-health-guidance-app-in-swahili/.

46. Heather Landi, "Digital health player Babylon Health to go public via $4.2B merger with blank check company," *FierceHealthcare*, June 3, 2021, https://www.fiercehealthcare.com/digital-health/another-day-another-spac-digital-health-player-babylon-health-to-go-public-via-4-2b.

47. Piotr Wnuk, "Will Babylon's app ever speak doctors' language?," *pharmaphorum*, July 12, 2018, https://pharmaphorum.com/views-and-analysis/will-babylons-app-ever-speak-doctors-language/.

48. "Virtual care: the next step in the evolution of healthcare," *Investing News Network*, 2021, https://investingnews.com/innspired/virtual-care-healthcare-evolution/.

49. Eribake and Ring, "Covid-19."

50. Andrew Jack, "Rwanda venture tests digital health potential in developing world," *Financial Times*, January 17, 2021, https://www.ft.com/content/4fe33c92-cbd5-459a-8df6-20d0d1f57ec8.

Some Criticisms

Babylon's system in the UK has been found to be of questionable effectiveness. A study found that its online symptom checkers "lack the functions to support the whole diagnostic process of an offline medical visit."[51] The system also focuses on particular diseases. Another study found that digital symptom checkers globally were used primarily by younger and more educated people. There was no sufficient evidence as to whether people take up the medical advice given.[52] The consulting firm Dalberg's evaluation of Babyl in 2018 found the solution can cut costs, which also includes the development of more efficient EHRs. It was also noted that with face-to-face consultations, Babyl has a higher "risk of fraud through false impersonation" by callers. The consulting firm also emphasized the importance of adjusting symptom-checking algorithms to "local health and disease patterns and to language and communication practices." The company's costs have increased due to a requirement to store all the data on a local cloud server hosted in Rwanda.

A review of the quality watchdog Care Quality Commission (CQC) found that Babylon did not provide a safe service in relation to prescribing regulations and best practice.[53] The CQC found that information was not always shared with a patient's primary doctor to ensure that prescribing was safe and appropriate. It also lacked a system to ensure that patients' conditions were being monitored appropriately.[54]

A further criticism Babyl's service faced in Rwanda is that it did not take local epidemiology into account. Rwanda's minister of health claimed that the Babyl app had no questions about

51. Yue You and Xinning Gui, "Self-diagnosis through AI-enabled chatbot-based symptom checkers: User experiences and design considerations," *AMIA 2020 Annual Symposium*, 2020.

52. Duncan Chambers, Anna J. Cantrell, Maxine Johnson, Louise Preston, Susan K. Baxter, Andrew Booth, and Janette Turner, "Digital and online symptom checkers and health assessment/triage services for urgent health problems: Systematic review," *BMJ Open*, August 1, 2019, https://bmjopen.bmj.com/content/9/8/e027743.

53. Andrew McConaghie, "Online doctor service Babylon tried to gag critical report," *pharmaphorum*, December 22, 2017, https://pharmaphorum.com/news/online-doctor-service-babylon-tried-gag-critical-report/.

54. Richard Staines, "National roll out of Babylon's NHS online doctor could be delayed," *pharmaphorum*, January 16, 2018, https://pharmaphorum.com/news/national-roll-nhs-online-doctor-delayed/.

> malaria.[55] Babylon founder Ali Parsa also noted that in the initial phase, Rwanda-specific data were not included in the Babyl system: "People are hyping AI often because they want to get finance. The reality is we are in day one. It's really in early infancy. AI will utterly outperform our wildest imaginations in years to come and utterly disappoint us in the short term."

The role of AI tools is becoming especially apparent in specialized medical consultations in LMICs. This is because medical AI is already doing things that even the world's best medical professionals are not capable of doing. AI is likely to have a powerful impact on areas such as ophthalmology and radiology. For instance, AI algorithms can examine radiology images much faster than humans.[56] Likewise, several AI apps use images of human eyes that practitioners, technicians, or even patients can diagnose for possible diseases. They no longer need the expertise of ophthalmologists. This is especially important in LMICs that have a severe shortage of health specialists.[57]

Making Healthcare Services Affordable, Efficient, and Cost-Effective

4R technologies are helping patients in LMICs receive better and more affordable healthcare services. Locally developed solutions are making healthcare more affordable. Aravind Eye Hospital Smart Vision Spectacles (SVS) cost US$292, compared to more than US$5,000 for smart eyewear products in advanced countries.[58]

4R technologies affect healthcare services not only by reducing costs but also through other mechanisms. For instance, different fraud exists in the healthcare sectors. 4R technologies provide various mechanism to prevent fraudulent actions, which can increase the efficiency and decrease costs of healthcare services. Likewise, 4R technologies such as drones can help

55. W. D. Heaven, "Your next doctor's appointment might be with an AI," *MIT Technology Review*, 2018, https://www.technologyreview.com/2018/10/16/139443/your-next-doctors-appointment-might-be-with-an-ai.

56. Ben Dickson, "What is the difference between artificial and augmented intelligence?," *TechTalks*, December 4, 2017, https://bdtechtalks.com/2017/12/04/what-is-the-difference-between-ai-and-augmented-intelligence.

57. W. Nicholson Price II, "Risks and remedies for artificial intelligence in health care," *Brookings*, 2019, https://www.brookings.edu/research/risks-and-remedies-for-artificial-intelligence-in-health-care/.

58. "Good news for those with low or no vision," *The Hindu*, August 18, 2021, https://www.thehindu.com/news/cities/Madurai/good-news-for-those-with-low-or-no-vision/article35964002.ece.

deliver medicines, vaccines, and other health products to the populations that live in remote and inaccessible locations.

Fighting Fraud in the Insurance Sector

The pervasiveness of fraud in the insurance sector is a concern in the LMICs. One estimate suggested that false claims in the Indian healthcare insurance industry account for 10%–15% of total claims. The industry is estimated to lose about US$90 million to false claims annually.

Major fraud categories or schemes reported include misrepresented services, services not provided, and services provided to "rented" patients. Likewise, the Chinese insurance industry suffers from rampant abuses and malpractice committed by patients and medical staff. In Lipanshui city in Guizhou, fraud cases were found in 107 of the 135 hospitals and medical centers. All hospitals in Anshun were found to engage in mismanagement of medical insurance. Some medical staff had provided fake medical records to get payments for treatments that were not performed. Such practices can be prevented with blockchain.

Fraudulent practices in the insurance sector can be reduced by performing automated verifications of policyholder identities and contract validity using blockchain. Submission and registration of claims are done online and are auditable. Relevant data such as encrypted data on injured parties prepared by hospitals and medical centers are obtained from third parties, which are made accessible to the insurer to verify payment. Payouts for claims can be made via a blockchain-based infrastructure or smart contracts.

Fighting Fake Products

According to the WHO, 42% of all fake medicines reported to the organization between 2013 and 2017 were from Africa.[59] About 250,000 children in Africa die every year from the use of fake drugs to treat diseases such as malaria or pneumonia.[60] Medicines in Africa change hands as many as thirty times before reaching a pharmacy retail point.[61] The manufacturers of legiti-

59. World Health Organization, *Global surveillance and monitoring system for standard and falsified medical products*, 2017, https://apps.who.int/iris/bitstream/handle/10665/326708/9789241513425-eng.pdf.?ua=1.

60. "Fake medicine kills hundreds of thousands in Africa," *Deutsche Welle*, 2019, https://www.dw.com/en/fake-medicine-kills-hundreds-of-thousands-in-africa/av-51535450.

61. "Cure for counterfeit drugs?," *Forbes Africa*, 2020, https://www.forbesafrica.com/health/2020/01/21/cure-for-counterfeit-drugs/.

mate products find it difficult to track their products, and it is not possible to trace a medicine back to its origin.

4R technologies such as AI (In Focus 6.4) and blockchain are used in LMICs to fight fake products. Blockchain can be used to track the entire supply chain from the raw materials used in pharmaceutical production to finished products sold by drug manufacturers to hospital and to end users could be the most effective way to fight counterfeit and substandard drugs (see In Focus 6.5 and 6.6). Afghanistan's Ministry of Health teamed up with smart contract platform Fantom to fight substandard or falsified medicines in the county. In July 2020, the ministry announced a trial of a blockchain-based smart medicine pilot program on Fantom's Opera main net, which is permissionless and open-source blockchain.

In Focus 6.4: RxAll's AI-Based Handheld Device to Fight Fake Drugs

The Nigerian startup RxAll launched an AI-based handheld device to fight fake drugs, which according to the World Health Organization (WHO) is a US$200 billion industry. It assesses a drug's compounds by connecting the device to a cloud-based database. The database contains information related to what the drugs should contain. The information is sent back to an app, and the database updated using AI. It has been used in Myanmar. The company also plans to enter into other developing countries such as Ghana, Cambodia, and Kenya.

According to the plan, pharmaceutical products of two Indian companies—Bliss GVS and Nabros Pharma—would be attached with labels provided by Fantom. Fantom's plan was to work with the Nigerian health startup Chekkit, which uses ID labels—secured on the ethereum blockchain and unstructured supplementary service data (USSD) or QR-code scanning to verify a product's authenticity.[62] The pharmaceutical distributor Royal Star Pharma would scan labels at each step of the distribution process. The scanned data is recorded in the Fantom blockchain using cryptographic encryption. The Royal Star and the Ministry of Public Health can verify products' authenticity by comparing the hash of the data stored in the blockchain to the hash printed on the label. The plan was to use blockchain

62. Samuel Nwite, "Chekkit partners with fantom to provide anti-counterfeit technology for Afghan's pharmaceutical industry," *Tekedia*, September 3, 2020, https://www.tekedia.com/chekkit-partners-with-fantom-to-provide-anti-counterfeit-technology-for-afghans-pharmaceutical-industry/.

to track supply chains of 80,000 products, including hand sanitizers, joint cream, kofanol chewable tablets, and diacare foot cream.[63]

Fantom's DAG-based smart contract platform has a high transactional throughput, which provides fast confirmation times at the required scale to power a country's entire healthcare system.[64] Fantom claims that its Lachesis algorithm confirms transactions in one to two seconds.[65]

In Focus 6.5: Indian Government Uses Blockchain to Fight Counterfeit Drugs

According to the WHO, India accounts for 35% of counterfeit drugs distributed worldwide.[66] The Indian government aims to fight this problem with blockchain.[67] In 2018, the Indian policy think-tank Niti Aayog and Oracle signed an agreement to use blockchain and IoT to conduct a pilot to track India's domestic pharmaceutical supply chains.[68] Among participants in the pilot were India's hospital chain Apollo Hospitals and the drug manufacturer Strides Pharma Science. Oracle reported that the pilot was successful. Oracle's blockchain registers a drug's record such as serial number and labeling in the manufacturer's drug supply chain.

Drugs were tracked during the supply chain journey from manufacturer to logistics to distributors to hospital or pharmacy to consumers. If a fake drug enters the system, the software detects the irregularity and notifies each node. The IoT tracked information such as the drug's chemical ingredients and the maintenance of acceptable temperature for some drugs or vaccines.[69]

---

63. Fitzpatrick, "Fantom's blockchain."

64. Nwite, "Chekkit."

65. Fantom, *What is Lachesis?*, https://fantom.foundation/lachesis-consensus-algorithm/.

66. Government of India, *Report of the expert committee on a comprehensive examination of drug regulatory issues, including the problem of spurious drugs* (New Delhi: Ministry of Health and Family Welfare, 2003).

67. A. Sharma, "AI spending to grow 43% in Middle East and Africa in 2019," *The National*, 2019, https://www.thenational.ae/business/technology/ai-spending-to-grow-43-in-middle-east-and-africa-in-2019-1.917352.

68. Oracle, *NITI Aayog, Oracle, Apollo Hospitals and Strides Pharma Sciences come together to end India's growing battle against fake drug distribution*, September 28, 2018, https://www.oracle.com/in/corporate/pressrelease/niti-aayog-oracle-pilot-real-drug-supply-chain-with-blockchain-iot-2018-09-28.html.

69. Sohini Bagchi, "Blockchain ensures greater transparency between brand and customer," *CXO Today*, July 24, 2020, https://www.cxotoday.com/news-analysis/blockchain-ensures-greater-transparency-between-brand-and-customer/.

In Focus 6.6: Uthabiti's Blockchain Solutions for Traceability in Kenya's Pharmaceutical Supply Chain

Kenya's web and mobile platform Uthabiti, used to consult doctors and other health auxiliaries online and to order prescribed or nonprescribed medicines from pharmacies, launched a blockchain solution to address counterfeit drugs.[70] After procuring medicines from manufacturers, Uthabiti performs a quality test at its own laboratory. The medicines are labeled with the product's safety report. Each medicine also has a unique blockchain ID. The medicines are then sent to partnering retailers, which allows Uthabiti to know the location of its products.[71]

Patients buying the retail medicines can verify authenticity through a mobile app or via SMS service.[72] Uthabiti's system checks against the database provided by the manufacturers to verify the product's legitimacy and replies to the query automatically.

As of August 2020, Uthabiti had partnered with four manufacturers and distributors of sexual health products. It had completed 2,300 deliveries and verified 5,000 scans.[73]

Overcoming Barriers to Accessibility

The challenge of overburdened healthcare systems in many LMICs is exacerbated by the remoteness and inaccessibility of the population. The remoteness of a location negatively affects the delivery and quality of healthcare services.[74] For instance, half of Ghana's population lives in rural or remote areas, where transporting refrigerated vaccines by road is a big challenge. In remote places with difficult terrain, a journey can take hours by car or truck.

Such barriers can be overcome through effective utilization of drones. Drones deliver packages directly straight to doctors in minutes without any direct human interaction. Drones fly close to 100 miles round-trip on a

70. International Telecommunications Union, *WSIS Team Report on the WSIS stocktaking 2019*, 2019, https://www.itu.int/net4/wsis/forum/2019/Files/Outcomes/DRAFT-WSISStocktakingReport2019-en.pdf.

71. Khumo Theko, "Medical counterfeit," March 8, 2020, http://iseeafrica.co.za/2020/03/08/medical-counterfeit/.

72. Forbes Africa, "Cure for."

73. Civicus, *Goalkeepers Youth Action Accelerator Towett Ngetich Kenya*, 2020, https://www.civicus.org/index.php/what-we-do/innovate/actionaccelerator/towett-ngetich/.

74. Gary McLean, Bruce Guthrie, and Matt Sutton, "Differences in the quality of primary medical care services by remoteness from urban settlements," *BMJ Quality & Safety* 16, no. 6, https://qualitysafety.bmj.com/content/16/6/446.short.

single battery charge, traveling up to 80 miles per hour.[75] The American medical product delivery company Zipline has drones (Zipline Zips) that deliver about four pounds (1.80 kg) of payload in under an hour over a 20,000-square-kilometer area surrounding a distribution center.[76] As of 2021, Zipline was delivering vaccines, medical supplies, and equipment to remote Ghana and Rwanda. The ultra-cold-chain technology needs a drop site the size of about two standard parking spaces.

Rwanda built an aerial network of drones to deliver medical supplies to remote villages. Consequently, drone delivery services have provided access to life-saving treatments. Zipline partnered with the Rwandan government to develop commercial drone delivery services for medical supplies to hospitals located in remote areas. The project was launched in 2016.

Zipline's drones now deliver blood products such as red blood cells, platelets, and plasma. Zipline operated two distribution centers in Rwanda, providing coverage to 80% of Rwanda.[77] Every day, about 150 deliveries are made by each distribution center, half a ton of freight. By September 2019, more than 13,000 deliveries of blood products had been made.[78] By September 2021, millions of units of blood, medical products, and vaccines had been delivered.[79]

In 2018, the government of Ghana approved a four-year contract worth US$12.5 million for Zipline to deliver blood and other medical supplies by drone. Zipline started its operation in Ghana in 2019. The plan is to make between 100 and 150 deliveries per day at an estimated cost of US$17 per delivery.[80] By September 2021, Zipline Ghana had completed over 105,000 deliveries, distributing more than 1 million medical commodities and over 3

---

75. Grace Dean, "Drones in Ghana deliver COVID vaccines to rural communities," *Business Insider*, March 5, 2021, https://www.businessinsider.com/covid-vaccine-ghana-drones-covax-who-coronavirus-zipline-rural-communities-2021-3.

76. Melissa Rusanganwa, "Five years later, the Rwandan Government, Zipline International have created an instant logistics network," *New Times*, September 1, 2021, https://www.newtimes.co.rw/opinions/five-years-later-rwandan-government-zipline-international-have-created-instant-logistics.

77. Elyse Graham, "Zipline and its implications on data colonialism," *Urge to Help*, August 19, 2021, https://theurgetohelp.com/articles/zipline-and-its-implications-on-data-colonialism/.

78. Ady Namaran Coulibaly, "Drone delivery of medical supplies in remote areas in Rwanda and Ghana and the realisation of SDG 3," *Modern Ghana*, October 1, 2019, https://www.modernghana.com/news/958501/drone-delivery-of-medical-supplies-in-remote-areas.html.

79. Rusanganwa, "Five years later."

80. Evan Ackerman, "In the air with Zipline's medical delivery drones," *IEEE Spectrum*, 2019, https://spectrum.ieee.org/in-the-air-with-ziplines-medical-delivery-drones.

million vaccine doses (including for COVID-19 and childhood diseases).[81] Zipline Ghana had four distribution centers that served over 610 centers and made an average of 150 flights a day.[82] Switzerland's healthcare company Novartis teamed up with Zipline to deliver medicines for sickle cell disease by drone to rural Ghana.[83]

Local 4R Innovations in LMICs

In recent years, firms in LMICs have introduced a number of healthcare innovations. In this section, I discuss a few useful solutions based on 4R technologies to treat various types of illness.

The prevalence of disabling hearing loss in children and seniors is highest in the developing economies in South Asia, Asia Pacific, and Sub-Saharan Africa. According to the WHO, prevalence in children is negatively related to a parent's literacy rate.[84] The South African startup hearX Group has developed a mobile app that provides a low-cost, clinically validated means to screen for hearing loss. People need only minimal training to use the app, which is thus accessible to poor people. The average test time is less than a minute. The app is becoming popular in multiple settings, including schools.[85] Also, hearX Group's dbTrack personal in-ear sound-tracking solution uses sound sensor technology in earphones to track and measure in-ear sound levels. The app provides feedback on sound exposure and safe listening time in real time.[86] It is described as a health tracker for the ears.[87]

Another example is Kenyan nonprofit organization inABLE's solution

---

81. "Zipline expands drone delivery of UK-donated AstraZeneca Covid-19 vaccines to health facilities in Ghana," *Modern Ghana*, 2021, https://www.modernghana.com/news/1103529/zipline-expands-drone-delivery-of-uk-donated-astra.html.

82. "MTN business reaffirms commitment to support Zipline with uninterrupted internet services," *News Ghana*, 2021, https://newsghana.com.gh/mtn-business-reaffirms-commitment-to-support-zipline-with-uninterrupted-internet-services.

83. Patrice Matchaba, "How to start a digital healthcare revolution in Africa in 6 steps," *World Economic Forum*, August 26, 2019.

84. WHO, *Global estimates on prevalence of hearing loss*, 2012, https://www.who.int/pbd/deafness/WHO_GE_HL.pdf.

85. Shujaat Ali, "IoT aids healthcare: 8 examples from around the world," *IoT Now*, April 23, 2020, https://www.iot-now.com/2020/04/23/102387-iot-in-healthcare-8-examples-from-around-the-world/.

86. Hearx Group, *New hearing prevention technology wins world*, 2018, https://www.hearxgroup.com/blog/NEW-HEARING-PREVENTION-TECHNOLOGY-WINS-WORLD-ITU-AWARD.html.

87. Julian Horsey, "Westone dbTrack sound monitoring earphones," *Geeky Gadgets*, December 4, 2018, https://www.geeky-gadgets.com/westone-dbtrack-sound-monitoring-earphones-04-12-2018/.

to support people with visual disabilities through assistive technologies. The organization teamed up with the Indian nonprofit I-STEM to develop technical solutions to help students with such disabilities to take exams. An AI-based solution has been developed: a virtual assistant.[88]

A final example is an AI system developed by Nigeria's Ubenwa that analyzes a baby's cries to predict asphyxia, the third most frequent cause of death among infants.[89] In a trial using 1,400 recorded baby cries, the solution's predictions were reported to be 95% accurate.[90]

4R Technologies to Fight COVID-19

4R technologies have also been used to fight the COVID-19 pandemic. For instance, in December 2021, South Africa's prompt genomic analysis played a key role in alerting the world about an unusual genome profile present in samples tested for coronavirus.[91] The Network for Genomic Surveillance in South Africa (NGS-SA), created in May 2020 in response to the COVID-19 pandemic,[92] first spotted the new variant in sequencing data from Botswana. National laboratories and the Africa Centers for Disease Control and Prevention (Africa CDC) teamed up to increase the level of Africa's sequencing response.

In October 2020, Africa CDC, the WHO, and other public, private and nonprofit organizations launched The African Pathogen Genomics Initiative, which is an Africa-wide network for viral sequencing. The aim of the initiative goes beyond COVID-19 surveillance to build technology infrastructure, expertise, and resources required in next-generation sequencing to confront future epidemic threats. It also helps to fight AIDS, tuberculosis, malaria, cholera, and other endemic diseases. The Initiative, which has centers in South Africa, Nigeria, Senegal, Ghana, Kenya, Democratic Republic

---

88. Fabio Moioli, "Council post: Artificial intelligence for social inclusion: Technologies and necessary steps," *Forbes*, 2021, https://www.forbes.com/sites/forbestechcouncil/2021/08/12/artificial-intelligence-for-social-inclusion-technologies-and-necessary-steps/?sh=3edc45be61ec.

89. Jonathan M. Spector and Subhash Daga, *Preventing those so-called stillbirths* (World Health Organization, 2008).

90. Akindare Okunola, "Artificial intelligence in Nigeria is an infant space with huge potential," *TechCabal*, 2018, https://techcabal.com/2018/08/08/artificial-intelligence-in-nigeria-is-an-infant-space-with-huge-potential.

91. Paul Adepoju, "African coronavirus surveillance network provides early warning for world," *Nature*, January 20, 2022, https://www.nature.com/articles/d41587-022-00003-3.

92. Nokukhanya Msomi, Koleka Mlisana, and Tulio de Oliveira, "A genomics network established to respond rapidly to public health threats in South Africa," *Lancet Microbe* 1, no. 6 (2020): E229–E230, https://doi.org/10.1016/S2666-5247(20)30116-6.

of the Congo, and Uganda, had sequenced over 70,000 viral genomes by January 2022. The network has also taken initiatives to expand pandemic surveillance into rural areas. The network of sequencing laboratories also covers remote areas.[93]

Likewise, drones equipped with cameras were used to slow the spread and lethalness of COVID-19 in Rwanda. Drones complemented efforts by radio and TV messages, community health workers, and other community leaders. Footage recorded by drones allowed local authorities to closely monitor areas in need of intervention or evacuation. Rwanda introduced high-tech robots at COVID-19 treatment centers to perform activities such as administering temperature checks, monitoring patient status, and keeping medical records of COVID patients. Robots minimized risk of infections among healthcare workers.[94]

Vodafone's South African subsidiary Mezzanine uses IoT in its Stock Visibility Solution (SVS), a network-agnostic mobile application for healthcare facilities' pharmacies. SVS helps track goods such as gloves, masks, and sanitizers using a smartphone barcode reader or wirelessly. As goods move through the supply chain, a SIM card can collect GPS location data and send it via the Vodafone cellular network. Vodafone's platform captures that information and transmits it to Mezzanine's cloud-based software. Updates are provided on the goods' location and status. Mezzanine can share that data with its customers and identify and address supply chain issues in real time. Mezzanine's IoT solutions were used in South Africa, Ghana, Nigeria, Mozambique, Zambia, Uganda, Kenya, Ethiopia, Malawi, and Tanzania.[95] During the COVID-19 pandemic, the SVS was deployed at 350 public hospitals in South Africa.[96]

4R technologies have also helped administer COVID-19 vaccines. Most vaccines require cold-chain or temperature-controlled supply chain during transportation and storage. About 3 billion of the world's 7.8 billion people

---

93. Adepoju, "African coronavirus."

94. WHO, *Rwanda: Drones for community awareness and nation-wide measures in COVID-19 response*, 2020, https://www.who.int/news-room/feature-stories/detail/rwanda-drones-for-community-awareness-and-nation-wide-measures-in-covid-19-response.

95. "Satellite Data—a solution for African farmers," *Space in Africa*, 2021, https://africanews.space/satellite-data-a-solution-for-african-farmers.

96. Claire Swedberg, "IoT companies track goods aimed at COVID-19 treatment, vaccinations," *Mezzanine*, April 4, 2021, https://mezzanineware.com/iot-companies-track-goods-aimed-at-covid-19-treatment-vaccinations/?utm_source=rss&utm_medium=rss&utm_campaign=iot-companies-track-goods-aimed-at-covid-19-treatment-vaccinations.

live in places that lack the temperature-controlled storage needed for an immunization campaign. They include most of Africa and Central Asia, much of India and Southeast Asia, and Latin America.[97]

Ghana, though, took advantage of 4R in administering COVID-19 vaccines in remote areas, and drones have played a critical role in delivering COVID-19 vaccine in the country. In March 2021, Zipline started delivering COVID-19 shots. On March 2, 2021, the first drone was launched from Zipline Distribution Centre at Mpanya in the south-central Ashanti. After 34 minutes of flying, the drone arrived in Asuofua in the Atwima Nwabiagya North District, which 70 kilometers (43 miles) from Mpanya. An insulated box containing 25 vials of vaccine parachuted down. Within five hours, 250 people were vaccinated. Another 35 drone deliveries were made to the Asuofua Health Center the same day.[98] As of March 2022, Zipline had delivered 1 million COVID-19 vaccine doses in the country.[99] The plan is to deliver about 2.5 million doses using the drones.[100]

Chapter Summary and Conclusion

The 4R is rapidly transforming the healthcare industry in LMICs. 4R technologies such as AI and blockchain are also being used to fight counterfeit drugs. AI is playing an important role in the democratization and decentralization of medical knowledge and excellence, and is enhancing access to healthcare services in countries characterized by an overburdened healthcare system.

Because of the remoteness and inaccessibility of large proportion of geographic areas, it is challenging to deliver healthcare services. Drone delivery

97. Lori Hinnant and Sam Mednick, "Vaccine storage issues could leave 3B people without access," *AP News*, October 19, 2020, https://apnews.com/article/virus-outbreak-pandemics-immunizations-epidemics-united-nations-fc4c536d62c5ef25152884adb1c14168.

98. Rich Haridy, "Drones deliver COVID-19 vaccines to remote African regions," *New Atlas*, March 8, 2021, https://newatlas.com/drones/coronavirus-vaccine-drone-delivery-zipline-ghana-covax/.

99. "Ghana, Zipline, deliver 1m COVID-19 vaccine doses via drone," *af24news*, March 30, 2022, https://af24news.com/2022/03/30/ghana-zipline-deliver-1m-covid-19-vaccine-doses-via-drone-service/.

100. Grace Dean, "Drones in Ghana deliver COVID vaccines to rural communities," *Business Insider*, March 5, 2021, https://www.businessinsider.com/covid-vaccine-ghana-drones-covax-who-coronavirus-zipline-rural-communities-2021-3.

of time-sensitive vaccines and medical supplies in countries such as Ghana and Rwanda helps us see how such barriers are being addressed.

Overall, impressive achievements have already been made in the utilization of 4R technologies in the delivery of healthcare services in LMICs. These technologies' potential can be more fully realized by developing appropriate physical and technological infrastructures, creating the enabling environment for more local healthcare innovations to thrive and facilitating the widespread diffusion of smartphones or other devices to access healthcare services.

CHAPTER 7

Agriculture

The agricultural sector in LMICs is characterized by lower productivity than in high-income countries. For instance, soybean yields are four times higher in the U.S. than in LMICs such as Indonesia, India, and the Philippines.[1] Low agricultural productivity is an important factor contributing to rural poverty in these countries.[2]

4R technologies provide an opportunity to increase agricultural productivity (In Focus 7.1). Recent evidence suggests that several pathways are capable of increasing smallholder farmers' productivity. For instance, being able to correctly identify crop diseases is critical in controlling such diseases. A study conducted in the Democratic Republic of Congo and in Benin achieved overall accuracy of 97% in detecting major diseases in banana plants in an analysis of aerial images using ML methods.[3] The findings are highly relevant because bananas and plantains are a primary food source for 90 million people in East, West, and Central Africa.[4] It is possible to send

1. Andrew D. Foster and Mark R. Rosenzweig, "Are there too many farms in the world? labor-market transaction costs, machine capacities and optimal farm size," *NBER Working Paper Series*, October 2017, https://www.nber.org/system/files/working_papers/w23909/w23909.pdf.

2. McKinsey Global Institute, *Nigeria's renewal: delivering inclusive growth in Africa's largest economy* (London: McKinsey, 2014).

3. Michael Gomez Selvaraj, Alejandro Vergara, Frank Montenegro, Henry Alonso Ruiz, Nancy Safari, Dries Raymaekers, Walter Ocimati, Jules Ntamwira, Laurent Tits, Aman Bonaventure Omondi, and Guy Blomme, "Detection of banana plants and their major diseases through aerial images and machine learning methods: A case study in DR Congo and Republic of Benin," *ISPRS Journal of Photogrammetry and Remote Sensing* 169 (2020): 110–124 https://www.sciencedirect.com/science/article/pii/S0924271620302410.

4. International Center for Tropical Agriculture, "Aerial images detect and track food

information about the nature of the specific threat and the patterns with which it is spreading to organizations or government agencies that can take actions.

In Focus 7.1: Intelligent Sensors, Data Analysis, and Automated Digital Greenhouse Increase Productivity

In May 2020, China Agricultural University and the Chinese e-commerce platform Pinduoduo organized a smart agriculture competition. The strawberry-growing competition took place in Yunnan province, and the FAO provided technical support.[5] Three teams of top strawberry growers ("traditional" teams) and four teams of scientific AI experts ("technology" teams) participated.[6]

The traditional teams relied on agricultural experience; technology teams employed internet-enabled devices such as intelligent sensors, data analysis, and an automated digital greenhouse.

In December 2020, the organizers announced the winner. The average production of the four technology teams was 196% more than that of the three traditional teams. Moreover, on average, the return on investment of the technology teams was 75.5% higher than the traditional teams.[7]

Smallholder farmers in LMICs have limited access to agricultural financial services, such as loans and crop insurance. For instance, Africa has an estimated 33 million farms,[8] and 63% of the population's livelihood depends on farming. Yet less than 1% of outstanding bank loans go to the agricultural sector.[9] The unavailability of credit means that most farmers cannot buy the machinery and seed stock needed to cultivate their lands. The agricultural

security threats for millions of African farmers," *EurekAlert!*, 2020, https://www.eurekalert.org/pub_releases/2020-10/icft-aid102220.php.

5. "China hosts its own AI cultivation challenge," *Horti Daily*, 2020, https://www.hortidaily.com/article/9223029/china-hosts-its-own-ai-cultivation-challenge/.

6. Winston Ma, "AI strawberries and blockchain chicken: how digital agriculture could rescue global food security," *World Economic Forum*, January 26, 2021, https://www.weforum.org/agenda/2021/01/china-digital-agriculture-global-food-security/.

7. "Technology beats humans at growing strawberries in Pinduoduo smart agriculture competition," *Globenewswire*, 2020, https://tinyurl.com/7tz3fnvk.

8. Grace Matheka, "OKO secures Ksh. 129 million seed funding to expand services to farmers," *HapaKenya*, April 29, 2021, https://hapakenya.com/2021/04/29/oko-secures-ksh-129-million-seed-funding-to-expand-services-to-farmers/.

9. Peter Wells, "Gro Intelligence looks to reap rewards from crop data," *Financial Times*, March 14, 2019, https://www.ft.com/content/e6530830-2b9f-11e9-9222-7024d72222bc.

and household-related financial needs of small-scale farmers are at approximately US$240 billion per year globally and to US$132 billion per year in Sub-Saharan Africa.[10] They received only US$10 billion in 2018 globally.[11] In recent years, digital technologies are playing an important role in servicing the financial needs of smallholder farmers in LMICs (In Focus 7.2 and 7.3).

High costs and lack of availability of insurance is also a concern. Insurers lack a clear understanding of a farm's operations and are reluctant to write policies. Such policies, however, tend to have high costs for smallholder farmers. Critics have been concerned that data from private sources is packaged and sold without any understanding of the quality. Such a tendency creates a challenge in determining "actuarially sound premiums for insurance."[12] 4R technologies can help create better insurance products for smallholder farmers. A more detailed treatment of such products appears in chapter 8. Additional challenges faced by the agricultural sector include poor or no transport links to uncultivated land, lack of communications, infrastructure, and issues with property rights.[13]

In Focus 7.2: Agrilife Connects Smallholder Farmers with Financial Institutions and Suppliers

Kenya-based mobile payment solution and service provider MobiPay's cloud-mobile platform Agrilife connects farmers with value chain partners such as dairy processors who purchase milk, credit appraisers, and local input and agro-dealers. In this way, Agrilife is a "one-stop-virtual-agri-info-shop"[14] that provides financial institutions and suppliers "near-real-time information" on farmers' ability to pay for services.[15] Agrilife reduces transaction costs by linking the various value chain actors. A farmer can make credit requests via mobile phone. The credit appraiser uses a range of data to assess creditworthiness. The input provider makes decisions on credit.

In the past, a farmer's needs may have been known to a field

10. CPI, *Examining the climate finance gap for small-scale agriculture*, 2020, https://www.climatepolicyinitiative.org/publication/climate-finance-small-scale-agriculture/#_ftn1.

11. Global Citizen, *How funding for farmers can help end hunger and poverty*, https://www.globalcitizen.org/en/content/why-smallholder-farmers-need-funding/.

12. Wells, "Gro Intelligence."

13. "AfricaMe-Team Agricultural data is becoming big business in Africa," *Africa Middle East*, April 30, 2020, https://africa-me.com/agricultural-data/.

14. *Agrilife: Bringing the agriculture value chain finance ecosystem to life*, Technical Centre for Agricultural and Rural Cooperation ACP-EU (CTA), July 14, 2014.

15. *Agrilife Platform*, 2013, www.capacity.org/capacity/opencms/en/topics/gender-and-social-inclusion/agrilife-web-platform-kenya.html.

officer or a program manager of a development organization only if the farmer was part of a household survey. Survey data take a relatively long time to analyze. By analyzing the data of many farmers through mobile payments and other transactions, a credit appraiser assesses farmers' creditworthiness. For example, based on milk sold to a dairy processor, a farmer receives an SMS code that can be used to access credit to purchase inputs, so use future production of milk as collateral. When a farmer sells milk, a Bluetooth-enabled digital scale is used. The transaction's "weight details" are transferred to the platform, which minimizes data discrepancies.[16] Digitization therefore minimizes the governing or transaction costs associated with opportunistic behaviors and uncertainty.[17]

As of September 2013, Agrilife had facilitated over US$2 million in revolving credit to about 120,000 small farmers in Kenya and Uganda. Century Microfinance Bank, while using the Agrilife platform, increased its outstanding loan portfolio from KSH 25.2 million to KSH 88.6 million.[18]

Platforms such as Agrilife also provide the opportunity to reach new markets for commercial banks. As of 2018, three banks were part of the AgriLife platform. Century Microfinance Bank provided finance to individual farmers by obtaining a loan guarantee from farmer cooperatives and other aggregators. It was reported that within four to five months, Century's outstanding loan portfolio increased from KSH 25.2 million to KSH 88.6 million with minimal extra costs.[19]

Some technology analysts have suggested that given economies of scale, smallholder farmers are not in a position to utilize 4R technologies. These technologies are for big farmers with high-tech offices.[20] On the contrary,

16. *Agrilife: Bringing.*

17. Aric Rindfleisch and Jan B. Heide, "Transaction cost analysis: Past, present, and future applications," *Journal of Marketing* 61, no. 4 (1997): 30–54.

18. J. Grossman and M. Tarazi, *Serving smallholder farmers: recent developments in digital finance*, Focus Note 94 (Washington, DC: CGAP, June 2014), www.cgap.org/sites/default/files/Focus-Note-Serving-Smallholder-Farmers-Jun-2014.pdf.

19. Elaine Tinsley and Natalia Agapitova, *Private sector solutions to helping smallholders succeed social enterprise business models in the agriculture sector*, March 2018, World Bank, https://documents1.worldbank.org/curated/en/851711521095180329/pdf/124304-WP-PUBLIC-AgriBookMar.pdf.

20. Peter Wells, "Gro Intelligence looks to reap rewards from crop data," *Financial Times*, March 14, 2019, https://www.ft.com/content/e6530830-2b9f-11e9-9222-7024d72222bc.

there are some effective examples of initiatives that have increased smallholders' access to 4R solutions, which can help surmount many of the challenges they face.

In Focus 7.3: Agryo Connects Lenders to Farmers

Brazil's Agryo is an agricultural technology company that connects lenders to farmers by combining data and AI modeling.[21] Agryo aims to connect over 500 million farmers to global financial institutions. It operates a chainlink node to make agri-finance data available on-chain, which can be used to develop environmentally focused blockchain-based financial products. It incentivizes sustainable farming practices through a land record monitoring system, which analyzes crop fields by collecting hundreds of datasets and processing them using specialized algorithms.[22] It documents every step of the process and methodology of producing data on-chain. Thus, anyone can verify Agryo's data to at any time.[23]

Chainlink's secure blockchain middleware provides unified services to users by bridging gaps between other applications, tools, and databases. It allows existing APIs to become compatible with leading blockchain networks. Its Agryo chainlink node can broadcast agricultural datasets directly within blockchain environments. The data can be sold to smart contract applications. It can also cryptographically sign data to prove it came from the Agryo API, in order to make it more trustworthy for automating data-driven financial transactions on blockchains.[24]

Smallholders' adoption of the 4R technologies has been recognized as a key mechanism to produce enough food to feed the world. For instance, some practitioners have argued that in high-risk farming areas or regions that face a shortage of skilled workers, robotic automation and AI enhancement of agricultural machinery should be the highest priority for all governments to respond to the global food-security challenges.[25]

21. Finextra Research, *Holt Fintech Accelerator announces 2020 cohort*, 2020, https://www.finextra.com/pressarticle/83586/holt-fintech-accelerator-announces-2020-cohort.

22. Isaque Eberhardt, "Agryo to launch a Chainlink node, bringing agricultural data on-chain to support," Agryo, June 3, 2021, https://www.agryo.com/Agryo_launch_risk_intelligence_oracle_in_ChainLink.html.

23. "Chainlink data providers explain why hybrid smart contracts demand high-quality data," *Chainlink Today*, 2021, https://chainlinktoday.com/chainlink-data-providers-explain-why-hybrid-smart-contracts-demand-high-quality-data/.

24. Eberhardt, "Agryo."

25. Olga Uskova, "An army of grain-harvesting robots marches across Russia," *IEEE Spectrum*, August 25, 2021, https://spectrum.ieee.org/robotic-farming-russia#toggle-gdpr.

Despite many benefits of agricultural tech solutions, their adoption rate has been slow in LMICs. For instance, only 12% of youth engaged in agriculture in Ghana are reported to use some form of technology to improve agricultural yields and quality.[26]

This chapter gives an overview of the roles of 4R technologies in increasing smallholder farmers' agricultural productivity, expanding their access to agricultural finance and facilitating the participation in agricultural markets. It also discusses how these technologies can promote sustainability in the agricultural value chain.

Improvements in Smallholder Productivity

New technologies play a key role in increasing agricultural productivity. For instance, while agricultural productivity in SSA has been low and is falling farther behind other world regions, some countries have been able to increase agricultural productivity. A wider adoption of new technologies has been a key factor in the enhanced productivity in more successful countries.[27]

A wide range of 4R technologies such as remote sensing, machine learning, drones, and IoT have facilitated agricultural productivity (In Focus 7.4). For instance, AI can predict the weather and other conditions affecting agricultural productivity, such as land quality, groundwater, crop cycles, and pests. Farmers can take action to enhance production.[28] Plantheus, developed in Nigeria, uses AI and image recognition to help farmers diagnose crop diseases and then recommends best practices for most crop diseases.[29]

Likewise, drones are used in agriculture to monitor crop health and irrigation equipment, herds and wildlife, and weed and disaster management.[30] As of 2019, Ghana's drone company Acquahmeyer was working with 8,000 farmers. Each farmer pays US$5 to US$10 per acre about six times a year.

26. "Only 12% of youth in agriculture use technology—Report," *GhanaWeb*, July 19, 2021, https://www.ghanaweb.com/GhanaHomePage/business/Only-12-of-youth-in-agriculture-use-technology-Report-1312489.

27. Keith Fuglie and Nicholas Rada, "Resources, policies, and agricultural productivity in Sub-Saharan Africa," USDA-ERS Economic Research Report No. 145 78, May 19, 2013, https://papers.ssrn.com/sol3/papers.cfm?abstract_id=2266459.

28. Tanha Talaviya, Dhara Shah, Nivedita Patel, Hiteshri Yagnik, and Manan Shah, "Implementation of artificial intelligence in agriculture for optimisation of irrigation and application of pesticides and herbicides," *Artificial Intelligence in Agriculture* 4 (2020): 58–73, https://doi.org/10.1016/j.aiia.2020.04.002.

29. "Charting Africa's digital future," *NewsDay Zimbabwe*, August 26, 2021, https://www.newsday.co.zw/2021/08/charting-africas-digital-future.

30. Talaviya et al., "Implementation of artificial intelligence."

The drones assess crops and soil and help in the application of pesticides. Each drone costs US$5,000 to US$15,000 and can spray 10,000 acres per year. The company started in mid-2018 with two drones and then increased that to ten. The company makes an annual net profit of US$15,000 to US$30,000 per drone.[31]

In Focus 7.4: 4R Technologies in Colombia's Cacao Industry

Cacao is grown across remote areas of Colombia. It is a smallholder crop and the farmers lack education. With a yield of 200 to 300 kilos per hectare, Colombian cacao growers have production far below the potential two to four metric tons per hectare (1,800 to 3,600 pounds per acre). About 60% of production is lost to pests and disease, for a cost of about US$1 billion annually. Colombia's cacao exports are less than a third of Ecuador's.[32]

Technology companies are developing solutions utilizing satellite data, IoT applications, and other apps to increase productivity and enhance transparency and traceability.[33] Two notable projects are the Ecological Productivity Management Information System (EcoProMIS), led by Agricompas, and COLCO, led by Satellite Applications Catapult.[34]

The EcoProMIS Project

The EcoProMIS project provides information on optimal conditions for free to farmers, who can use the information to apply fertilizer at the right moment. The project sells anonymous data to external partners in the service industry, government agencies, NGOs, and food processors. As of 2019, Agricompas had three employees in the UK and two in Colombia, and was piloting the app with ten cacao and palm growers in Colombia. Its goal is to reach 56,000 rice, cacao, and palm growers across the country by working with farmer associations.[35]

EcoProMIS estimates farmers' yields and greenhouse-gas emis-

31. Emma Reynolds, "How technology is helping African farms to flourish," *CNN*, October 18, 2019, https://www.cnn.com/2019/10/18/business/smart-farming-africa-tech-intl.

32. Aurora Solá, "Technology innovations look to change the cacao landscape in Colombia," *Mongabay*, November 10, 2020, https://news.mongabay.com/2020/11/technology-innovations-look-to-change-the-cacao-landscape-in-colombia/.

33. Solá, "Technology."

34. Solá, "Technology."

35. Agri-Innovation Den, *Meet our 2019 Agri-innovation Den Finalists*, 2019. https://www.aginnovationden.com/roelof-kramer-meet-our-2019-agri-innovation-den-finalists/.

sions on a near real-time basis. Using space technology, EcoProMIS tracks crop performance in various conditions. The aerial data is combined with data from farmers to assess farms' crop production and biodiversity.[36]

The Colombian Cocoa Control System (COLCO) Project

The COLCO project, a two-year project, was funded by the UK government's public body UK Research and Innovation's Innovate UK through the Newton-Caldas Fund. The project aims to improve farm productivity, product consistency, quality, and post-harvest processes through IoT integration. To achieve these objectives, the firms involved will increase monitoring, certification, and localized processing along value and supply chains.[37]

COLCO partners include UK technology and innovation Center Satellite Applications Catapult, the High Value Manufacturing Catapult's Manufacturing Technology Centre (which focuses on advanced manufacturing), CABI, and Cervest in the UK, and CCC, Croper, Dextera, Fedecacao, Nutresa, Netux, and Ruta-N in Colombia. The project is supported by the British Embassy in Bogotá.[38]

Activities in the production phase affect cacao volume. Quality depends on post-harvest activities. As of February 2021, about US$4,937,300 was funded in COLCO by Innovate UK. COLCO will provide a data repository that includes production and post-harvest process information, bean-quality assessment, pest and disease management capabilities, yield prediction, and fermentation and drying monitoring. It will also build a framework to develop integrated production and post-harvest technologies.[39]

The second phase of the project started in June 2020 with a goal to commercialize services trialed in phase 1. The second phase also involves developing and implementing a smart-trading system to

36. Katherine Walla, "These 21 projects are democratizing data for farmers," *GreenBiz*, November 25, 2019, https://www.greenbiz.com/article/these-21-projects-are-democratizing-data-farmers.

37. Manufacturing Technology Centre, *Colco: Project overview*, http://www.the-mtc.org/our-projects/colco.

38. Mark Jarman, "Project to deliver change for cacao in Colombia to be extended by 12 months," *Catapult*, https://sa.catapult.org.uk/news/colco-project-extended-by-12-months/.

39. UK Research and Innovation, *COLCO—Colombian Cocoa control system*, 2021, https://gtr.ukri.org/projects?ref=104429.

increase traceability and transparency, get a better price for farmers, and develop a more climate-resilient, secure value chain.[40]

In November 2020, Mark Jarman, in charge of COLCO's pilot projects, noted that farmers were getting advice from an extension service every two weeks and emphasized the importance of using the service daily to improve their decision-making.[41]

Some telecommunication companies have developed IoT-based solutions for smallholder farmers. To leverage the power of the IoT, MTN, which is present in 19 African countries, has partnered with Aotoso Technology in a proof-of-concept process to provide connected collars for cattle in the Sudanese market. Farmers can use SIM cards on the collar and their cell phones to get vital information about the cattle, such as to inform feeding and breeding strategies as well as prevention of cattle theft.

Likewise, Orange, which is present in 16 countries in Africa, codeveloped Wazihub, an IoT project that helps mainly informal sector farmers monitor their livestock and crops for better disease detection and irrigation. Vodacom, present in seven countries in Africa, has MyFarmWeb, which collects data from multiple IoT sensors across a farm. The data is collated centrally on an app and can be used to make crucial decisions. There are 77 farms in South Africa's eastern Free State province that use MyFarmWeb and its precision farming technology to gather data on farming corn, soybeans, and sugar beans.[42] As of 2020, the platform was deployed at more than 4,000 farms, covering over 1 million hectares of land.[43]

Many such solutions use multiple 4R technologies to amplify the effects on agricultural productivity. For instance, agriculture software platform Dimitra uses satellites to evaluate crop performance. It supplements that data with farmer observations and IoT soil sensors. The information is fed into a machine-learning algorithm to help farmers make better decisions about how to prepare, sow, care for, and harvest crops, and then get them to market. For instance, it is not possible to perform these functions by relying

40. Jarman, "Project to deliver change.

41. Warp Institute, *Apps support peace crop*, 2020, https://warpinstitute.org/%F0%9F%87%A8%F0%9F%87%B4-apps-support-peace-crop/.

42. "How telecommunication companies are using technology in agriculture and healthcare in Africa," *CNBC Africa*, 2021, https://www.cnbcafrica.com/2021/how-telecommunication-companies-are-using-technology-in-agriculture-and-healthcare-in-africa/.

43. Vodafone, *MyFarmWeb: A cloud-based IoT solution for farmers*, 2020, https://www.vodafone.com/news/digital-society/myfarmweb-cloud-based-iot-solution-farmers.

on just human analysis. For instance, Dimitra notes that it is a daunting task to send soil specialists to 1.3 million farms using its solution.[44]

Access to Agricultural Finance

As noted earlier, assessing poor people's and smallholder farmers' creditworthiness is difficult without established credit history. Thus, only a small proportion of agricultural and household-related financial needs of small-scale farmers are satisfied; banks find it unattractive to provide loans to such farmers. Likewise, the loan portfolios of most microfinance institutions (MFIs) in developing countries are typically concentrated in urban areas. Systemic risks associated with droughts, floods, cyclones, and other extreme weather events tend to make agricultural loans less attractive and to hinder the ability and enthusiasm of MFIs to expand their services to rural farmers. Smallholder farmers' lack of access to agricultural finance is associated with low crop productivity.[45]

New technology and fintech startups are developing and implementing emerging technologies such as AI and ML to assess farmers' creditworthiness and provide loans. For instance, Myanmar-based MFI Maha Agriculture started combining harvesting data based on weather monitoring with its credit-scoring model, which is expected to improve predictive capabilities and increase the number of borrowers.[46] Some are using technologies such as blockchain (In Focus 7.6), which is expected to help build the creditworthiness of smallholder farmers.

In Focus 7.5: BanQu and Anheuser-Busch Work Together to Promote Supply Chain Transparency

According to BanQu, of the 2.7 billion unbanked and underbanked people, about 1 billion supply 5,000 global brands.[47] One such example

44. A. Ahmad, "AI-driven technology finds solution to global food security," *Technology Times*, 2021, https://www.technologytimes.pk/2021/08/13/al-driven-technology-finds-solution-to-global-food-security/.

45. Betty Kibaara and James K. Nyoro, "Expanding the agricultural finance frontier: A Kenyan case," *African Association of Agricultural Economists (AAAE), 2007 Second International Conference*, August 20–22, 2007, Accra, Ghana, https://ageconsearch.umn.edu/record/52109/.

46. Emily Bary, "How artificial intelligence could replace credit scores and reshape how we get loans," *MarketWatch*, October 29, 2018, www.marketwatch.com/story/ai-based-credit-scores-will-soon-give-one-billion-people-access-to-banking-services-2018-10-09.

47. Ben Paynter, "This digital ledger helps small farmers get a fair deal," *Fast Company*,

of a global brand is Anheuser-Busch InBev, which works with large commercial farmers in the U.S. and Argentina and small landholder farmers in countries such as Zambia, Uganda, and India.[48] In June 2018, BanQu teamed up with Anheuser-Busch InBev to promote supply chain transparency and traceability in Zambia. The BanQu system is also referred to as the Chembe Cassava Online Buying Project in Zambia. The partnership started with the cassava crop value chain, with the aim of providing economic empowerment to small-scale farmers. Using BanQu solutions, Anheuser-Busch's local business, Zambian Breweries, can track products throughout the supply chain, from farmer to local businesses to aggregated buyers and retailers.[49]

Farmers require only a text-capable mobile phone and a national identity card.[50] A farmer supplying Anheuser-Busch receives a digital payment through BanQu's platform. The crypto tokens can be redeemed for cash or applied for payment to other transactions, such as energy bills.[51]

Farmers may benefit from the immutable records of economic activities linked with their digital profiles. For instance, with this record, farmers can connect with NGOs, local cooperatives, MFIs, and banks to receive loans, grants, and training. The program started with 2,000 farmers. The system is projected to track 2,000 tons of cassava, a high-quality starch used in beer. Zambian Breweries is expected to add 2,500 additional cassava farmers by the end of 2019.[52]

Since 2018, BanQu has had a presence in Uganda. The goal is to reach 7,000 barley farmers there by the end of 2019.[53] As of June 2019, Nile Breweries, a subsidiary of Anheuser-Busch, implemented the system to track over 5,000 barley farmers in the Sebei region of eastern Uganda. A farmer receives an SMS message that shows the quality, quantity, and price of the crop sold to Nile Breweries. The farmer can access the payment by presenting the code received in the SMS to a partner bank or mobile telecom.[54]

April 8, 2019, https://www.fastcompany.com/90328012/this-digital-ledger-helps-small-farmers-get-a-fair-deal?partner=rss&utm_source=rss&utm_medium=feed&utm_campaign=rss+fastcompany&utm_content=rss.

48. BanQu, *How a startup is working with corporates to solve inequality and climate change*, June 9, 2021, https://banqu.co/press-coverage/how-a-startup-is-working-with-corporates-to-solve-inequality-and-climate-change/.

49. Paynter, "This digital ledger."

50. Paynter, "This digital ledger."

51. Paynter, "This digital ledger."

52. Paynter, "This digital ledger."

53. Paynter, "This digital ledger."

54. "How blockchain technology is changing lives of farmers in Eastern Uganda,"

In 2018, BanQu joined AB InBev's Accelerator program to scale the idea in other locations. The company announced a plan to start a new program with 1,000 Indian barley farmers.

Malaysia's Poladrone was reported to be developing AI algorithms that process farms' aerial images to assess crop performance, which would provide data to predict farmers' ability to repay.[55] Similarly, Singapore's Adatos processes satellite imagery of farmland to produce harvesting data. Among its clients is a Thai company that accounts for 45% of farm machinery in Southeast Asia. Adatos was reported to be helping the company develop new precision agriculture products.[56] All these are encouraging developments because general consumer and business finance companies and microcredit organizations have had limited success in serving the needs of economically active low-income families and microenterprises cost-effectively and sustainably.

With the proliferation of low-cost satellites, it is becoming increasingly easy and affordable to get data and information to gain a better understanding of various economic, social, and environmental indicators.[57] In the near future, it will be possible to use satellite imagery of the roof of a farmer's house and plot for the identification of a farmer and assessment of ability to repay a loan.

Smallholder Farmers' Participation in Agricultural Markets

Smallholder farmers' market access is a key factor influencing the performance of agriculture in developing countries and enhancing and diversifying their livelihoods.[58] Yet smallholder farmers' market participation in

Equatornews, 2019, https://equatornews.today/business/how-blockchain-technology-changing-lives-farmers-eastern-uganda.

55. Paul Voutier, "Digital payments key to rural credit scoring in S-E Asia," *Business Insider*, September 12, 2019, https://www.businesstimes.com.sg/garage/digital-payments-key-to-rural-credit-scoring-in-s-e-asia.

56. "The retired ex-US-intel officer who started an AI business in Southeast Asia," *Tech in Asia*, September 2, 2018, https://www.techinasia.com/exusintel-officer-retirement-start-ai-business-southeast-asia.

57. Nir Kshetri, *Big data's big potential in developing economies: impact on agriculture, health and environmental security* (Oxford: Centre for Agriculture and Biosciences International, 2016).

58. Christopher B. Barrett, "Smallholder market participation: Concepts and evidence from eastern and southern Africa.," *Food Policy* 33, no. 4 (2008): 299–317.

many developing is severely constrained by poor access to agricultural and market information.[59] This lack of access results in problems such as moral hazard and adverse selection, which lead to an increase in transaction costs and prevent participation in markets by smallholder farmers.[60]

Smallholder producers from LMICs, especially in Africa, are engaged in subsistence or semisubsistence agricultural activities. These activities are often characterized by low productivity, low marketable surplus, and low investment. This situation of low returns is known as low equilibrium poverty trap.[61]

Economic Exchange and Threshold Costs

Three stages of transactional relationships have been identified: contact, contract, and control.[62] Each of these phases has threshold costs, which put smaller transactions associated with smaller firms as suppliers as well as buyers at a disadvantage. For instance, an appointment needs to be made and executed with a potential transaction partner. A party needs to judge the perspective of the other party.[63] The parties then need to establish a contract and channels of communication.[64] Traditional contracts face various challenges in incorporating distributive justice.[65] Moreover, the systemic injustice that has persisted in society makes it illogical to impose the burden of redistribution on a party to contract.[66] Thus, smallholder farmers are likely to exploited by powerful supply chain actors. A study conducted in Punjab, India, found that contracting firms decide the quality norms and reject a crop in part or in full, which puts contracting farmers at a disadvantage.

---

59. Christopher B. Barrett and Michael R. Carter, "The economics of poverty traps and persistent poverty: Empirical and policy implications," *Journal of Development Studies* 49, no. 7 (2013): 976–990.

60. Bekele Shiferaw, G. Obare, and Geoffrey Muricho, "Rural institutions and producer organizations in imperfect markets: Experiences from producer marketing groups in semi-arid eastern Kenya," *Natural Resources Forum* 32, no. 1 (2006): 25–38.

61. Barrett, "Smallholder."

62. Bart Nooteboom, "Information technology, transaction costs and the decision to make or buy," *Technology Analysis and Strategic Management* 4, no. 4 (1992): 339–350.

63. B. Nooteboom, "Firm size effects on transaction costs," *Small Business Economics* 5 (1993): 283–295.

64. Nooteboom, "Firm size."

65. A. Baggchi, "Distributive justice and contract," in *Philosophical Foundations of Contract Law* (Oxford: Oxford University Press, 2014).

66. J. Coleman and A. Ripstein, "Mischief and misfortune," *McGill Law Journal* 41, no. 1 (1995): 91–130.

Thus, when there is abundant supply on the open market, farmers get a higher price than the market offers. The price is, however, only marginally higher; rejection of crops based on quality is far more common.[67]

The parties engaged in a transaction also need to set up a scheme for control in the form of guarantees and controls in bilateral governance or an arbitrator in trilateral governance.[68] Control-related costs also affect small farmers adversely. For instance, a form of trilateral mode is a neoclassical contract in which third-party participation is used to manage transactions with high uncertainty and asset specificity and also low frequency. Some examples include verification of wine grades or certification of special products (e.g., eco, fair trade, origin).[69] In the coffee industry, for instance, costs related to paperwork and physical inspection are estimated to be as high as US$0.91 per pound.[70] There are persuasive arguments that 4R technologies can reduce the costs associated with contact, contract, and control (Table 7.1).

Contact

Technology companies are providing loan services to farmers without ever meeting the borrower personally, which has decreased the cost of contact. Kenya's Apollo Agriculture, for instance, utilizes machine learning, remote sensing, and mobile payments to provide input finance and agronomic advice to smallholders at a lower cost than current solutions. Farmers sign up with their mobile phones. Apollo gathers data about each farmer from different sources, including satellite, to build a detailed profile of the farmer's agricultural and economic life. Data related to farm and house size, crops planted, and yields are combined using machine learning to assess credit risk.[71]

---

67. Sharanjit S. Dhillon, Navchetan Singh, and Sharangit S. Dhillon, "Contract farming in Punjab: An analysis of problems, challenges and opportunities," *Pakistan Economic and Social Review* 44, no. 1 (2006): 19–38.

68. Nooteboom, "Firm size."

69. Hrabrin Bachev, "Needs, modes and efficiency of economic organizations and public interventions in agriculture," *Munich Personal RePEc Archive*, October 2010, https://mpra.ub.uni-muenchen.de/25979/1/MPRA_paper_25979.pdf.

70. Nir Kshetri, "Blockchain and the economics of customer satisfaction," *IEEE IT Professional* 21, no. 1 (2019): 93–97.

71. Fatah Iftin, *Solving the credit impasse: How big data and AI are generating funding opportunities for smallholder farmers in Africa*, Next Billion, April 24, 2020, https://nextbillion.net/credit-big-data-ai-farmers-africa/.

Table 7.1. 4R technologies effect on contact, contract, and control for smallholder farmers

Phase	Some mechanism to reduce costs	Example
Contact	4R technologies make it possible for farmers to engage in transactions without a face-to-face meeting,	Kenya's Apollo Agriculture's use of ML, remote sensing and mobile money to provide input finance and agronomic advice.
Contract	Smart contracts can reduce costs	Etherisc and ACRE utilize blockchain-based smart contracts to offer crop insurance for smallholder farmers in Kenya.
Control	4R technologies can reduce the costs to prove the quality of agricultural products.	Demetria's AI-based apps for coffee farmers to help them track bean quality.

Contract

Traditional contracts often disadvantage smallholder farmers. For instance, while contract farming can play a key role in providing agricultural services to smallholders, two main problem areas result in high costs, which undermine viability: contract default and scale of operations. A buyer may break a contract with a farmer, for instance, by failing to deliver inputs and services at the correct time. They may refuse to receive produce or raise quality standards arbitrarily. If such things happen, the farmer is often at a disadvantage. Farmers also break contracts due to production failure. In some cases, they may sell product to competing buyers and fail to repay credit. Factors such as lack of effective legal systems, lack of collateral held by smallholders, and a weak insurance sector create risks for companies entering into contracts. Due to a small scale of farmer operations, smaller farmers are not in a position to benefit from contract farming.[72]

4R technologies can reduce contract-related costs. For instance, smart contracts can lower costs for farmers and for those supplying finance, insurance, and other products. For instance, insurance policies can take advantage of smart contract features, which can be programmed to occur if certain outcomes or conditions are met.[73] In Kenya, for instance, Etherisc and

72. Jonathan Coulter, Andrew Goodland, Anne Tallontire, and Rachel Stringfellow, *Marrying farmer co-operation and contract farming for agricultural service provision in sub-Saharan Africa*, Le Hub Rural, November 1999, http://www.hubrural.org/IMG/pdf/contractingoutzambia.pdf.

73. World Bank Group, *Smart contract technology and financial inclusion*, May 1, 2020, https://openknowledge.worldbank.org/handle/10986/33723.

ACRE utilize blockchain-based smart contracts to offer crop insurance for smallholder farmers.

Control

Threshold costs of control-related actions, such as ensuring that small farmers engage in sustainable practices and do not use pesticides, also put smallholders at a disadvantage. The costs increase because it is difficult to reach small farmers because they are dispersed, which increases the costs of service delivery and monitoring. The small scale of production, characterized by low volumes of input and output, leads to high per-unit transaction costs. In Zimbabwe, an estimate suggested that to break even, a horticultural exporter needs to pay smallholder suppliers less than 30% of the price per kilogram paid to commercial farmers who deliver directly to a packing plant.[74]

There are also high costs of control to ensure that smallholder farmers comply with quality standards. Research in information economics and cognitive science views classifications of products as based on objective characteristics.[75] However, in many markets, the perception of a product's quality cannot be derived objectively. The quality in these markets is a function of a social process of qualification. In such a process, a wide range of actors—producers, sellers, experts, journalists, trade and industry associations, marketing specialists, and consumers—may participate to construct the qualities of products.[76,77,78] In such situations, a dense network of relationships forms among various actors, which is likely to reduce incentives for opportunism[79] Dense relationships and interactions are also likely to generate a lot of "gossip," which helps enhance trust. In the absence of such institutions, some individuals are willing to engage in opportunistic behaviors, which increases the roles of 4R technologies to enhance trust in business relationships. For instance, technologies such as AI (In Focus 7.7) and blockchain (In Focus

74. Coulter et al., *Marrying farmer*.

75. Douglas L. Medin and Cynthia Aguilar, "Categorization," in *The MIT Encyclopedia of the Cognitive Sciences* (Cambridge: MIT Press, 1999), 104–106.

76. François Eymard-Duvernay, "Les qualifications des biens," *Sociologie du Travail* 44 (2002): 267–272.

77. Lucien Karpik, *Valuing the unique: The economics of singularities* (Princeton: Princeton University Press, 2010).

78. Jörg Rössel and Jens Beckert, "Quality classifications in competition, price formation in the German wine market," MPIfG Discussion Paper 12/3, 2012, Max Planck Institute for the Study of Societies, Cologne.

79. Michael D. Cohen and Robert Axelrod, *Harnessing complexity: Organizational implications of a scientific frontier* (New York: Basic Books, 2000).

7.8), machine vision, and IoT (In Focus 4.1) are being used to assess the quality of products.

In Focus 7.6: Demetria's AI-Based Solution to Determine Coffee-Bean Quality

Coffee-bean quality is mostly determined by cupping, which is used to evaluate various aspects of a coffee sample. A cupping session involves taste, smell, and feel to differentiate coffees from one another.[80] The process is manual and carried out by the industry's certified tasting experts. It is an expensive and time-consuming process.

Cupping is not affordable for most of the world's 12.5 million smallholder farmers, who account for 60 percent of the world's coffee beans. They cannot determine or manage the quality of their crop.[81]

AI and Coffee Quality

The Israeli and Colombian agriculture tech startup Demetria aims to transform the cupping process using AI. Instead of using human experts, Demetria's sensors read biochemical markers of taste.[82] The process involves scanning green coffee with a handheld device. The collected data is sent to the cloud for analysis using AI. The device uses near-infrared (NIR) sensors to shine different light wavelengths onto green coffee. Each wavelength interacts differently with organic compounds inside the bean. The interactions are measured and interpreted by the AI system and a "sensory fingerprint" is created. That fingerprint is sent to a cloud-based intelligence platform, which matches the bean profile to the industry standard coffee flavor wheel. The wheel, first developed in the mid-1990s, helps coffee professionals taste and evaluate different coffee flavors. It also monitors biochemical markers such as size, shape, and color.

To train its ML program, Demetria gathers quality analysis from human Q Graders, who are certified by the Coffee Quality Institute

80. John Carrier, "Coffee cupping—A sensory experience," *I Need Coffee*, July 27, 2012, https://ineedcoffee.com/coffee-cupping-a-sensory-experience/.

81. Hannah Brown, "Israeli-Colombian food-tech startup Demetria helps industry wake up & smell coffee," *Jerusalem Post*, March 2, 2021, https://www.jpost.com/israel-news/israeli-colombian-food-tech-startup-demetria-helps-industry-wake-up-and-smell-coffee-660681.

82. Howard Bryman, "With AI-powered green coffee analysis, Demetria closes $3 million round," *Daily Coffee News*, March 3, 2021, https://dailycoffeenews.com/2021/03/03/with-ai-powered-green-coffee-analysis-demetria-closes-3-million-round/.

(CQI) to do cupping for arabica coffee.[83] Demetria compiles the data into what it calls a "quality data cloud." It claims that the process accurately predicts coffee taste. Demetria's tools are expected to complement rather than replace the work of cuppers. The technology helps identify samples that do not require cupping. In this way, Demetria aims to utilize cuppers' time more efficiently. Demetria argues that coffee farmers can increase their income by using the system to sort out high-quality lots, which otherwise are mixed with lower-quality beans.

Demetria completed a pilot project with Carcafe, the Colombian coffee division of agricultural commodity traders Volcafe/ED&F Man. Demetria's solution determines which green beans suit a distinct high-value cupping profile identified by Carcafe.

Additional Funding

In March 2021, Demetria closed a US$3 million round of seed funding, led by the international investor Celeritas and a group of private investors. The company's plan is to use the fund to launch a suite of SaaS solutions to replace manual processes for assessing bean quality. The solutions are expected to deliver taste assessment and profile beans, measure quality, and ensure traceability throughout the supply chain efficiently and accurately.[84]

App for Farmers

The company is also working with the Colombian National Federation for Coffee Growers, Federación Nacional de Cafeteros (FNC), to develop apps for farmers to help them track bean quality and price coffee.[85]

83. Angie Katherine and Molina Ospina, "What is the Q grader program & should you do it?," *Perfect Daily Grind*, May 24, 2017, https://perfectdailygrind.com/2017/05/what-is-the-q-grader-program-should-you-do-it/.

84. D. Foxwell, "Startup Demetria secures funding for AI-powered coffee quality system," *STiR*, March 15, 2021, https://stir-tea-coffee.com/coffee-report/startup-demetria-secures-funding-for-ai-powered-coffee-quali/.

85. "Demetria, a Colombian—Israeli startup, uses AI to determine taste and quality of coffee beans," *Comunicaffe*, March 3, 2021, https://www.comunicaffe.com/demetria-a-colombian-israeli-startup-uses-ai-to-determine-taste-and-quality-of-coffee-beans/.

In Focus 7.7: E-Livestock Global Brings Visibility to Zimbabwe's Cattle Industry

Zimbabwe was first in the Middle East and Africa to use the Mastercard blockchain-based solution E-Livestock Global to bring visibility to the cattle industry. Commercial farmers and dipping officers tag the head of each cattle with an ultra-high frequency RFID tag, as mandated by the Ministry of Agriculture. The cattle and owner are registered onto the solution.[86] The E-Livestock Global app aims to provide end-to-end visibility to the cattle supply chain. When an animal is dipped or vaccinated or receives medical treatment, the tag records the event onto the blockchain traceability system. In this way, it maintains a secure and tamper-proof trail of each animal's history. Farmers can use the record to prove ownership, which supports sales and exports. They can also obtain a loan using cattle as collateral. E-Livestock Global's system can help them gain access to global markets due to the visibility captured and recorded on the blockchain. Buyers can efficiently manage operations and guarantee product quality to customers.[87] An outbreak of tick-borne disease in 2018 killed 50,000 cattle. The lack of a traceability system has prevented Zimbabwe from exporting beef to Europe and the Middle East. This reduced earnings from beef exports, which is important to the country's economy.[88]

Sustainability in the Agricultural Value Chain

4R technologies help promote sustainable practices in the agricultural value chain (In Focus 7.8). For instance, blockchain can facilitate sustainable farming by allowing complete visibility of activities such as seed purchase and crop management. Smart contracts can match demand and supply,

---

86. "E-Livestock Global launch Mastercard blockchain-based solution to bring visibility to the cattle industry in Zimbabwe," *MasterCard Social Newsroom*, 2021, https://newsroom.mastercard.com/mea/press-releases/e-livestock-global-launch-mastercard-blockchain-based-solution-to-bring-visibility-to-the-cattle-industry-in-zimbabwe/.

87. Rachel Wolfson, "Fruits of the land: Blockchain traceability gives farmers a competitive advantage," *CoinTelegraph*, July 15, 2021, https://cointelegraph.com/news/fruits-of-the-land-blockchain-traceability-gives-farmers-a-competitive-advantage.

88. "E-Livestock Global launches blockchain-based MasterCard solution to bring visibility to Zimbabwe's livestock industry," *NNN News*, June 17, 2021, https://nnn.ng/e-livestock-global-launches-blockchain-based-mastercard-solution-to-bring-visibility-to-zimbabwes-livestock-industry/.

which can make the market more transparent and cut out intermediaries. These lead to increased profitability and reduction in uncertainty. Farmers who reduce greenhouse-gas emissions can receive carbon credits in the form of tokens.[89]

Blockchain can also help farmers take socially responsible actions: those that contribute to social and distributive justice, improve fairness in the allocation of resources (e.g., by increasing the price paid to commodity producers and farmers in developing countries), or promote the sustainability of community.[90] One such use has been in fighting tackle child labor.

A barrier to develop "green" smart contracts has been the inability of blockchains to interact meaningfully with data about the state of the real world, including the natural environment. Oracles can "shepherd" data about the world onto blockchains. Oracles are becoming production-ready. They are broadcasting agricultural datasets directly onto blockchains, which allows smart-contract developers to build applications around crop yields, soil quality, weather reports, and carbon offsets. As datasets related to weather patterns, IoT sensor readings, and other environmental indicators are being fed onto blockchains, developers are beginning to write a wide range of environmentally conscious smart contracts.[91]

In Focus 7.8: Japan Tobacco International Integrates Blockchain into Its ARISE Initiative

The international tobacco division of Japan Tobacco, Japan Tobacco International (JTI), has integrated BanQu's solution into its ARISE initiative, which aims to tackle child labor in communities.[92] When JTI Leaf Technicians visit farmers, they input farmer-specific information into a database via a mobile app. The data entered includes

89. Fiorenzo Manganiello and Nessim Sariel-Gaon, "How blockchain and other digital technology can help China achieve its agriculture and climate change goals," *South China Morning Post*, March 24, 2021, https://www.scmp.com/comment/opinion/article/3126554/how-blockchain-and-other-digital-technology-can-help-china-achieve.

90. Nicola Dempsey, Glen Bramley, Sinéad Power, and Caroline Brown, "The social dimension of sustainable development: Defining urban social sustainability," *Sustainable Development*, September 27, 2011, 289–300.

91. "Blockchain can help us tackle climate change: Here's how," *World Economic Forum*, 2021, https://www.weforum.org/agenda/2021/06/blockchain-can-help-us-beat-climate-change-heres-how/.

92. "JTI uses BanQu's non-crypto Blockchain to protect children, improve workers' livelihoods, and fulfill its social responsibility in their supply chain," BanQu, 2021, https://banqu.co/use-cases/jti-uses-banqus-non-crypto-blockchain-to-protect-children-improve-workers-livelihoods-and-fulfill-its-social-responsibility-in-their-supply-chain/.

family kinship, household data, and schooling details of children. JTI collected such information in Brazil for many years due to regulatory requirements. It started collecting the information in Malawi, Tanzania, and Zambia for ARISE communities.

Farmers consent to providing information. They control their own identity in the system. They also get access to "growth enablers" such as crop insurance, soil and irrigation management, and crop rotation guidance. The plan is to allow farmers to interact with JTI through the app. JTI uploads the data collected via the mobile app to a customized blockchain system supported by BanQu.[93]

BanQu's system tracks if a child is receiving a service from ARISE. It allows school attendance to be verified. Winrock tracks school attendance and progress at other ARISE activities. The system is immutable and the information cannot be changed.[94] JTI has launched the program in Tanzania, Brazil, Malawi, and Zambia. BanQu's Child Labor Monitoring System (CLMS) provides watchdog groups with more insight into the problem and creates data, which can help eliminate child labor. The CLMS project was tested in Brazil and Malawi for six months. The pilot was conducted by JTI Leaf Technicians, ARISE implementation partner Winrock International and BanQu. The project was launched in Tanzania and Zambia.[95]

Measures to Increase the Use and Impact of 4R Technologies

It is important to increase the use and impact of ag tech. Some possible mechanisms are outlined in this section.

The Roles of Extension Services

Agricultural extension services may provide an opportunity to increase the use and impact of 4R technologies. In many developing countries, agricul-

93. ARISE, *Closer together annual review 2019*, 2019, http://ariseprogram.org/files/4915/9177/5043/ARISE_AR_2019.pdf.

94. Brij Kothari and Tathagata Bandyopadhyay, "Lifelong reading for a billion people," *Stanford Social Innovation Review* 18, no. 3 (2020): https://ssir.org/articles/entry/lifelong_reading_for_a_billion_people#.

95. BanQu, *ARISE: How BanQu and JTI are working together to end child labor*, 2020, https://banqu.co/blog-posts/arise-how-banqu-and-jti-are-working-together-to-end-child-labor/.

tural extension agents travel to provide industry information to farmers. For instance, COLCO's pilot projects provides an extension service every two weeks.[96] An analysis of improved mangrove rice varieties in Guinea showed that the number of times a farmer was visited by extension agents and participated in extension-related workshops were both significant in influencing farmers' adoption decisions of the rice.[97] This implies that the researcher-extension-farmer linkages were extremely weak in Guinea. Most mangrove varieties grown by farmers were driven by farmer-farmer contacts.[98]

Importance of Multiple Technologies and Data Sources

Farmers benefit from data collected from as many platforms as possible—including satellite, farm sensors, marketing, internet and cloud—and delivered on a single platform such as a mobile phone application.[99] A 2020 *Journal of Satellite Precipitation Measurement* study in Rwanda found that rain gauge instruments installed on the ground have limitations and that satellites can more accurately forecast weather.[100] Combining satellite data with machine-learning and deep-learning techniques makes it possible to improve the prediction of weather patterns and reduce uncertainty among farmers.

Locally Developed Solutions

Outsiders lack wide legitimacy in local areas and can do little to bring changes in technology-related and other practices in LMICs.[101] Many farmers do not tolerate being told what to do by Western experts.[102] Locally

96. Warp Institute, *Apps support*, https://warpinstitute.org/%F0%9F%87%A8%F0%9F%87%B4-apps-support-peace-crop/.

97. A. A. Adesina and J. B. Baidu-Forson, "Farmers perceptions and adoption of new agricultural technology: evidence from analysis in Burkina Faso and Guinea, West Africa," *Agricultural Economics* 13 (1995): 1–9.

98. M. M. Zinnah, J. L. Compton, and A. A. Adesina, "Research-extension-farmer linkages within the context of the generation, transfer and adoption of improved mangrove swamp rice technology in West Africa," *Quarterly Journal of International Agriculture*, 32 (1993): 201–211.

99. Kshetri, *Big data's*.

100. "Satellite data."

101. G. Almond and S. Verba, *The civic culture revisited* (Boston: Little, Brown and Co., 1980).

102. Million Belay and Bridget Moambe, "Bill Gates should stop telling Africans what kind of agriculture Africans need," *Scientific American*, July 6, 2021, https://www.scientif

developed solutions targeting local farming communities are likely to gain higher legitimacy. There has been an encouraging development on this front in recent years. African tech companies raised more than KSH 6.4 billion in 2020. They represent 8.6% of all startups in the continent. Kenya's GRO Intelligence raised US$85 million.[103] In December 2020, SunCulture received additional US$14 million in funding to expand its off-grid technology business (chapter 5) across Africa.[104]

Among LMICs, China is on the forefront of the development of AgTech. Alibaba was reported to be developing voice recognition technology for pigs to detect whether the animals are in pain or trouble. Online retailer JD.com has been working on an AI-powered system to develop optimum feeding plans for pigs.[105]

Likewise, Brazil has emerged as a global power in AgTech. According to the study "Radar Agtechs Brasil 2019," the country had 1,125 AgTech companies.[106] The Brazilian startup Solinftec is a highly visible example of a Brazilian AgTech company facilitating digital transformation. Solinftec's AI assistant Alice integrates and processes data from machines, people, climate stations, and other sources.[107] To use Alice, farmers embed smart black boxes in their machinery and deploy IoT devices in fields. Alice calculates farmers' needs and provides real-time recommendations.[108]

The initial solution was developed for the sugarcane industry. It also offers solutions for other crops. As of August 2019, Solinftec was used on

---

icamerican.com/article/bill-gates-should-stop-telling-africans-what-kind-of-agriculture-africans-need1.

103. Efayomi Carr, "Agritech is the next frontier in enhancing food security," *This Day* July 28, 2021, https://www.thisdaylive.com/index.php/2021/07/28/agritech-is-the-next-frontier-in-enhancing-food-security/.

104. Jonathan Shieber, "SunCulture wants to turn Africa into the world's next breadbasket, one solar water pump at a time," *TechCrunch*, December 4, 2020, https://techcrunch.com/2020/12/04/sunculture-wants-to-turn-africa-into-the-worlds-next-bread-basket-one-solar-water-pump-at-a-time/.

105. Anna Fifield, "Orwell's nightmare? Facial recognition for animals promises a farmyard revolution," *Washington Post*, August 24, 2020, https://www.washingtonpost.com/world/asia_pacific/facial-recognition-china-animals-farms-agriculture/2020/08/23/9808c710-d6fb-11ea-b9b2-1ea733b97910_story.html.

106. Daniel Azevedo, "Brazilian agtech boom produces 1.125 start-ups," *Future Farming*, September 10, 2019.

107. Nanalyz, "7 Brazilian."

108. "Growmark teams up with Solinftec to increase farm efficiency," *CropLife*, 2019, https://www.croplife.com/precision/growmark-teams-up-with-solinftec-to-increase-farm-efficiency/.

more than 6.5 million hectares, monitoring 20,000 pieces of equipment and managing 100,000 active daily users.[109]

Chapter Summary and Conclusion

4R technologies hold significant promise in enhancing smallholder farmers' welfare in LMICs. Several mechanisms and processes play a role, including these technologies' contributions in increasing agricultural productivity and their roles in meeting the needs of smallholder farmers and helping them to participate in economic exchanges.

Measures at various levels can help increase the use and impact of agricultural technologies. Agricultural technologies that have a proven record are more likely to be accepted by farmers. The questionable effectiveness of agricultural extension agents and farmers' lack of participation in extension-related workshops make it difficult to provide useful knowledge and practical solutions that farmers can use and benefit. This situation needs to be improved to increase the benefits of 4R technologies to farmers. It is also important to use multiple technologies and data sources.

Locally developed solutions can be perceived as having a higher degree of legitimacy than foreign solutions. There has been rapid growth of locally developed solutions targeting local farming communities, such as SunCulture's off-grid technology in Kenya. It is important to learn from countries such as Brazil and Kenya, which have successfully built the local AgTech industry.

109. Nanalyz, "7 Brazilian."

CHAPTER 8

Finance, Banking, and Insurance

4R technologies hold great potential to transform the way finance, banking, and insurance industries operate in LMICs. A study conducted by the global management company McKinsey & Company found that widespread adoption and use of digital finance in these economies could increase their GDP by US$3.7 trillion or 6% by 2025.[1] For instance, digital finance increases access for unbanked people and enterprises to finance and allows them to engage in entrepreneurial activities. Digital finance is also likely to reduce public spending leakage and increase tax collection.[2]

These technologies' role in addressing information opacity problems is the key mechanism with potential to contribute to the transformation. In general, digitization has addressed information asymmetry problems in the financial system and the labor market. The information flow has enhanced efficiency, certainty, and security.[3]

Another barrier that needs to be broken down in this sector concerns power asymmetries, which usually benefit intermediaries such as notaries, banks, and brokers. Google's chief economist Hal Varian noted that "new contractual forms due to better monitoring" are a key benefit of computer-

---

1. McKinsey.com, *FinTechnicolor: the new picture in finance*, 2016, www.mckinsey.com/~/media/mckinsey/industries/financial%20services/our%20insights/bracing%20for%20seven%20critical%20changes%20as%20fintech%20matures/fintechnicolor-the-new-picture-in-finance.ashx.

2. McKinsey.com, *FinTechnicolor*.

3. Njuguna Ndung'u and Landry Signé, "The Fourth Industrial Revolution and digitization will transform Africa into a global powerhouse," *Brookings*, January 8, 2020, https://www.brookings.edu/research/the-fourth-industrial-revolution-and-digitization-will-transform-africa-into-a-global-powerhouse/.

mediated transactions.[4] Especially decentralized networks such as blockchain can eliminate such asymmetries.[5]

A further point worth emphasizing is that supply chain finance, insurance, and consumer credit have been identified as key application areas likely to be transformed by smart contracts, especially in LMICs. For instance, smart contracts can help close the gap in demand and supply of financing for small businesses. Such contracts can lower costs for these entities and cut risks for financial institutions. For instance, insurance policies can take advantage of smart contracts' features.[6]

Financial institutions are more confident providing credit to individuals and organizations if information about collateral is verified with and stored in blockchains. This technology also facilitates SMEs' access to trade financing (In Focus 8.1). In West Africa and Kenya, blockchain-based efficient verification of property records and transactions has increased access to credit, which operated informally in the past.[7] Blockchain is thus likely to bring tremendous benefits in LMICs with large informal economies. For instance, in Nigeria, the informal economy represents more than half of GDP and about 95% of transactions take place in cash.[8]

In Focus 8.1: Interswitch's Blockchain System for End-to-End Visibility

The African digital payments and commerce company Interswitch partnered with Microsoft to launch the Interswitch Blockchain Service, built on Microsoft Azure. The service connects entrepreneurs, financial institutions, and corporate organizations by providing end-to-end visibility. It ensures fast and seamless trade financing in supply chain operations. SMEs can access funding in a shorter time

4. Shoshana Zuboff, "Big other: Surveillance capitalism and the prospects of an information civilization," *Journal of Information Technology* 30 (2015): 75–89.

5. Claus Dierksmeier and Peter Seele, "Blockchain and business ethics," *Business Ethics: A European Review*, December 16, 2019, https://onlinelibrary.wiley.com/doi/full/10.1111/beer.12259.

6. World Bank, *Smart contract technology and financial inclusion*, 2020, https://openknowledge.worldbank.org/handle/10986/33723.

7. S. Gebre, "Blockchain opens up Kenya's $20 billion informal economy," *Bloomberg*, June 13, 2018, https://www.bloomberg.com/news/articles/2018-06-14/blockchain-is-opening-up-kenya-s-20-billion-informal-economy.

8. N. Munshi, "Explosion in electronic payments powers start-up boom in Nigeria," *Financial Times*, August 19, 2021, https://www.ft.com/content/5fa49678-9eed-45e8-9c3d-6e19a2237b81.

(up to three weeks) from participating banks such as United Bank for Africa (UBA), Guaranty Trust Bank (GTB), and Zenith Bank.[9] The Microsoft Azure Blockchain was launched in 2018. It integrates Microsoft Flow and Logic Apps, which "offer hundreds of connectors to thousands of applications." It made the creation of blockchain apps easier. Blockchain enables a single, authentic dataset shared across counterparties.[10]

Data from diverse sources is a game changer for the finance, banking, and insurance industry in LMICs. In the insurance industry, for instance, by using satellite information and data collected on the ground, it is possible to plan a response (ex ante) rather than react to disasters after they occur (ex post). This process also speeds up response to disaster. For instance, data from satellites can provide the immediate first impression of situation. There is no need to conduct a lengthy needs assessment. For instance, if satellite data indicates a drought event is taking place, immediate response measures can be launched to help those in urgent need. One example of this is a crop insurance scheme. If satellite data indicates that a threshold agreed on the contract is reached (e.g., degree of modeled crop loss), an insurance payout can be made immediately to the farmer.[11]

This chapter provides an account of how 4R technologies are transforming the finance, banking, and insurance sectors in LMICs. Key topics covered include 4R technologies' roles in promoting financial inclusion and facilitating the adoption of crop insurance among farmers.

The 4R and Financial Inclusion

Current Challenges

Prior researchers have identified two main problems that contribute to low penetration of financial services among low-income families and microenterprises in LMICs: inefficiency and informational opacity. First, traditional

9. Technology Record, *Interswitch and Microsoft launch blockchain-based service*, 2018, https://www.technologyrecord.com/Article/interswitch-and-microsoft-launch-blockchain-based-service-73302.

10. "How Interswitch blockchain service eases pressure on low SME financing," *BusinessDay* November 15, 2018, https://businessday.ng/technology/article/interswitch-blockchain-service-eases-pressure-low-sme-financing/.

11. Saito and Lung, "Using satellite."

banks have been unwilling and reluctant to serve small-scale borrowers, such as low-income people and small businesses, due to high transaction costs and inefficient processes associated with making small loans to these borrowers.[12] Second, informational opacity is a barrier to access.[13] Traditionally, a national credit bureau collects and distributes reliable credit information, increasing transparency and minimizing a bank's lending risks. Many LMICs are characterized by the lack or poor performance of credit-rating agencies to provide information about the creditworthiness of SMEs. This puts SMEs at a disadvantage in the credit market, as SMEs are more informationally opaque than large corporations because they often lack certified audited financial statements. Thus, it is difficult for banks to assess or monitor the financial conditions of SMEs.[14]

Informational opacity may lead to moral hazard and adverse selection problems. For instance, it is possible that a borrower has provided false and misleading information to a lender about assets, skills, and credit capacity, but the lender provided a loan given the inability to verify the information. This is a problem of adverse selection associated with the information opacity related to ability to repay loans. Likewise, the borrower may have capacity to pay but may not have entered a contract with the lender in good faith. This is a problem of moral hazard due to the information opacity related to willingness to repay loans.

Role of 4R Technologies in Addressing Challenges

Addressing Informational Opacity

The implementation of 4R technologies in a financial company can bring a number of benefits. The use of big data, AI, and ML in assessing, evaluating, and refining the creditworthiness of potential borrowers has been a crucial mechanism in the transformation of the finance and banking sector of LMICs. These technologies have helped address challenges related to informational opacity and to moral hazard and adverse selection problems (In Focus 8.2, In Focus 8.3, and In Focus 8.4).

---

12. B. Rogaly, "Micro-finance evangelism, 'destitute women,' and the hard selling of a new anti-poverty formula," *Development in Practice* 6, no. 2 (1996): 100–102.

13. Adverse selection (anti-selection or negative selection) arises from information asymmetry between buyers and sellers prior to a deal. In such a case, one party is unable to determine if the other party is lying. Likewise, moral hazard is the problem of not being able to determine if the other party is cheating or acting dishonestly following a deal.

14. Nir Kshetri, *Global entrepreneurship: environment and strategy* (New York: Routledge, 2019).

Analysts have suggested the importance of paying greater attention to three categories of data that can be used as reliable proxies for creditworthiness of low-income people and microenterprises: verifying identity, assessing ability to repay, and assessing willingness to repay.[15] Various categories of personal financial and nonfinancial information are used as proxy measures for these risk factors.

Identity

Identity-related information helps ensure that statements or facts about a potential borrower are the same as what has been provided or described by the borrower. Such information helps reduce fraud. Various approaches and data sources are used to evaluate identity. A South African mobile payments provider was reported to be piloting the use of location data as a low-cost mechanism to validate self-reported addresses. To do so, it looks at the user's cellphone's nightly location patterns.[16] Some fintech companies use data from government agencies, allowing the company to quickly confirm the validity and authenticity of a potential borrower's identification. Tencent's WeBank verifies a person's identity using data from the Chinese Ministry of Public Security.

Ability to Repay

To determine ability to repay a loan, possession of means needs to be assessed. For instance, it is suggested that individuals living in rural areas who top up their cellphones on the same day every week and pass by more than two mobile phone masts during the week are likely to have more reliable financial habits than those who top up irregularly and do not travel.[17] Ecuador's fintech firm Kullki provides loans to consumers without prior credit history. Its credit-ranking system uses the KSOCIAL score, which involves data mining from the borrower's mobile devices. Information analyzed includes

15. Tobias Baer, Tony Goland, and Robert Schiff, *New credit-risk models for the unbanked*, April 1, 2013, http://www.mckinsey.com/insights/risk_management/new_credit-risk_models_for_the_unbanked.

16. Tilman Ehrbeck, "Driving financial inclusion at 4G speed," *Huffington Post*, June 7, 2015, http://www.huffingtonpost.com/tilman-ehrbeck/smartphones-driving-finan_b_7011282.html.

17. "A pionful of dollars: the world's poor need the stability and security that banks have traditionally offered, but increasingly they do not need banks to provide it," *The Economist*, November 15, 2014, http://www.economist.com/news/briefing/21632441-worlds-poor-need-stability-and-security-banks-have-traditionally-offered.

how people organize their lives. For instance, those who organize contacts by both first and last names are viewed as more creditworthy.[18] Also, attention to detail, such as grammar and punctuation in text messages, is taken into account.[19] The assumption is that a borrower's ability to pay is tightly linked to the digital footprint generated by their virtual presence.

Willingness to Repay

The willingness to pay is different from the ability to pay. For instance, paying a loan may disrupt usual household consumption patterns or deplete assets, putting the borrower at risk of poverty.[20] Possession of means thus may not be a sufficient condition to pay a loan. It is also important to assess whether the borrower has a strong disposition or inclination toward paying down debt. Past credit history and payment behavior are used to assess willingness to repay. Credit history information, such as a review of utility, rent, telephone, insurance, and medical bill payments, is valuable. Since such information is not available for most borrowers in LMICs, fintech companies have developed alternative approaches. For instance, the German company Monedo, which has operations in many LMICs, considers consumers who spend time online gambling as credit risks. Potential borrowers who have friends who have already repaid a loan are viewed as less risky.[21]

Reducing Transaction Costs

4R technologies such as big data and AI can drastically reduce operating costs for financial institutions, since machines perform most of the work that human agents otherwise need to perform. Such systems provide a faster response (In Focus 8.2). For instance, the financial affiliate of China's Ant Group's AI-based customer service chatbot handles up to 3 million queries per day. The costs associated with approving a small business loan are reported to be 2 RMB (US$0.32), compared to more than 2,000 yuan

18. Catherine Cheney, "How alternative credit scoring is transforming lending in the developing world," *Deveximpact*, September 8, 2016, https://www.devex.com/news/how-alternative-credit-scoring-is-transforming-lending-in-the-developing-world-88487.

19. Cheney, "How alternative."

20. S. Russell, "Ability to pay for health care: concepts and evidence," *Health Policy and Planning* 11, no. 3 (1996): 219–237.

21. Jeevan Vasagar, "Kreditech: A credit check by social media," *Financial Times*, January 1, 2016, www.ft.com/content/12dc4cda-ae59-11e5-b955-1a1d298b6250.

(US$318) at a traditional bank.[22] Overall, by using complex and sophisticated rules compared to those of traditional credit-scoring systems, AI can provide a more accurate assessment of a borrower faster than human agents and at a lower cost.

In Focus 8.2: Mexico's Konfio's Online Financial Services for SMEs

Mexico's Konfio, which provides online financial services for SMEs, makes loan disbursement in about 24 hours to small and midsize companies compared to the months taken by traditional banks. According to a Bloomberg article published on September 6, 2019, its interest rates are half of those charged by traditional banks and the delinquency rate was 4.8% in 2018, compared with 5.4% for the banking industry in general.

Konfio utilizes alternative data sources, artificial intelligence, and data science in its lending decisions. A December 3, 2019, article published by the e-commerce and online payment news outlet PYMNTS.com reported that Konfio borrowers can complete the application process in about eight minutes. Loans average US$12,000, compared to US$40,000 for traditional banks in Mexico. Companies such as Konfio use systems with enormous processing power that allow them to handle large amount of data in a short time.

AI algorithms can also analyze and flag risk cases. Such systems can reduce fraudulent transactions. Fraud detection systems can analyze customers' behavior and other information to trigger a cybersecurity mechanism when anomalous activities occur. Banks also employ AI to prevent money laundering. ML can recognize many suspicious activities that human beings are not capable of. The costs of investigation in transactions involving money laundering can be cut drastically using ML. For instance, Ant Group uses deep-learning technology to detect fraud.[23] AI-based systems can address problems related to high costs of due diligence, which has been a serious concern in many developing economies. For instance, due primarily

22. Shu Zhang and Ryan Woo, "Alibaba-backed online lender MYbank owes cost-savings to home-made tech," *Reuters*, January 31, 2018, www.reuters.com/article/us-china-banking-mybank/alibaba-backed-online-lender-mybank-owes-cost-savings-to-home-made-tech-idUSKBN1FL3S6.

23. Perez and Soo, "China a fast."

to fraud, bad loans account for about 20% of bank loans in India.[24] Loan fraud in the country is estimated at US$2 billion annually, leading to low trust and high interest rates.[25]

AI-based systems such as chatbots also reduce operating and transaction costs. For instance, Nigeria's United Bank for Africa (UBA) has a banking chatbot called Leo that helps customers with transactions such as transferring money, paying bills, buying mobile airtime, and checking account balance.[26] Customers can chat with Leo on WhatsApp, Facebook messenger, and Apple business chat. The chatbot responds immediately. UBA has 18 million account holders in 20 African countries. One year after launch on Facebook and WhatsApp messengers, Leo recorded over 195 million transaction volume on the platform.[27]

In Focus 8.3: MyBucks's AI-Based Financial Services to Refugees in Malawi

Luxembourg-based company MyBucks started a branch of its New Finance Bank subsidiary in Malawi's Dzaleka Refugee Camp in Dawa, 70 kilometers from the capital Lilongwe, in April 2018. By November of that year, 4,200 accounts had been opened and the branch was already profitable.[28] By November 2019, more than 7,500 accounts had been opened for Dzaleka residents.[29]

MyBucks's subsidiary in Dawa is arguably the first of its kind. The services offered by the bank include mobile banking, financial literacy training, professional training, group loans, and personal loans.

The bank uses an Oracle-core banking system. On top of the

24. William Suberg, "Indian bank wants joint effort to share data on blockchain," *CoinTelegraph*, February 16, 2017, https://cointelegraph.com/news/indian-bank-wants-joint-effort-to-sha///re-data-on-blockchain.

25. Abhishek Pitti, "Why India can become the global center for blockchain innovation," *Nasdaq*, July 17, 2018, www.nasdaq.com/article/why-india-can-become-the-global-center-for-blockchain-innovation-cm992358.

26. mTransfersHQ, "Banking chatbots in Nigeria," *Medium*. 2018, https://medium.com/mtransfers/banking-bots-in-nigeria-21a3e6c8600e.

27. Omobayo Azeez, "How we changed banking into lifestyle with Ai technology, by UBA," *business a.m.*, September 23, 2019, https://www.businessamlive.com/how-we-changed-banking-into-lifestyle-with-ai-technology-by-uba/.

28. Penny Crosman, "Fintech MyBucks brings banking to African refugee camp," *American Banker*, November 21, 2018, https://www.americanbanker.com/news/fintech-mybucks-brings-banking-to-african-refugee-camp.

29. "World's first bank in refugee camp brings hope to residents," *African Business*, February 26, 2020, https://african.business/2020/02/economy/worlds-first-bank-in-refugee-camp-brings-hope-to-residents/.

Oracle system is MyBucks's AI-based banking and lending software (Haraka). The software uses facial recognition and helps predict borrowers' ability to repay.[30] Haraka (which means "swiftly" in Swahili) utilizes AI to analyze social, behavioral, and physical data (https://corporate.mybucks.com/haraka). Haraka predicts credit risk to make decisions to offer an applicant small loans.[31] After the app is downloaded and the applicant grants permissions to access data, the app can provide such loans within minutes. Haraka utilizes relevant data to come up with a credit score.[32] A full automation of the process is the key to being able to provide nano-loans.

In Focus 8.4: Migo Uses Machine Learning in Lending Decisions

Nigeria's Migo is one of 12 apps that promise loans to Nigerians with "zero documentation and collateral."[33] It employs machine learning to make lending decisions for middle- and low-income consumers. Migo's partner banks also use its technology to reduce risks (Idris, 2020). Potential borrowers provide personal information when they apply for a loan. Based on the information, the loan amount varies from N500 to N500,000 (US$1.30–US$1,300). The company normally starts with a small loan. Borrowers are eligible for larger amounts after trust is built. For instance, borrowers who pay loans on time are then viewed as more trustworthy.[34] Migo's API is plugged in by its partners, such as banks, telecommunications operators, and merchants. By 2019, over 3 million loans had been offered to more than 1 million customers in Nigeria.[35]

Migo also uses AI to determine best fit for consumers. Its products are offered to banks or other financial services providers, such as mobile operators. It earns a percentage of each transaction. Migo harvests data from those transactions, which helps it create better credit products. Migo's partners in Nigeria include 9Mobile.[36]

30. Crosman, "Fintech."

31. MyBucks, *MyBucks issues their 2019 fintech forecast*, 2019, https://corporate.mybucks.com/articles/mybucks-issues-their-2019-fintech-forecast/4l4sKU1pPgtMzbIodPOPBH.

32. Vincent Nwanma, "Serving the average African," *Global Finance Magazine*, May 1, 2018, https://www.gfmag.com/magazine/may-2018/serving-average-african.

33. Olumuyiwa Olowogboyega, "In search of quick loans, Nigerians give up privacy," *Techcabal*, October 14, 2019, https://techcabal.com/2019/10/14/in-search-of-quick-loans-nigerians-give-up-privacy/.

34. Migo, "FAQ," 2020, www.migo.money/en-ng/faq/.

35. Olowogboyega, "In search of."

36. "How Migo's credit-as-a-service innovation attracted clients in Brazil," *TechCabal*,

Open Banking and Financial Inclusion

Open banking is a system under which banks, with customers' consent, securely share financial data and services with third-party providers (TPPs) such as fintech companies. APIs—a set of codes and protocols that decide how different software components interact and allow different applications to communicate with one another—facilitate such sharing.[37]

By expanding open banking to include telecommunications, social media, digital wallets, and other types of data, consumers may be able to engage in banking activities with a simple SIM card registration. Prohibitive cost barriers facing unbanked customers due to cumbersome sign-up and KYC procedures can thus be overcome.

Smartphone penetration is low in many LMICs. However, with only feature phones consumers can take advantage of many open-banking-enabled products, such as automatic savings sweeps and smart loan repayment. These services use data on the back end, so consumers are not required to own a smartphone.[38] Another open-banking product is decentralized KYC, which allows users to verify their identities to satisfy regulatory needs and maintain control over their information.

Users who need identity verification are often required to go through laborious and repetitive process (e.g., filling out the same documents repeatedly). By using blockchain-based solutions, they go through the verification process only once. For instance, the AI-powered identity verification provider Civic offers "on-demand, secure, and low-cost access to identity verification services."[39] Civic has teamed up with public blockchain platform Solana to provide low-cost, scalable identity management solutions in emerging economies.[40] To improve blockchain scalability, Solana uses a combination of proof of stake (PoS) consensus instead of proof of work (PoW). It combines PoS with so-called proof of history (PoH). In a PoH

January 14, 2020, https://techcabal.com/2020/01/14/how-migos-credit-as-a-service-innovation-attracted-clients-in-brazil/.

37. Alicia Phaneuf, "How open banking and bank APIs are boosting fintech growth," *Business Insider*, January 13, 2021, https://www.businessinsider.com/open-banking-api-trends-explained.

38. "The promise of open banking in driving financial inclusion in Africa," *CNBC Africa*, July 27, 2021, https://www.cnbcafrica.com/2021/the-promise-of-open-banking-in-driving-financial-inclusion-in-africa/.

39. WAXEX, *Waves African Exchange Pty Ltd partners with Civic Technologies for KYC verification*, 2018, https://waxex.africa/2018/03/19/waves-african-exchange-pty-ltd-partners-with-civic-technologies-for-kyc-verification/.

40. *Civic*, https://solhack.com/squads/civic/forum/discussion/civic-blog-updates/.

approach, a leader node timestamps blocks, which provides cryptographic proof of the passed since the last proof. The new block is then shared with validator nodes, which verify the proofs.[41] Solana claims that its systems support 50,000 transactions per second.[42] In this way, a reusable identity is created that can be trusted across any platform.[43]

An unintended consequence of open banking, however, is that it may lead to an increased exclusion since low-credit-quality customers may be identified more easily. Also, some customers could be technically excluded.[44]

The Insurance Industry

Agricultural insurance can help farmers manage risks that are too large for them to tackle on their own, and thus it is likely to lead to significant welfare gains. Estimates suggest that by removing risk through crop insurance, smallholder farmers' investment and income can be increased by 20% to 30%.[45]

Yet smallholder farmers have dismally low rates of adoption of agricultural insurance. Only 20% of smallholder farmers in developing countries are reported to have agricultural insurance.[46] In Sub-Saharan Africa, this proportion is only 3%.[47]

The main reason behind the adoption of farm insurance products is that such products are difficult to develop and offer.[48] Key challenges from an insurer's perspective include moral hazard, adverse selection, and high

41. Solana, *Synchronization*, https://docs.solana.com/cluster/synchronization.

42. Solana, https://www.coinbase.com/price/solana.

43. Civic Technologies, *4 key takeaways: Decentralized KYC for ICOs and token sales*, 2018, https://www.civic.com/blog/4-key-takeaways-decentralized-kyc-for-icos-and-token-sales/.

44. PWC, *The case for open banking in Nigeria*, https://www.pwc.com/ng/en/assets/pdf/case-open-banking-nigeria.pdf.

45. Jane Bird, "Smart' insurance helps poor farmers to cut risk Blockchain will enable cheaper and simpler cover," *Financial Times*, December 4, 2018, https://www.ft.com/content/3a8c7746-d886-11e8-aa22-36538487e3d0.

46. "Etherisc to develop a blockchain-based crop insurance for Kenyan farmers," *Ledger Insights*, November 16, 2020, https://www.ledgerinsights.com/etherisc-blockchain-parametric-crop-insurance-kenya-chainlink/.

47. "Chainlink awards grant to support the joint venture between ACRE Africa and Etherisc," *Chainlink*, November 14, 2020, https://blog.chain.link/chainlink-awards-grant-to-support-the-joint-venture-between-acre-africa-and-etherisc/.

48. T. Maestro, A. Garrido, and M. Bielza, "Drought insurance," in *Drought: Science and Policy* (New York: Wiley, 2018), 147–162.

administrative costs.[49] First, farmers that face high risk of loss are more likely to invest in insurance. Such farmers may not provide all relevant information to the insurer. The insurer lacks the ability to determine whether the farmer is lying. This state of information asymmetry, in which farmers may strategically withhold information from the insurer leads to an adverse selection problem. Second, farmers who are insured might behave in more risky ways. For instance, the farmer may not undertake sufficient effort to increase farming activities. This leads to moral hazard since the insurer in not able to determine whether the farmer is cheating or acting dishonestly.[50] Third, there are high administrative costs for some types of insurance plans. For instance, in a type-specific insurance, the insurer needs to identify various types of farmers and then design individual contracts for each type. Each type of farmer will be offered a contract that reflect the risks facing their type of farms. A drawback of this type of contract is that it results in high administrative costs since detailed information is required about the farmer and farm operation to design an optimal contract.

Several technological innovations have been developed to address these issues (In Focus 8.5). More recently attention has been focused on the development of blockchain-based crop insurance products for smallholder farmers.

In Focus 8.5: OKO's Agricultural Insurance

As national and international space programs are expanding satellite networks and operating increasingly powerful instruments, more frequent and higher-resolution remote-sensing data are becoming available to provide insurance services.[51] Insurance startup OKO makes use of such data to provide agricultural insurance. OKO has developed data-science expertise to do so.

OKO uses satellite data and mobile payments to create automated insurance products for farmers affected adversely by droughts and floods.[52] The satellite resolution has improved, and powerful

49. Maestro et al., "Drought insurance."

50. Maestro et al., "Drought insurance."

51. GSMA, *Agricultural insurance for smallholder farmers: Digital innovations for scale*, May 22, 2020, https://www.gsma.com/mobilefordevelopment/resources/agricultural-insurance-for-smallholder-farmers-digital-innovations-for-scale/.

52. Charlotte Tucker, "Luxembourg-based OKO raises around €1 million to bring innovative insurance to smallholder farmers across Africa," *EU-Startups*, April 23, 2021, https://www.eu-startups.com/2021/04/luxembourg-based-oko-raises-around-e1-million-to-bring-innovative-insurance-to-smallholder-farmers-across-africa/.

algorithms analyze data.[53] Farmers only need a feature phone to connect to OKO.[54]

OKO field agents go to markets to explain the product. The rest of the process can be completed remotely, which has the potential for additional automation. Farmers who have an Orange SIM card dial a code into a basic mobile phone to a free call back from OKO to provide a quote. The quote is based on crop type, field size, and location. Farmers then finalize their registration. The premium is typically 3%–6% of insured value for around 0.5 hectare to 2 hectares. OKO expects that it will become profitable when it scales up to 30,000–50,000 insured farmers, and renewal costs could be as low as €1-€1.50 per farmer.[55]

The Market

Its first official product was launched in Mali in January 2020.[56] It later expanded to Uganda. The company had about 7,000 paying customers in Mali in 2021. It compensated more than 1,000 farmers affected by floods in 2020.[57]

Strategic Partnership

OKO has built a strategic partnership with a number of value-chain partners. OKO's insurance service has been integrated in Orange's USSD menu in Mali. It has teamed up with Orange Mali to conduct a voice SMS campaign to educates farmers on its insurance service. Mobile money agents help customers set up their accounts and make transactions. It has started own call center to help farmers and supplement their insurance knowledge in local languages.[58]

53. Archyde, *Everything is done with my phone": In Mali, farmers take out insurance by satellite*, May 20, 2021, https://www.archyde.com/reportage-everything-is-done-with-my-phone-in-mali-farmers-take-out-insurance-by-satellite/.

54. Grace Matheka, "OKO secures Ksh. 129 million seed funding to expand services to farmers," *HapaKenya*, April 29, 2021, https://hapakenya.com/2021/04/29/oko-secures-ksh-129-million-seed-funding-to-expand-services-to-farmers/.

55. Helen Castell, *Making index-based insurance profitable*, Technical Centre for Agricultural and Rural Cooperation ACP-EU (CTA), August 27, 2019, https://spore.cta.int/en/finance/all/article/making-index-based-insurance-profitable-sid07179486f-209e-4e9f-877a-66c78d9a0c88.

56. Startup Luxembourg, *KO, a high-tech insurance for African farmers*, April 30, 2020, https://www.startupluxembourg.com/news-and-insights/fit-4-start-ensuring-the-product-market-fit-1.

57. Tucker, "Luxembourg-based."

58. GSMA, *Agricultural insurance*.

Insurance payout can be collected from an Orange Money collection point in the village. OKO has a partnership with a microfinance company. OKO insurance can serve as a guarantee for farmers to take out microcredit.[59] It has teamed up with Allianz in underwriting and reinsuring crop insurance products.[60]

A tool provided for free by the University of Reading, England, is used to calculate the amount of rain in areas of 16 square kilometers.[61]

Crop insurance is not the only type of insurance facilitated by 4R technologies. More than 600 million people in Africa who commute every day lack insurance protection in case of accidents during their trips. The Kenyan startup MotiSure has launched an IoT-based insurance tech platform to provide personal accident coverage for commuters and riders.[62] The platform is API powered. IoT sensors are attached to vehicles such as motorcycle, car, bus, or other assets. In case of an accident, an alert is triggered to MotiSure's telematics platform, which monitors vehicles and other assets with GPS and on-board diagnostics (OBD) using a computerized map. The first notice of loss (FNOL) and claims process starts immediately. For instance, an ambulance could be dispatched and payouts could be made.[63]

Blockchain-Based Crop Insurance for Smallholder Farmers

Many of these problems can be largely overcome through the use of a parametric insurance model (also known as index-based cover)—one way to close the protection gap in developed and developing markets.[64]

In such models, payouts are made based on the occurrence of clearly defined events. The parties have an agreed-upon payout in case of the event's occurrence.[65] For example, if an earthquake takes place within a five-mile

59. Archyde, *Everything is done.*

60. *Allianz partners with OKO for development of index-based crop insurance in Africa*, August 6, 2019, https://chronicle.lu/category/innovation/30018-allianz-partners-with-oko-for-development-of-index-based-crop-insurance-in-africa.

61. Archyde, *Everything is done.*

62. "Kenya's MotiSure launches API-powered insurtech platform," *Disrupt Africa*, September 1, 2021, https://disrupt-africa.com/2021/09/01/kenyas-motisure-launches-api-powered-insurtech-platform/.

63. Twitter, https://twitter.com/motisure/status/1433665270799900673, 2021.

64. SP Global, *NGL revenues surge in Q2; economic rebound underway; Lumen's dilemma*, 2021, https://www.spglobal.com/marketintelligence/en/news-insights/blog/ngl-revenues-surge-in-q2-economic-rebound-underway-lumens-dilemma.

65. Adelyn Zhou, "How blockchain smart contracts are reinventing the insurance

radius of a policyholder's home, the insurer will automatically pay out US$100,000.

Examples of such schemes include crop insurance based on area yield and weather. In an area yield scheme, a farmer will receive insurance payout only if there is a general loss in the area. In a weather insurance scheme, an insured farmer is paid in the occurrence of a weather-related loss. In these schemes, every farmer pays the same premium. A farmer is not required to prove a loss. Everybody gets the same compensation in case of a low farming yield or bad weather.

Recent developments in blockchain technology and related areas have made it possible to design smart contracts, which hold a promising potential to offer affordable index-based crop insurance products to smallholder farmers in developing countries. Of special relevance is the development of so-called oracles, which provide data related to real-world conditions that are needed to enforce smart contracts.

To take an example, decentralized oracle network company Chainlink, which provides a secure connection between smart contracts and off-chain data and services, has teamed up with several players in the insurance sector such as New York–based parametric insurance startup Arbol, East Africa's agricultural microinsurance provider for smallholder farmers Agriculture and Climate Risk Enterprise Ltd. (ACRE), and smart contract developer Etherisc. Chainlink provides verifiable data about weather conditions for insurance products targeted at smallholder farmers.[66] Chainlink was launched on ethereum in 2019. However, it is agnostic and can be used with other blockchains such as bitcoin, Hyperledger, Polkadot, Cosmos, and Ava.[67] Chainlink's services are used in decentralized finance (DeFi) applications, social networks, nonfungible token (NFT) platforms, and interoperability projects.[68] Table 8.1 presents some examples of blockchain-based crop insurance schemes in various stages of development in three LMICs.

---

industry," *Nasdaq*, January 29, 2021, https://www.nasdaq.com/articles/how-blockchain-smart-contracts-are-reinventing-the-insurance-industry-2021-01-29.

66. Adriana Hamacher, "2020 has been Chainlink's year: Here's why," *Decrypt*, December 25, 2020, https://decrypt.co/51015/2020-has-been-chainlinks-year-heres-why.

67. "Easily sell your APIs and data to any blockchain via Chainlink," *Chainlink*, October 1, 2020, https://blog.chain.link/easily-sell-your-apis-and-data-to-any-blockchain-via-chainlink/.

68. Marcel Pechman, "Chainlink (LINK) looks for momentum while pro traders target $40," *Cointelegraph*, August 26, 2021, https://cointelegraph.com/news/chainlink-link-looks-for-momentum-while-pro-traders-target-40.

Kenya

In 2014, the social enterprise Kilimo Salama—which had insured over 233,000 farmers in Kenya, Tanzania, and Rwanda[69]—and had become the largest insurance project in Africa in 2011, transitioned to a commercial business with the new name Agriculture and Climate Risk Enterprise (ACRE).[70] The number of insured farmers fell in 2014, which also triggered the transition. ACRE has directed its focus toward developing an insurance policy utilizing blockchain-based smart contracts that is indexed to local weather. The occurrence of an extreme weather event defined in the contract automatically triggers payments. It facilitates timely and fair payouts through mobile money. This new initiative involved new partners such as smart contract developer Etherisc and digital insurance platform provider Sprout Insure. Sprout Insure builds transparent, smart, weather-indexed crop insurance contracts in a distributed ledger system.[71]

To improve product quality, ACRE uses localized measurements produced by Global Index Insurance Facility–sponsored weather stations, which validate satellite data (also referred to as ground-truthing).[72] GIIF is a program launched by the World Bank Group to facilitate the adoption of index-based insurance among smallholder farmers and provide other catastrophic risk-transfer solutions in developing countries.

In November 2020, the collaboration between Etherisc and ACRE Africa received a Chainlink Community Grant to develop a decentralized insurance protocol to deliver sustainable, blockchain-based reliable and affordable parametric crop insurance solutions. The goal is to serve up to 250,000 smallholder farmers in Kenya. Farmers can pay premiums in small installments, which could be as low as US$0.50.[73] The insurance is built on top of

---

69. T. Rosenberg, "Doing more than praying for rain," *New York Times*, May 9, 2011, thttp://opinionator.blogs.nytimes.com/2011/05/09/doing-more-than-praying-for-rain/.?_r=0.

70. A. Patt, N. Peterson, M. Carter, M. Velez, U. Hess, and P. Suarez, "Making index insurance attractive to farmers," *Mitigation and Adaptation Strategies for Global Change* 14, no. 8 (2009): 737–753.

71. Blended Finance, *Sprout Insurance blockchain climate risk crop insurance*, 2020, https://www.blendedfinance.earth/nature-linked-insurance/2020/11/16/sprout-insurance-blockchain-climate-risk-crop-insurance.

72. Global Index Insurance Facility (GIIF), *ACRE Africa: Protecting rural Africa through creative partnerships and technology*, January 28, 2020, https://www.indexinsuranceforum.org/news/acre-africa-protecting-rural-africa-through-creative-partnerships-and-technology.

73. "Chainlink awards."

Etherisc's Generic Insurance Framework (GIF), which is an ethereum-based decentralized insurance protocol. It is indexed to local weather parameters. Chainlink connects to different external weather data sources and broadcasts the data within Etherisc's smart contracts.[74]

As of July 2021, Etherisc and ACRE Africa had covered more than 17,000 smallholder farmers from 17 regions in Kenya under the collaboration. About 6,000 were expected to be compensated for lost or affected crops before the end of the season in 2021.[75] A payment is made directly to the farmer's mobile phone through an API that connects the Etherisc software solution to mobile phone money-transfer service, M-PESA. All payments are notarized on-chain using Etherisc's GIF for increased transparency and auditability.[76]

Cambodia

Arbol has teamed up with financial and technical assistance facilitator Agribee Cambodia to offer blockchain-based crop insurance. Chainlink provides decentralized weather data for Arbol.[77] A pilot for a parametric extreme weather solution for the Cambodian agricultural sector was launched in October 2020.[78] Based on results of the pilot study, the plan is to scale the program to 20,000 farmers in 2021.[79] Global Parametrics, a specialty provider of parametric solutions for natural hazard and extreme weather risk

74. Etherisc, *Etherisc teams up with Chainlink to deliver crop insurance in Kenya*, November 14, 2020, https://blog.etherisc.com/etherisc-teams-up-with-chainlink-to-deliver-crop-insurance-in-kenya-137e433c29dc.

75. "Etherisc onboards 17K Kenyan farmers covered by blockchain-based crop insurance," *Cointelegraph*, July 21, 2021, https://cointelegraph.com/news/etherisc-onboards-17k-kenyan-farmers-covered-by-blockchain-based-crop-insurance.

76. Etherisc, *Etherisc update: Etherisc and Acre Africa announce first payouts through blockchain based platform*, August 16, 2021, https://blog.etherisc.com/etherisc-update-etherisc-and-acre-africa-announce-first-payouts-through-blockchain-based-platform-a0c5194214f4.

77. William Foxley, "Chainlink to provide data for farming insurance startup arbol: Chainlink will provide climate data for insurance startup Arbol's weather derivatives. Small to medium sized firms can use the Ethereum-based dapp for hedging natural events," *Coindesk*, September 14, 2021. https://www.coindesk.com/chainlink-to-provide-data-for-farming-insurance-startup-arbol.

78. Emerging Risks, *Parametric pilot targets rice farmers*, https://emergingrisks.co.uk/parametric-pilot-targets-rice-farmers/.

79. "Cambodian farmers to get crop protection insurance cover, thanks to Global Parametrics," *Khmer Times*, October 2, 2020, https://www.khmertimeskh.com/50769142/cambodian-farmers-to-get-crop-protection-insurance-cover-thanks-to-global-parametrics/.

in emerging economies, and the reinsurance group Hannover Re supported the initiative.[80]

Arbol's drought coverage charges a premium per hectare of US$13, with a maximum payout per hectare of US$150. The trigger in the insurance contract is 89% of average cumulative rainfall, and the exit is 50% of average cumulative rainfall.[81] If the rainfall total during an agreed period exceeds the trigger, no payout will occur. If the rainfall total is less than the exit, the policyholder receives the maximum payout.

Arbol's application is built on ethereum smart contracts. It secures data via the peer-to-peer protocol Interplanetary File System (IPFS).[82] IPFS allows users to download web pages and content stored across multiple servers. It also provides "historical versioning" to show any manipulation in documents. This is different from the current paradigm of Hypertext Transfer Protocol (HTTP), which sends requests for online content to a single server that stores information and means that, if anything is changed or blocked, there is no reliable way to know or access the content.[83]

Sri Lanka

In 2018, Etherisc teamed up with Aon and Oxfam to develop a blockchain-based crop insurance system in Sri Lanka. Etherisc and Oxfam reported that they were working with one of the world's largest insurance brokers to expand the market for their services.[84] The initial goal was to automate existing policies and reduce operating costs.[85] In July 2019, a blockchain-based microinsurance platform was launched, and about 200 smallholder

80. Steve Evans, "Global Parametrics & Hannover Re back another Arbol-powered weather risk pilot," Artemis October 2020, https://www.artemis.bm/news/global-parametrics-hannover-re-back-another-arbol-powered-weather-risk-pilot/.

81. AGRIBEE, *Protection scheme*, https://agribee.co/home/category?parent_id=22&id=75&menu_id=0.

82. Foxley, "Chainlink to provide."

83. Daniel Kuhn, "InterPlanetary File System is uncensorable during coronavirus news fog digital citizens are turning to decentralized protocols like IPFS to share news during the coronavirus outbreak," *Coindesk*, March 18, 2020, https://www.coindesk.com/tech/2020/03/18/interplanetary-file-system-is-uncensorable-during-coronavirus-news-fog/.

84. Ian Allison, "Oxfam in Sri Lanka will use ethereum to deliver microinsurance," *Coindesk*, October 23, 2018, https://www.coindesk.com/global-charity-oxfam-will-use-ethereum-to-deliver-microinsurance.

85. Bird, "Smart' insurance."

Table 8.1. Some examples of blockchain-based crop insurance in LMICs

Country	Insurers	Blockchain technology providers	Implementation status
Kenya	ACRE Africa	Etherisc and Chainlink	July 2021: covered more than 17,000 smallholder farmers from 17 regions in Kenya; 6,000 were expected to be compensated in 2021
Sri Lanka	Aon	Etherisc	July 2019: about 200 smallholder paddy-field farmers enrolled November 2019: first payouts to policyholders
Cambodia	Global Parametrics and Hannover Re	Arbol, Chainlink	October 2020: pilot launched Plan to scale up to 20,000 farmers in 2021 based on pilot

paddy-field farmers were enrolled.[86] In November 2019, the system made first payouts to the policyholders.[87]

Increasing Investments in Productive Entrepreneurial Businesses

Not all investments are created equal and will pay off. Investments in productive entrepreneurial businesses should be encouraged. However, in some cases, the borrower may use the loans in unproductive activities and get stuck in a debt trap because of easily available loans. The following equations show the relationship between financial production and real production and how these two indicators are related to capital and labor. First, the equation for financial production is $F = G(Kf, Lf)$, where Kf and Lf are capital

86. Aon, *Aon, Oxfam, and Etherisc launch first blockchain-based agricultural insurance policies for smallholder farmers in Sri Lanka*, July 1, 2019, https://aon.mediaroom.com/2019-07-01-Aon-Oxfam-and-Etherisc-launch-first-blockchain-based-agricultural-insurance-policies-for-smallholder-farmers-in-Sri-Lanka.

87. Etherisc, "Oxfam, Etherisc, and Aon deliver pay-outs with first blockchain-based agricultural insurance policies for smallholder farmers in Sri Lanka," *PR Newswire*, November 04, 2019, https://www.prnewswire.com/news-releases/oxfam-etherisc-and-aon-deliver-pay-outs-with-first-blockchain-based-agricultural-insurance-policies-for-smallholder-farmers-in-sri-lanka-300949728.html.

and labor used to produce output in the financial sector, F.[88] The equation for real production is $R = \mathrm{R}(K^r, L^r, F)$, where K^r and L^r are capital and labor used in production in the real sector, R. The real production function is a function of the output in the financial sector. Note that the real sector of an economy comprises the production of goods and services (e.g., agriculture, industry, construction, wholesale, retail and services sectors). The real production equation indicates that financial externalities exist in the real sector of an economy. However, conflicting findings have been reported concerning the nature of such externalities. A study of 71 LMICs, for instance, found a positive effect of financial intermediation on economic growth in about 85% of the countries.[89] Another study, however, found "negligible or negative association between financial development and growth."[90]

Various approaches, such as financial services offered (e.g., liquidity or payment services, wealth management) and products or commodities provided by banks (e.g., loan and deposit products), are employed to define and measure bank output.[91] Financial outputs such as credit availability do not necessarily contribute to economic growth. The use of credit from banks to purchase productive resources (e.g., farm assets such as tractors) can lead to economic growth.

Financing and Delivering Welfare Programs

4R technologies are playing key roles in financing and delivering welfare programs to reach the poverty market. As discussed in chapter 4, during the COVID-19 lockdown, the governments of Peru and Togo turned to AI, satellite imagery, and remote-sensing data for help locating households in poverty or extreme poverty to deliver cash payments. Blockchain has also helped in financing and delivering welfare programs (In Focus 8.6).

---

88. P. J. Dawson, "Financial development and economic growth in developing countries," *Progress in Development Studies* 8, no. 4 (2008): 325–331.

89. M. O. Odedokun, "Alternative econometric approaches for analysing the role of the financial sector in economic growth, time series evidence from LDCs," *Journal of Development Economics* 50 (1996): 119–146.

90. R. Ram, "Financial development and economic growth, additional evidence," *Journal of Development Studies* 35 (1999): 164–174.

91. Mark Ruddock, *NAICS 5221—depository credit intermediation—industry synopsis*, Food and Agriculture Organization of the United Nations, June 21, 2008, http://www.fao.org/fileadmin/templates/ess/documents/meetings_and_workshops/pricesmarch/IndustrySynopsis_NAICS5221_Canadian_Depository_Credit_Intermediation.pdf.

In Focus 8.6: Sarafu in Kenya

A community inclusion currency (CIC) is a blockchain-based open-source solution to facilitate disaster response and recovery through mobile networks, which enables fragile and vulnerable communities to exchange their digital tokens for goods and services. One such example is Sarafu ("coins"), which was launched by the U.S.-based engineering firm BlockScience and Grassroots Economics in 2019 to help low-income families in Kenya. It was developed with funding from global government donors. Anyone with a Kenyan mobile phone line, including a feature phone, can use it.[92]

Donors' financial contributions are put into a community fund that is used to create the community credit.[93] The Kenya Red Cross teamed up with the Danish Red Cross and Grassroots Economics to launch a basic income system called Sarafu Network. To use the service, a person dials *384*96# (Safaricom) or *483*46# (Airtel) in Kenya, chooses a preferred language, and enters a PIN. The users then receive 50 Sarafu (KES 50 or about US$0.50) and can immediately start using it.[94] Those that do not need a basic income are reported to support others by giving theirs to people in need.

The system relies on Unstructured Supplementary Service Data (USSD) to enroll users and complete transactions. Thus, no internet connectivity is needed to engage in transactions and there is an easy user interface. Blockchain is at the back end of the application. The existing knowledge of using mobile money technology M-PESA in Kenya makes it simple to use.[95]

---

92. CryptoGuru, "How the Sarafu community inclusion cryptocurrencies are used in Kilifi County, Kenya," *BitKE*, July 30, 2021, https://bitcoinke.io/2021/07/sarafu-in-kilifi-kenya/.

93. Mohammed Yusuf, "Kenyans turn to community initiative currency to stay afloat," *Voice of America*, June 25, 2020, https://www.voanews.com/africa/kenyans-turn-community-initiative-currency-stay-afloat.

94. "The Kenya Red Cross launches a token-driven basic income system after the success of the Sarafu blockchain community currency," *BitcoinKE*, March 5, 2021, https://bitcoinke.io/2021/03/the-sarafu-basic-income-system/.

95. Edna Laboso, "How Sarafu, a community inclusion currencies is keeping families in Nairobi's informal settlements afloat," *Impact Hub*, https://impacthubmedia.com/14649/how-sarafu-a-community-inclusion-currencies-is-keeping-families-in-nairobis-informal-settlements-afloat/.

Usefulness in the Pandemic

Sarafu helped Kenyan slum dwellers pay for basic necessities such as food, water, and sanitary items during the COVID-19 pandemic. Each week, families were given virtual vouchers worth 400 Kenyan shillings (US$4) to use to buy essential goods. Vendors then sent the vouchers to Nairobi-based social enterprise Grassroots Economics to redeem for cash.

In Mukuru kwa Njenga, a slum in eastern Nairobi, more than a third of the vendors had signed up for the project by July 2021. Whereas cash aid can be spent on anything, Sarafu can be used to pay for only essential items, such as food, health supplies, and educational resources. Since the platform runs on blockchain, all transactions can be tracked, so it is transparent and receivers can spend aid only to buy necessities. In July 2021, the platform made an average of 1 million Kenyan shillings (US$9,0000) in daily transactions.[96]

Stimulation of Economic Activity

Local cryptocurrencies such as Sarafu are aimed at facilitating "loops" of economic activity and creating a network effect. The more businesses and individuals that use it, the more valuable it becomes. One way to increase the usefulness of a cryptocurrency is to increase its circulation in the community and make it possible for participants to cover as many of their expenses as possible.[97]

Sarafu can be used to buy goods in an area. In Kenya's coastal city of Kilifi, farmers are reported to use Sarafu to sell their vegetables and to buy supplies such as seeds and fertilizer.[98] To buy things not available in the local community, Sarafu points can be converted into local currency.[99]

---

96. Kagondu Njagi, "Kenya slum dwellers battle COVID-19 downturn with virtual currency," *Reuters*, January 19, 2021, https://www.reuters.com/article/kenya-tech-slums-idUSL8N2JG326.

97. Jackie Brown, "The power of local money for a thriving local economy," Yes Magazine, July 9, 2021, https://www.yesmagazine.org/economy/2021/07/09/local-money-equity.

98. Ruud Elmendorp, "Cryptocurrency booming among Kenyan farmers," *Voice of America—English*, July 26, 2021, https://www.voanews.com/africa/cryptocurrency-booming-among-kenyan-farmers.

99. Elmendorp, "Cryptocurrency."

Creating Belongingness

Local cryptocurrencies such as Sarafu can also encourage a feeling of community belongingness. Kilfi vendors think that when they accept Sarafu, they help their customers earn their way out of debt.[100]

Sarafu can be earned if people are engaged in income-generating activities. There are weekly Sarafu market days to sell goods and services using basic income (Sarafu ya Jamii) in Kilifi. Such markets are an ideal place to meet other people who support community-driven basic income. These markets provide an opportunity to buy and sell with Sarafu to support the community.[101]

A study of intermediary-less markets in Athens where food products were sold directly to consumers found that this model helped fight the food industry's oligopoly of intermediaries and move to a more economically sustainable model.[102] Cryptocurrencies such as Sarafu can play an even bigger role in promoting economically sustainable models since institutions such as banks and credit card companies are not needed to create trust between buyers and sellers.

Fraud, corruption, discrimination, and mismanagement block some money that is intended to reduce poverty and improve education and healthcare from actually helping people. An estimate suggested that about 30% of development funds do not reach the intended recipients due to problems such as third-party theft and mismanagement.[103]

Blockchain can improve humanitarian assistance by fighting these problems (In Focus 8.6). For instance, the WFP's Innovation Accelerator started its Building Blocks pilot in early 2017 to distributes food vouchers to refugees without having to rely on the services of financial institutions. In the first stage, food and cash assistance was provided to needy families in Pakistan's Sindh province. In 2017, the WFP started distributing food vouchers in Jordan's refugee camps by delivering cryptographically unique

100. I. Khabuqwi, "Kilifi Kenya—a hub for community driven basic income," *Grassroots Economics*, February 10, 2021, https://www.grassrootseconomics.org/kilifi-kenya.html.

101. Khabuqwi, "Kilifi Kenya."

102. N. Papacharalampous, "A new rural in the city: A no-middlemen markets' ethnography," *Journal of Rural Studies* 86 (2021): 702–710.

103. Ben Paynter, "How blockchain could transform the way international aid is distributed," *Fast Company*, September 18, 2017, www.fastcompany.com/40457354/how-blockchain-could-transform-the-way-international-aid-is-distributed.

coupons to participating supermarkets. Supermarket cashiers are equipped with iris scanners to identify the beneficiaries and settle payments. UN databases verify biometric data about refugees. Building Blocks' ledger records the transactions on a private version of the ethereum blockchain: the Parity Ethereum. No banks are involved and beneficiaries thus receive goods directly from the merchants.

The Parity Ethereum used in the system employs four nodes to validate transactions. This means that transactions cannot be seen by actors that are not a part of the authorized peer nodes. An additional benefit is that cryptocurrency mining is not needed to validate the transactions. This feature removes a key bottleneck to the processing speed and transaction capacity. The system is designed to scale.

The WFP reported that by October 2017, it had distributed US$1.4 million in food vouchers to 10,500 Syrian refugees in Jordan. As of early 2019, 106,000 refugees in Jordan's Azraq and Za'atari camps received cash transfers in cryptocurrencies. By that time, 1.1 million cryptocurrency transactions transferred more than US$ 23.5 million to refugees.[104] In 2019, the WFP disbursed more than US$64 million through the pilot program, which helped the organization save 98% in financial transaction fees.[105] The WFP expects that blockchain-based solutions could reduce overhead costs from 3.5% to less than 1%.[106]

Processing speed and efficiency can be increased with cryptocurrencies. For instance, once refugees are registered, the WFP's blockchain system encrypts their data and transfers vouchers almost instantaneously. In this regard, a key appeal of blockchain-based fintech products and cryptocurrencies is the ability to intervene quickly and efficiently in situations that face difficult environments. For example, when remote places lacking financial infrastructure such as ATM machines and banks face disasters such as earthquakes or storms, blockchain can help humanitarian organizations provide cash assistance faster than other available means.[107] Cryptocash (e.g., tokens)

---

104. "The World Food Program: Fighting hunger with blockchain," *Food Tank*, January 2019, https://foodtank.com/news/2019/01/the-world-food-program-fighting-hunger-with-blockchain/.

105. Damian Radcliffe, "From mobile money to blockchain: How this UN agency's tech stops people starving," *ZDNet*, December 10, 2020, https://www.zdnet.com/article/how-the-nobel-prize-winning-world-food-programme-is-using-ict-to-hack-hunger-in-the-middle-east-and-beyond.

106. Nir Kshetri and Jeff Voas, "Blockchain-enabled e-voting," *IEEE Software* 35, no. 4 (2018): 95–99.

107. World Food Programme, *What is "blockchain' and how is it connected to fighting*

can represent local currencies such as Pakistani rupees and Jordanian dinar that can be traded outside the banking system. Recipients can use the cryptocash to buy goods and services in participating shops. Cryptocurrencies can even replace scarce local cash, which allows aid organizations, residents, and merchants to engage in trannsactions[7].

When the applications reach a more advanced development stage in the future, more benefits can be realized. For instance, the WFP expects that refugees may be able to access funds by controlling their own cryptographic keys. This would also allow them to incorporate and integrate personal data from diverse sources. For instance, their medical records could be with the WHO, academic credentials with the UN Children's Fund (UNICEF), and nutritional data with the WFP. In this way, they can build an economic identity.

Chapter Summary and Conclusion

4R technologies can address concerns related to informational opacity. These technologies also reduce transaction costs and make it profitable for financial institutions to provide small-scale loans to poor borrowers.

Technologies such as AI, blockchain, IoT, and satellite are also expected to increase adoption of various types of insurance products. Among the most notable of these is that blockchain-based smart contracts offer low-cost crop insurance to smallholder farmers in LMICs. Some insurance companies have started offering index insurance products for smallholder farmers in LMICs that combine on-chain codes with off-chain oracle solutions. In case of an extreme weather event, policies are automatically triggered. While not all challenges farmers encounter can be resolved, smart contracts enhance transparency and help farmers receive insurance payouts in a timely manner. Adoption among smallholder farmers can be further increased by creating more relevant data and facilitating the diffusion of smartphone and other electronic devices and e-payment systems.

hunger?," March 6, 2017, https://insight.wfp.org/what-is-blockchain-and-how-is-it-connected-to-fighting-hunger-7f1b42da9fe.

PART 4

The Way Forward

CHAPTER 9

Enablers, Opportunities, Barriers, and Challenges

4R technologies offer significant possibilities for improving the welfare of the B4B. For instance, these technologies have made it possible to adopt sustainable and environmentally friendly solutions in many LMICs. In many African cities, for instance, electricity is expensive, and piped gas supply is limited. Bottled LPG is often the best available alternative to charcoal.[1] 4R technologies have helped increase adoption of LPG. For instance, Nigeria's renewable energy startup Nupe Energy's solution involves an LPG cylinder's smart meter that controls gas flow, monitors fuel level remotely, and facilitates gas purchase. The solution is compatible with different cellular networks (2G/GSM/GPRS).[2]

The digital transformation in LMICs is facilitated by the emergence of innovative local companies such as Nupe Energy, PayGo Energy, M-Gas, and KopaGas in the energy sector; by M-PESA in the payment sector; and by RxAll in the healthcare sector. Even more encouraging, there are some world-class companies in LMICs' digital sector. In Africa, for instance, Nigeria's Flutterwave and Kenya's Gro Intelligence made *Time*'s 2021 global list "100 Most Influential Companies." The Gro Intelligence platform "aggregates, normalises and models complex data to illuminate the interrelationships between food, climate, trade, agriculture, and macro-economic conditions." It has over 40,000 datasets and processes more than 650 trillion

1. M. Rowling, "Smart gas cooking seeks to break African cities' dirty charcoal habit," *Reuters*, August 6, 2019, https://www.reuters.com/article/us-africa-energy-cooking-feature-idUSKCN1UW161.

2. "Nupe Energy is turning gas cylinders into connected devices," *TechCabal*, May 12, 2021, https://techcabal.com/2021/05/12/nupe-energy-metered-gas-cylinders/.

datapoints for different customer segments such as governments, financial institutions, agricultural input companies, retailers, and food and beverage companies.[3] Gro Intelligence works with governments, food companies like Unilever and Yum!, and banks like BNP Paribas to provide analysis and forecasts.[4]

Digital technologies' increased affordability has opened up further possibilities in utilizing these technologies to enhance social and economic development. In September 2021, ten African countries offered charges of less than US$1 for a gigabyte of data compared to the global average of US$4.07. The cost of mobile internet in Sudan was US$0.27 per gigabyte of data, the cheapest in the world.[5]

Despite all the promises, however, there are many hurdles to overcome to realize the full potential of these technologies. Key barriers are related to lack of technological and infrastructural facilities and lack of political will. Many LMICs are also facing problems of slow internet connectivity and lack of infrastructure. For instance, as of November 2019, the mobile broadband penetration rate in Africa was 47%, and fourth-generation (4G) cellular penetration rate was only 10%. By 2025, the proportion of 2G users is expected to drop from 46% to 12%.[6]

Even more important, despite rapidly falling costs of ICT products and services, many products targeted at the poor are beyond the reach of most of LMIC's population. For instance, many LMICs do not have widespread diffusion of smartphones or other appropriate devices and internet access to benefit from available healthcare solutions. For instance, as of 2021, only 7.5 million Tanzanians own a smartphone capable of downloading Ada Health's app.[7] Such challenges are also currently preventing more widespread adoption of other advanced solutions such as blockchain-based crop insurance. Many farmers in Sri Lanka lacked electronic devices and internet access,

3. Tom Jackson, "2 African startups on TIME's 100 most influential companies list," *Disrupt Africa*, May 3, 2021, https://disrupt-africa.com/2021/05/03/2-african-startups-on-times-100-most-influential-companies-list/.

4. Kate Barlett, "Tech in Africa is taking off," *OZY*, August 19, 2021, https://www.ozy.com/the-new-and-the-next/tech-in-africa-is-taking-off/439675/.

5. Conrad Onyango, "Africa: Are internet data plans affordable?," *Africa Report*, September 11, 2021, https://www.theafricareport.com/125706/africa-are-internet-data-plans-affordable/.

6. "Why African mobile networks must invest in 4G," *EABW News*, November 15, 2019, https://www.busiweek.com/why-african-mobile-networks-must-invest-in-4g/.

7. Ben Turner, "Tanzania's digital doctor learns to speak Swahili," *Financial Times*, May 16, 2020, https://www.ft.com/content/7ed03336-6a0a-11ea-a6ac-9122541af204.

which hindered their attempt to register for the policy.[8] Likewise, in 2019, only 10% to 20% of Cambodian farmers had smartphones. Some farmers are illiterate or lack internet access and a mobile data plan.

Small companies in LMICs also face barriers to adopt 4R technologies. Like other technologies, deployment of 4R technologies tends to diffuse from larger to smaller organizations, commonly known as the rank effect.[9] Blockchain systems are expensive to implement and manage. For instance, despite some smaller companies' access to blockchain-based solutions, due to cost and complexity, this technology is out of reach for many organizations. For instance, as of August 2020, the costs per month of using IBM's enterprise blockchain platform, based on Hyperledger Fabric,[10] included a membership fee of US$1,000 and per-peer fee of US$1,000. This translates to an annual peer fee of US$12,000 for each additional member added to IBM enterprise blockchain platform.

On the entrepreneurship front, while some countries and technology hubs have reported high levels of entrepreneurial activities, a concern has been questionable quality of such activities. For instance, although many companies describe themselves as blockchain companies, few real use cases have emerged. For instance, among China's 262 public companies that categorized themselves as blockchain companies, as of September 2020, only 23 had mentioned blockchain use cases that had gone live.[11]

Among the major obstacles is lack of political will. For instance, some blockchain projects were reported to face opposition from key actors due to the technology's potential to increase transparency. For example, the Honduran government's plan to transfer land registry in blockchain was stopped due to political issues.[12]

---

8. Etherisc, "Oxfam, Etherisc, and Aon deliver pay-outs with first blockchain-based agricultural insurance policies for smallholder farmers in Sri Lanka," *PR Newswire*, November 4, 2019, https://www.prnewswire.com/news-releases/oxfam-etherisc-and-aon-deliver-pay-outs-with-first-blockchain-based-agricultural-insurance-policies-for-smallholder-farmers-in-sri-lanka-300949728.html.

9. G. Gotz, "Monopolistic competition and the diffusion of new technology," *Rand Journal of Economics* 30, no. 4 (1999): 679–693.

10. Aran Davies, "How much does it cost to build a blockchain project?," *DevTeam.Space*, 2020, https://www.devteam.space/blog/how-much-does-it-cost-to-build-a-blockchain-project/.

11. Shuyao Kong, "Blockchain's been a Bust for China's 'Blockchain 50' Public Companies," *Decrypt*, September 30, 2020, https://decrypt.co/41657/blockchains-been-a-bust-for-chinas-blockchain-50-public-companies.

12. "Blockchain Land Title Project 'Stalls' in Honduras," *HondurasNews*, December 2015, https://www.hondurastoday.com/2015/12/blockchain-land-title-project-stalls-inhonduras/.

This chapter highlights the enablers and opportunities for low-income people to utilize 4R technologies in enhancing their economic and social welfare. It also analyzes barriers and challenges in adopting these technologies.

Key Enablers

Freely Available Data

Decision makers can make use of freely available data and open-source software. One such data source is Digital Earth (DE) Africa platform. Using DE Africa, it is possible to translate satellite images into actionable information. Decision makers in various fields—science, policy, agriculture, and industry—can access and use the data.[13] DE Africa makes current and historical satellite images available. According to the World Economic Forum, African economies could realize economic benefits of up to US$2 billion by utilizing data from DE Africa. Using the open-source technology Open Data Cube, which has access, management, and analysis functions for geospatial data, raw data can be processed to help decision makers.

Local Capacity Building

One the plus side, a number of capacity-building measures have been introduced in several LMICs that have the potential to improve the digital landscape. Many LMICs have demonstrated capabilities to create technological solutions to address local challenges.[14] Kenya's iHub (Innovation Hub) has brought together innovators and the world's leading organizations to accelerate startup activity.[15] In Kenya, iHub was the first innovation hub; it started in Nairobi in March 2010. By 2015, 152 companies had been

13. "Welcome to Digital Earth Africa (DE Africa)," *Digital Earth Africa*, https://www.digitalearthafrica.org.

14. Nir Kshetri, "Creation, deployment, diffusion and export of Sub-Saharan Africa-originated information technology-related innovations," *International Journal of Information Management* 36, no. 6 (2016, pt. B): 1274–1287.

15. Charles Mulungi, "Could tech 'lions' prowling Silicon Savannah hold key to East Africa's post-pandemic recovery?," *East African Business Week*, September 1, 2021, https://www.busiweek.com/could-tech-lions-prowling-silicon-savannah-hold-key-to-east-africas-post-pandemic-recovery/.

formed out of iHub.[16] Cofounder Erik Hersman noted that it opened as a "space for the tech community in Nairobi to gather, to call home and build connections to each other and work on ideas from."[17]

Rwanda's kLab provides an open space for collaboration and innovation. It brings together students, thinkers, and entrepreneurs to turn concepts into products and services. Similarly, Uganda's Outbox incubation and innovation space provides support to local entrepreneurs. Its kLab was started in 2012 and had more than 1,400 members by 2017. Over 80 companies had been created. It is open 24/7 and offers free space, free mentorship, and free internet to members.[18]

Likewise, a recent report has challenged the opinion that the Middle East and North Africa (MENA) region is a technology "desert." Instead, the region is a "fertile" technological environment with its own independent character and influenced by Western, Eastern, and local cultures.[19]

Some LMICs are developing more advanced capacities. To get access to more computing power to run sophisticated AI algorithms, many Chinese companies are designing and building their own AI-optimized chips.[20]

Skills and Exchange to Benefit from the 4R

Some efforts have also been reported on developing skills in areas such as data training and literacy. Some African universities are offering data science

16. Aubrey Hruby and Jake Bright, "The rise of silicon savannah and Africa's tech movement," *TechCrunch*, July 23, 2015, https://techcrunch.com/2015/07/23/the-rise-of-silicon-savannah-and-africas-tech-movement/.

17. Toby Shapshak, "Kenya's iHub enters a new chapter," *Forbes*, May 11, 2016, https://www.forbes.com/sites/tobyshapshak/2016/03/11/kenyas-ihub-enters-a-new-chapter/?sh=390c4fea4f6a.

18. Matthew Mercer, *Driving growth: How Rwanda's kLab is powering entrepreneurial change*, Centre for Public Impact, May 8, 2017, https://www.centreforpublicimpact.org/insights/driving-growth-rwandas-klab.

19. Kanta Dihal, Tomasz Hollanek, Nagla Rizk, Nadine Weheba, and Stephen Cave, *Imagining a future with intelligent machines a Middle Eastern and North African perspective imagining a future with intelligent machines* (Cambridge: Leverhulme Centre for the Future of Intelligence, 2021), https://static1.squarespace.com/static/5b2a5992fcf7fd5871d926dd/t/60d1fbdb30db914783l4cbdf/1624374282079/ImaginingAI-MENA.pdf.?fbclid=IwAR2HjMjP0AHwP_Zxpgb7ox8a0dtXCR0A5vJTYbwwOg_EF4vY8cAUukL5wy8.

20. C. Arkenberg, "China inside: Chinese semiconductors will power artificial intelligence," *Deloitte Insights*, December11, 2018, https://www2.deloitte.com/xe/en/insights/industry/technology/technology-media-and-telecom-predictions/chinese-semiconductor-industry.html.

degrees and providing shorter courses for policymakers and professionals.[21] In Rwanda, universities have started teaching 4R principles.[22] The Rwandan government's target is to achieve digital literacy for all youths aged 16 to 30 by 2024.[23]

Researchers have also started organizing scientific events such as conferences, workshops, and symposia about 4R technologies. To take an example, partly in response to the inaccessibility of prestigious AI conferences in the West, in 2017, a small group of African data scientists created their own conference, Deep Learning Indaba. Attendees at the conference range from experienced AI engineers to graduate students. The conference had 300 attendees in 2017 and more than 600 in 2019.[24]

Rapidly Falling Costs of 4R Technologies

Rapidly falling costs of ICTs are illustrated in chapter 1 with data comparing the prices of IoT sensors and hard drives for 2009 and 2017. Costs are also falling for solutions based on other 4R technologies. For instance, some drone models can be acquired for as little as US$650.[25]

Likewise, innovations around satellite technology have lowered the cost of devices. It is thus becoming possible for low-income nations to design or manufacture their own small satellites. Instead of relying on single-use rockets, which cost US$200 million to launch a satellite, reusable rockets reduce the cost to about US$60 million, and this has the potential to drop to US$5 million. Satellite mass production could decrease costs from US$500 million per satellite to US$500,000.[26] In many cases, it could be less expensive to launch and maintain a satellite than to purchase satellite imagery and

21. Natalie Donback, "Harnessing the potential of satellite data to leapfrog progress across Africa," *Devex*, June 23, 2020, https://www.devex.com/news/harnessing-the-potential-of-satellite-data-to-leapfrog-progress-across-africa-96757.

22. "4IR: The future of mining is now," *Mining Review Africa*, November 6, 2019, https://www.miningreview.com/gold/4ir-the-future-of-mining-is-now-and-you-better-be-on-board/.

23. "Charting Africa's digital future," *NewsDay Zimbabwe*, August 26, 2021, https://www.newsday.co.zw/2021/08/charting-africas-digital-future/.

24. IBM AI, "Building a community to spread AI across Africa," *Forbes*, May 21, 2020, https://www.forbes.com/sites/insights-ibmai/2020/05/21/building-a-community-to-spread-ai-across-africa/?sh=67674d7657a2.

25. Nathaniel Allen, *Africa's evolving cyber threats*, Africa Center for Strategic Studies, January 19, 2021, https://africacenter.org/spotlight/africa-evolving-cyber-threats/.

26. Morgan Stanley, *Space: Investing in the final frontier*, 2019, https://www.morganstanley.com/ideas/investing-in-space.

other services from commercial companies and foreign governments.[27] Over the past few decades, there has been significant advancement in space-based technologies in Africa.[28]

The rapidly falling costs of some technological solutions have changed the soundness of some arguments related to whether to acquire these technologies. For instance, regarding satellites, one view is that due to high costs of launching and maintaining satellite operations, it is more appropriate for developing countries to build capacity to use and interpret satellite data. The opposite argument is that such operations provide technological autonomy and offer the opportunity to balance consumption and production of satellite data and promote sustainability of resources and environment. All lower-orbit satellites, especially those over the poles, return to the same spot too infrequently. They thus do not adequately or effectively satisfy the needs of many countries. Such satellites often return to a spot every five days, or even every 10 or 20 days. Sometimes, it is not possible for the satellite to see through cloud cover. LMICs should thus develop their own space programs. Moreover, free images are sometimes old. If countries want certain images, they have to pay for a satellite to take it.[29] This indicates that decisions about launching a satellite may not be feasible today for a small developing country but may make economic sense a few years down the road when costs further decrease.

Major Opportunities

4R technologies provide opportunities in a number of areas. Relevance for the B4B of the four major 4R technologies discussed in chapters 2–5 and opportunities created are presented in Table 9.1. Likewise, three main applications of these technologies and their transformations of key industries were considered in chapters 6, 7 and 8. A few other key opportunities are discussed in this section.

27. Judd Devermont and Temidayo Oniosun, "Is the United States losing the African space race?," *War on the Rocks*, June 23, 2020, https://warontherocks.com/2020/06/is-the-united-states-losing-the-african-space-race/.

28. T. Woldai, "The status of earth observation (EO) & geo-information sciences in Africa—Trends and challenges," *Geo-spatial Information Science 23*, https://www.tandfonline.com/doi/full/10.1080/10095020.2020.1730711.

29. T. Anderson, "Launching your own satellite—The pros and cons," *SciDev.Net*, November 11, 2009, https://www.scidev.net/global/features/launching-your-own-satellite-the-pros-and-cons/.

Table 9.1. Opportunities created by major 4R technologies to the B4B

Technology	Essence of the technology	Relevance for the B4B and examples
AI	Machines can apply logic and reasoning, which would enable intelligent automation of tasks.	Due to decreasing costs and increasing user-friendliness AI applications are becoming accessible to the B4B (e.g., PlantVillage Nuru, chapter 2).
Blockchain	Decentralized and distributed ledger creates transparency of transactions	Opaque systems facilitate corrupt practices due to the lack of accountability to the public and stakeholders. Such practices are more likely to victimize poor people.[1] Blockchain can help fight opacity and hence corruption (e.g., the WFP's use of blockchain to distribute food vouchers) (chapter 8).
Remote sensing	Information about an object or a place can be acquired without any physical contact.	Digital Earth Africa offers free satellite data: has partnered with the Ghana Statistical Service (GSS) to build capacities to collect and analyze satellite data, which can be used to fight illegal mining.[2]
The IoT	Facilitates constant monitoring of the environment using information created by interconnected "things," and remotely manage various aspects of the environment by turning the information into actionable intelligence.	Rapidly declining costs of IoT devices and increasing connectivity have made it practical, feasible, and cost effective to deploy IoT. solutions to serve low-income populations (e.g., Sanku's Project Healthy Children (chapter 6)).

[1]Transparency International Bangladesh, Corruption in Bangladesh. 2002, TIB: Dhaka
[2]"Illegal mining depleting Apamprama Forest Reserve" November 17, 2021 https://www.ghanabusinessnews.com/2021/11/17/illegal-mining-depleting-apamprama-forest-reserve/.

Marginalized Groups' Participation in Economic Activities

4R technologies are facilitating the participation of low-income people in economic activities. One example is the BanQu platform discussed in chapter 1. With BanQu's software platform, a waste picker is linked with a company that sources recycled materials. Each time a person sells materials, such as a bottle, they receive a text message that validates their existence in the supply chain. The information about the identity of the waste picker and the amount of material they bring is provided to the company that buys the materials.[30]

There are also important indirect benefits that firms may experience from their participation in economic activities. For instance, since 4R technologies are making it easier to access foreign markets, it is worth noting that exporters all over the world are more productive than domestically oriented firms in the same location and industry. Unlike in other regions, African exporters are found to improve their relative performance after they enter foreign markets. They have higher post-entry productivity and rates of productivity growth.[31]

Solutions based on blockchain and other 4R technologies offer a number of opportunities for increasing smallholder farmers' access to crop insurance. First, prior research has suggested that information transparency can increase farmers' confidence in weather index insurance (WII) products.[32] Blockchain solutions can provide such transparency since all parties can independently verify relevant data and information in blockchain-based smart contracts. The data cannot be tampered with by the insurer or other actors.[33]

Blockchain systems can help make prompt insurance payouts in the case of yield-reducing weather events. For instance, in the previous system used by ACRE, the data from a weather station was aggregated and combined with satellite data at the end of every season to map out rain patterns. The company worked with agronomists to calculate the index and identify locations that experienced too much rain, too little rain, or rain at the wrong time. Farmer payouts were calculated based on crops, location, and the amount invested in seeds.[34] ACRE's experience showed that timely payout after weather adversities can contribute to a 30% increase in policy renewal.

30. Deonna Anderson, "People make the circular economy world go round," *Greenbiz*, March 5, 2021, https://www.greenbiz.com/article/people-make-circular-economy-world-go-round.

31. J. Van Biesebroeck, "Exporting raises productivity in sub-Saharan African manufacturing firms," *Journal of International Economics* 67, no. 2 (2005): 373–391.

32. A. Patt, N. Peterson, M. Carter, M. Velez, U. Hess, and Po Suarez, "Making index insurance attractive to farmers," *Mitigation and Adaptation Strategies for Global Change* 14, no. 8 (2009): 737–753.

33. Etherisc, *Etherisc teams up with Chainlink to deliver crop insurance in Kenya*, November 14, 2020, https://blog.etherisc.com/etherisc-teams-up-with-chainlink-to-deliver-crop-insurance-in-kenya-137e433c29dc.

34. Jonathan Kalan, "Tech fix for Africa's big farming challenge," *BBC*, April 7, 2013, http://www.bbc.com/future/story/20130408-tech-taps-africas-farm-potential.

ACRE's blockchain solution makes it possible to send payouts to eligible farmers' mobile accounts within seconds.[35] This can produce more trust among smallholder farmers and increase product demand.[36]

Third, smart contract–led automation is likely to reduce operating costs of insurers, which can result in lower premiums for farmers. A study conducted by Etherisc and the Global Innovation Lab for Climate Finance claimed that Etherisc's smart contract will reduce the costs required to issue a policy by up to 41%.[37] It translates into 30% premium reduction.[38]

Most WII in West Africa rely on precipitation data alone. In such models, rainfall acts as a proxy for drought. Recent research has found that combining precipitation data with reference evapotranspiration (RefET)—the combined rate of evaporation from Earth's surface and transpiration from the leaves of plants—could be more useful in guiding insurance payouts.[39] Some initiatives have been undertaken on this front. ACRE is adding more data to design a more optimal insurance policy. As of 2020, a soil moisture index was being tested for use in combination with the usual rainfall-based indexes. The company plans to install soil moisture sensors in the weather stations to complement remote-sensing data.[40]

Opportunity for South-South Cooperation

Many 4R-related innovations originated in LMICs and subsequently have been used in other LMICs. This process is facilitated by South-South cooperation (SSC), which involves the exchange of resources, technology, and knowledge among economies in the Global South. Indeed, SSC is a key force in the digitization of LMICs.

Many successful examples of internationalization of firms that have developed products based on 4R technologies have been reported. The Nigerian provider of commercial financing and mobile payment services

35. Global Index Insurance Facility, *ACRE Africa: Protecting rural Africa through creative partnerships and technology*, January 28, 2020, https://www.indexinsuranceforum.org/news/acre-africa-protecting-rural-africa-through-creative-partnerships-and-technology.

36. Etherisc, https://medium.com/@etherisc.

37. Etherisc, *Etherisc teams up with Chainlink.*

38. Global Index Insurance Facility, *ACRE Africa.*

39. Earth Resources Observation and Science (EROS) Center, *Satellite data show promise as guide for West African crop insurance*, September 16, 2020, https://www.usgs.gov/center-news/satellite-data-show-promise-guide-west-african-crop-insurance?qt-news_science_products=1#qt-news_science_products.

40. Global Index Insurance Facility, *ACRE Africa.*

Flutterwave is highly internationalized (In Focus 9.1). PayGo announced a plan to expand operations to Asia, starting with Vietnam and Bangladesh, in partnership with the Japanese company Saisan.[41] As of May 2021, the company had started two pilot projects of its LPG solution in Asia.[42] Likewise, Brazilian AgTech company Solinftec, which has launched the AI assistant Alice, opened North American offices in Indiana.[43] It is also expanding to other South American countries, Russia, and Ukraine.[44]

In Focus 9.1: Flutterwave's Digital Payment Processing

Flutterwave is a Nigerian provider of commercial financing and mobile payment services. Flutterwave's API integrates different payment systems and methods. The goal is to make it easy to process payments for banks and merchants in Africa. As of February 2020, the payment platform had processed more than 100 million transactions valued at over US$5.4 billion.[45]

As of December 2019, it served over 60,000 merchants in six African countries and the United Kingdom, supported over 150 currencies, and facilitated payment services from over 68 gateways, such as MasterCard, Visa, PayPal, and AliPay.[46] Its clients included Arik Air, Uber, and Wakanow.[47] Flutterwave operates in Ghana, Kenya, Nigeria, Rwanda, South Africa, and Uganda.[48] It plans to

41. Alexander Onukwue, "Kenya-based PayGo Energy is exporting its gas meter tech to Asia," *Techcabal*, June 16, 2020, https://techcabal.com/2020/06/16/paygo-energy-cylinder-gas-meter-saisan-japan/.

42. Shrikant Avi, "PayGo Energy: Striking strategic partnerships to scale up clean cooking," *Clean Cooking Alliance*, May 19, 2021, https://cleancookingalliance.org/news/05-19-2021-paygo-energy-striking-strategic-partnerships-to-scale-up-clean-cooking/.

43. "Brazil's Solinftec sets up shop in the U.S.," *Indianapolis Business Journal*, June 3, 2019.

44. Rob Leclerc, "The Road to Automated Agriculture Begins in Brazil," *AgFundernews*, June 10, 2019, https://agfundernews.com/the-road-to-automated-agriculture-begins-in-brazil.html.

45. Dolapo Aina, "Flutterwave launches operations in Rwanda," *The Guardian*, February 13, 2020, https://guardian.ng/technology/c55-technology/flutterwave-launches-operations-in-rwanda/.

46. Olumuyiwa Olowogboyega, "Countdown: The 10 most important African tech companies of the decade," *Techcabal*, December 16, 2019.

47. Partner, "Flutterwave pioneers instant settlement for Nigerian merchants," *TechCabal*, August 20, 2019, https://techcabal.com/2019/08/20/flutterwave-pioneers-instant-settlement-for-nigerian-merchants/.

48. Aina, "Flutterwave."

enter Cameroon, Egypt, Ethiopia, and Morocco, as well as China and India.[49]

In March 2020, Flutterwave teamed up with the Nigeria-based fintech Paga to allow Paga users to access products from merchants using Flutterwave. Paga's 14 million customers can make payments directly from their wallets to merchants such as airlines, hotels, ticketing companies, media outlets, and fashion retailers.[50] Increasing Paga users' access to useful financial services has promoted financial inclusion.

Flutterwave teamed up with e-commerce fraud prevention provider Forter, which uses AI and ML technology for merchants to accurately identify legitimate buyers from fraudsters. The technology makes it possible to approve more legitimate online transactions and block frauds.[51]

As of April 2021, Flutterwave served more than 290,000 businesses and processed over 140 million transactions worth over US$9 billion worldwide. It is Africa's largest payment service provider. The partnership with Forter provides increases merchants' confidence to accept more transactions. Online shoppers get the flexibility to use alternative payment methods and traditional options such as credit cards.[52]

Challenges, Barriers, and Threats

Shortage of Skills and Talent

Perhaps the most serious challenge is the lack of 4R skills in LMICs. Indeed, lack of skills is a global problem rather than one centered in LMICs. Accord-

49. Tage Kene-Okafor, "Flutterwave might be starting operations in Asia soon," *TechPoint Africa*, August 27, 2019, https://techpoint.africa/2019/08/27/flutterwave-expands-to-asia/.

50. "Paga, Flutterwave, links users to international merchants," *The Paypers*, March 6, 2020, https://thepaypers.com/online-payments/paga-flutterwave-links-users-to-international-merchants--1241015#.

51. R. Pavithra "Forter and Flutterwave partner to create friction free commerce in Africa and beyond," *International Banking Systems (IBS) Intelligence*, April 16, 2021, https://ibsintelligence.com/ibsi-news/forter-and-flutterwave-partner-to-create-friction-free-commerce-in-africa-and-beyond/.

52. "Forter and Flutterwave partner to reduce merchant fraud in African markets," *Finextra Research*, April 15, 2021, https://www.finextra.com/pressarticle/87128/forter-and-flutterwave-partner-to-reduce-merchant-fraud-in-african-markets.

ing to an IDC global survey of 2,473 organizations released in May 2019, 25% of organizations worldwide that were using AI solutions reported failure of 50% of their AI projects.[53] The lack of skilled staff and unrealistic expectations were among the top reasons for failure.[54] For instance, there has been a problem related to information silos. For instance, in developing AI solution for financial services, there are not many people who understand both AI and finance. Moreover, people with these skills often are from different institutions and departments.[55]

According to a study conducted by Tencent Research Institute, there were only about 300,000 "AI researchers and practitioners" in the world in 2017.[56] There is a demand for millions of such roles in companies and government agencies worldwide.[57] According to the Tencent report, the U.S., China, Japan, and UK account for most AI manpower. The next two countries with the concentration of AI manpower are Israel and Canada. Among these countries, the U.S. is reported to be far ahead of others, mainly because it has the highest number of universities that offer ML-related courses. The U.S. also has more AI startups than any other nation. It was estimated that in 2017 there were 2,600 AI startups worldwide, of which the U.S. had more than 1,000 and China had about 600.[58] Element AI, a Montreal-based lab, has argued that Tencent counted too many coders who contribute to projects but lack the expertise to create novel algorithms and applications from scratch.[59]

There is an even larger gap between demand and supply for highly skilled

53. Ritu Jyoti and Michael Shirer, "IDC survey finds artificial intelligence to be a priority for organizations but few have implemented an enterprise-wide strategy," *Business Wire*, July 8, 2019.

54. Gil Press, "This week In AI stats: Up to 50% failure rate in 25% of enterprises deploying AI," *Forbes*, July 19, 2019, https://www.forbes.com/sites/gilpress/2019/07/19/this-week-in-ai-stats-up-to-50-failure-rate-in-25-of-enterprises-deploying-ai/#1010ee7372ce.

55. Gillian Tett, "Artificial intelligence is reshaping finance," *Financial Times*, November 19, 2020, https://www.ft.com/content/c7d9a81c-e6a3-4f37-bbfd-71dcefda3739.

56. James Vincent, "Tencent says there are only 300,000 AI engineers worldwide, but millions are needed," *Verge*, December 5, 2017, https://www.theverge.com/2017/12/5/16737224/global-ai-talent-shortfall-tencent-report.

57. Cade Metz, "Tech giants are paying huge salaries for scarce A.I. talent," *New York Times*, October 23, 2017, https://www.nytimes.com/2017/10/22/technology/artificial-intelligence-experts-salaries.html.

58. Vincent, "Tencent says."

59. Jeremy Kahn, "In the war for AI talent, sky-high salaries are the weapons," *Bloomberg Law*, February 13, 2018, https://news.bloomberglaw.com/daily-labor-report/in-the-war-for-ai-talent-sky-high-salaries-are-the-weapons.

AI workforce. According to Element AI, in 2017, there were fewer than 10,000 people in the world with serious AI research skills.[60] Among them only about 3,000 were estimated to be looking for a job, and there were at least 10,000 related positions open in the U.S. only. LMICs find it difficult to attract this pool of talent and retaining them. Venture capitalist Kai-Fu Lee, who previously ran Google's business in China, noted that "Google is paying a million dollars for these superstars."[61]

LMICs other than China are facing a more severe shortage of AI skills. Finding highly skilled AI talent such as ML engineers has been a big challenge for companies in LMICs.[62] For instance, one estimate suggested that India had only about 50–75 AI researchers in 2019.[63] According to Aspiring Minds' Annual Employability Survey 2019, 80% of Indian engineers were unfit for an AI job.[64] India has also faced a severe shortage of qualified faculty members to teach AI courses in its universities.[65]

For blockchain, as of 2018, only 0.1% of 20 million software developers in the world knew about blockchain codes. No more than 6,000 coders were estimated to have the skill and experience to develop high-quality blockchain solutions.[66] Out of India's 2 million software developers, only 5,000 were estimated to have blockchain skills. Some speculate that about 80% of these developers may pursue job opportunities outside the country.[67]

The lack of relevant skills also constrains LMICs from using data from satellite and other sources. For instance, in addition to building and launching a satellite, the work on the ground to understand and make use of the data can be prohibitive for many countries.[68]

The shortage of technology skills is especially severe in rural and unde-

60. Metz, "Tech giants."

61. Kahn, "In the war."

62. Sunny Sen, "India moves to address AI talent supply gap, gets a leg-up from Google, Microsoft, Intel," *Factor Daily*, January 18, 2018.

63. Ruchi Gupta, "The state of artificial intelligence development in India," *ViaNews*, February 12, 2019.

64. "80% of Indian engineers not fit for jobs, says survey," *Business Today*, March 25, 2019, https://www.businesstoday.in/current/corporate/indian-engineers-tech-jobs-survey-80-per-cent-of-indian-engineers-not-fit-for-jobs-says-survey/story/330869.html.

65. S. Ravi and P. Nagaraj, "Harnessing the future of AI in India," *Brookings*, October 18, 2018, https://www.brookings.edu/research/harnessing-the-future-of-ai-in.

66. P. Suprunov, "How much does it cost to hire a blockchain developer?," *Medium*, 2018, https://medium.com/practical-blockchain/how-much-does-it-cost-to-hire-a-blockchain-developer-16b4ffb372e5.

67. Meha Agarwal, "Blockchain: India likely to see brain drain as 80% developers may move abroad," *Inc42*, July 23, 2018, https://inc42.com/buzz/blockchain-india-likely-to-suffer-brain-drain-as-80-developers-prepare-to-move-abroad/.

68. Anderson, "Launching your."

veloped areas. It is unreasonable to expect that blockchain solutions be sent into rural Africa for artisanal miners to use them.[69] Even if such systems are set up with outside help, small farmers and miners cannot perform technical tasks such as troubleshooting and maintenance.

Barriers to the free exchange of knowledge with the West affect LMICs' efforts to develop skills and talent. AI researchers in LMICs are facing various difficulties in accessing the AI knowledge that exists in the West. For example, Vancouver hosted the 2019 Conference on Neural Information Processing Systems (NeurIPS), the world's largest AI research conference. The Canadian government was reported to have denied visas from many AI researchers from Africa and Asia.[70] Fifteen of 44 people planning to attend the "Black in AI" workshop were denied entry. Many volunteers of the Masakhane project, which is working with groups such as Translators without Borders and academics to find language datasets to use machine learning to translate African languages, were also denied visas.[71] Immigration officials allegedly feared that the researchers would not return home.[72]

Data Divide

The data divide—or the gap in data availability for scientific research and decision-making, has adversely affected LMICs.[73] It is a major barrier to benefiting from innovations in agriculture, finance, and healthcare. It is difficult to assess the situation of poverty due to the lack of data. An analysis of poverty data using the World Development Indicators (WDI) database, which identifies 1,101 poverty datapoints between 1976 and 2013 based on household consumption surveys, found that 77 countries lacked adequate poverty data.[74]

The issue of limited datasets for AI and other projects are of concern. To

69. Catherine Early, "Can high-tech solutions take the risk out of artisanal mining?," *Reuters*, October 31, 2019, https://www.ethicalcorp.com/can-high-tech-solutions-take-risk-out-artisanal-mining.

70. Khari Johnson, "Canada is denying travel visas to AI researchers headed to NeurIPS—again," *Venture Beat*, November 9, 2019, https://venturebeat.com/2019/11/09/canada-is-denying-travel-visas-to-ai-researchers-headed-to-neurips-again/.

71. Johnson, "Canada is."

72. Johnson, "Canada is."

73. Jonathan Cinnamon and Nadine Schuurman, "Confronting the data-divide in a time of spatial turns and volunteered geographic information," *GeoJournal* 78 (2013): 657–674.

74. Umar Serajuddin, Nobuo Yoshida, and Hiroki Uematsu *Much of the world is deprived of poverty data: Let's fix this*, World Bank, April 30, 2015, https://blogs.worldbank.org/developmenttalk/much-world-deprived-poverty-data-let-s-fix.

learn patterns, machine learning and deep learning require massive amounts of relevant training data and huge computing power. If a country's healthcare system lacks such data and does not allow access to relevant health data to train the algorithm, AI systems cannot function well.[75]

In healthcare, one example is undeveloped clinical datasets, such as genomes. For instance, advances in human genomics make it possible to reconstruct population history and identify genes that are likely to make a person vulnerable to specific diseases. However, most genome studies so far have focused on people of European descent. In 2009, 96% of genomes sequenced belonged to the European ancestry.[76] The proportion was 19% in 2017,[77] which slightly increased to 22% in March 2018.[78] For instance, while India represents 20% of the world's population, only 0.2% of the world's fully mapped genomes are of Indian origin. Likewise, less than 3% of genetic material used in global pharmaceutical research is from Africa. Africans and people of African descent are reported to be more genetically diverse than any other population.[79] Taken together, people of African and Latin American descent and the indigenous population represented less than 4% in the Genome Sequence Database.[80] The lack of genome data hampers the B4B's ability to benefit from exciting advances in the area of human genomics in the past three decades.

In the agricultural sector, index-based crop insurance has been promoted widely among small-scale farmers. An index-based insurance scheme's effectiveness in reducing risk depends on how well the actual crop yield and profits correlate with the index.[81] Many measurement challenges need to be

75. Jeremy Hsu, "Medical advice from a bot: The unproven promise of Babylon Health," *Salon*, December 14, 2019, https://www.salon.com/2019/12/14/medical-advice-from-a-bot-the-unproven-promise-of-babylon-health_partner.

76. Nicholas Parry, "Gene databanks ignore non-Europeans," *Health Issues India*, August 30, 2017, https://www.healthissuesindia.com/2017/08/30/gene-databanks-ignore-non-europeans/.

77. Parry, "Gene databanks."

78. G. Sirugo, S. M. Williams, and S. A. Tishkoff, "The missing diversity in human genetic studies" *Cell* 177 (2019): 26–31.

79. Tage Kene-Okafor, "African genomics startup 54gene raises $25M to expand precision medicine capabilities," *TechCrunch*, September 16, 2021, https://techcrunch.com/2021/09/16/african-genomics-startup-54gene-raises-25m-to-expand-precision-medicine-capabilities.

80. Giorgia Guglielmi, "Facing up to injustice in genome science," *Nature*, April 16, 2019, https://www.nature.com/articles/d41586-019-01166-x.

81. J. W. Glauber, "Crop insurance reconsidered," *American Journal of Agricultural Economics* 86, no. 5 (2004): 1179–1195.

addressed. For instance, to design an effective weather insurance scheme, a high density of weather stations is a prerequisite.[82] Currently the payment to a policyholder relies on satellite images to detect whether extreme weather has affected a given area. A criticism ACRE faced was that its weather-monitoring stations are primarily installed in heavily farmed areas in Kenya and Rwanda. This means that areas without heavy farming lack data. In Kenya, a weather-monitoring station may cover an area with up to 1,000 farmers. Canada's International Development Research Centre found that the promised benefits of crop insurance have not materialized because a weather-monitoring station gathers data on wide areas and fails to capture the microclimate that a farmer experiences.[83] It provides general views about the effects of drought or floods but fails to accurately measure local rainfall. There have been cases in which satellite data indicated that an area had sufficient rainfall but farmers in microclimates experienced crop loss and were not offered payouts. Some discontinued their insurance schemes. Likewise, in Sri Lanka, existing data sources provided by weather stations were insufficient.[84]

Lack of Computing Power

Many 4R applications are computing intensive and data intensive. Training some ML algorithms requires expensive equipment and computing power.[85] Small LMICs are often unable to afford such luxuries. Because of the lack of sufficient processing capacity, many government agencies are unable to use the data they receive from their own satellites or from free sources.[86] Some

82. S. L. Jorgensen, M. Termansen, and U. Pascual, "Natural insurance as condition for market insurance: Climate change adaptation in agriculture," *Ecological Economics* (2020): 106489.

83. Kagondu Njagi, "Kenyan farmers snap crops with phones to improve insurance payouts," *Reuters*, October 11, 2019, https://www.reuters.com/article/us-climate-change-kenya-insurance/kenyan-farmers-snap-crops-with-phones-to-improve-insurance-payouts-idUSKBN1WQ0Q7.

84. Etherisc, "Oxfam, Etherisc, and Aon deliver pay-outs with first blockchain-based agricultural insurance policies for smallholder farmers in Sri Lanka," *PR Newswire*, November 4, 2019, https://www.prnewswire.com/news-releases/oxfam-etherisc-and-aon-deliver-pay-outs-with-first-blockchain-based-agricultural-insurance-policies-for-smallholder-farmers-in-sri-lanka-300949728.html.

85. Ahmed El Adl, "Debunking the myths and reality of artificial intelligence," *Forbes*, April 22, 2019, https://www.forbes.com/sites/cognitiveworld/2019/04/22/debunking-the-myths-and-reality-of-artificial-intelligence/#108bfa1543b5.

86. Anderson, "Launching your."

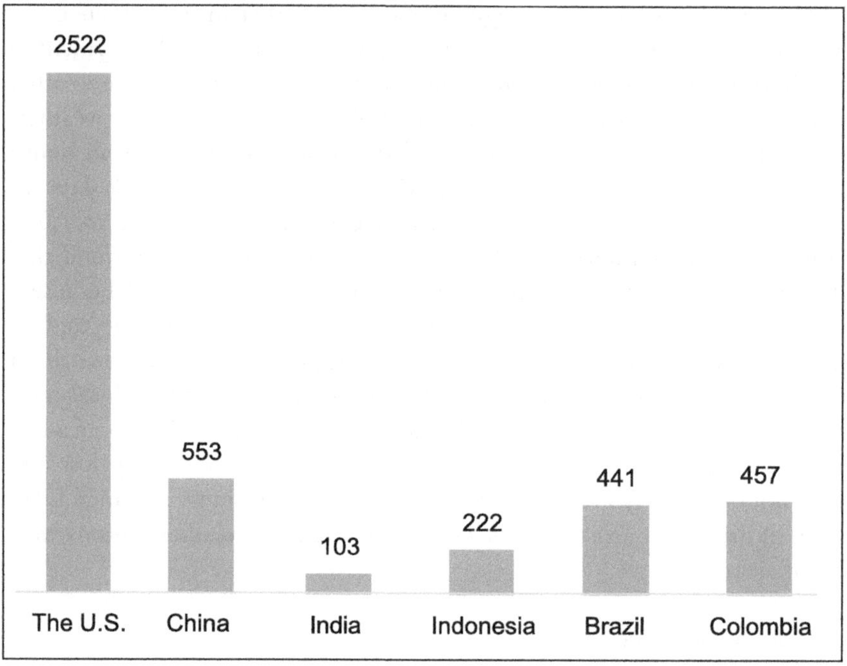

Figure 9.1. Per capita computing power in gigaflops (GFLOPS) (2020). Source: Liang Hua, *Ubiquitous computing power: The cornerstone of an intelligent society* (position paper), Huawei, February 2020, https://www-file.huawei.com/-/media/corporate/pdf/public-policy/ubiquitous_computing_power_the_cornerstone_intelligent_society_en.pdf.?la=en.

AI solutions can take as long as 12 hours to run on typical computers in Kenya.[87]

In general, except for China, most LMICs lack the required computing power to perform such functions (Figure 9.1). According to the TOP500 project, which ranked systems based on ability to solve a set of linear equations, China had the highest number of supercomputers in June 2021: 186 Chinese machines were listed, compared to 122 in the U.S.[88] In June 2019, India had fewer than five of the world's 500 most powerful supercomputers, compared to China's 219.[89]

87. David Pilling, "AI in Africa healthcare falls short of potential," *Financial Times*, May 18, 2020, https://www.ft.com/content/90fa8f44-6847-11ea-a6ac-9122541af204.

88. TOP500, *Fugaku holds top spot, Exascale remains elusive*, 2021, https://www.top500.org/news/fugaku-holds-top-spot-exascale-remains-elusive/.

89. Som Satsangi, "Why India is ready to become a supercomputer power," *World Eco-*

Low-Quality Infrastructure

Low-quality infrastructure also hinders the use of 4R technologies. As discussed in chapter 1, a large proportion of the population in LMICs lives in rural areas, which makes building infrastructure such as a 5G network difficult and costly.[90] The demand for Africa-based data centers to provide low-latency and high-bandwidth connectivity is increasing. Without cheap and reliable power, however, data centers are less likely to be built in Africa.[91] The lack of infrastructure has resulted in the high cost of data in SSA. Due to overburdened networks, data bundles offered are generally smaller. Consequently, the average cost per gigabyte (GB) is higher than in countries with unlimited packages.[92]

Lack of Research Focus

While some developing world companies are generating 4R technologies locally, they lack focus on fundamental and basic research. For instance, Chinese AI companies are reported to be more application focused. There has been less emphasis on advancing foundational technologies and research.[93] Likewise, regarding higher value-added innovative activities such as patents and research and development (discussed in chapter 3), while Chinese firms have filed a lot of blockchain patents, they lag the U.S. in granted patents.[94] According to the China Patent Protection Association, as of May 2020, there were 3,924 blockchain patents granted worldwide. China's share was less than half that of the U.S.

---

nomic Forum, September 23, 2019, https://www.weforum.org/agenda/2019/09/what-are-supercomputers-used-for-india/.

90. World Bank, *Rural population (% of total population)*, 2018, https://data.worldbank.org/indicator/SP.RUR.TOTL.ZS.

91. Rose Mutiso and Katie Hill, "Why hasn't Africa gone digital?," *Scientific American*, August 11, 2020, https://www.scientificamerican.com/article/why-hasnt-africa-gone-digital/.

92. Carmen Ang, "What does 1GB of mobile data cost in every country?," *Visual Capitalist*, July 3, 2020, https://www.visualcapitalist.com/cost-of-mobile-data-worldwide/.

93. Karen Hao, "Three charts show how China's AI industry is propped up by three companies," *MIT Technology Review*, January 22, 2019, https://www.technologyreview.com/s/612813/the-future-of-chinas-ai-industry-is-in-the-hands-of-just-three-companies/.

94. Eliza Gkritsi, "Alibaba is the top global blockchain patent holder," *Technode*, July 3, 2020, https://technode.com/2020/07/03/alibaba-leads-global-blockchain-patent-but-china-lags-behind-us-and-s-korea/.

Low Resource Languages

From the perspective of AI, natural language processing (NLP) is arguably the most mature and widely adopted application, having gained enough ML capabilities with real-world experience.[95] ML algorithms for NLP, however, are mainly developed for English, spoken by a small proportion of the population in LMICs. For instance, only 10% of the Indian population speaks English.[96]

The Indo-Aryan or Indic languages are a major language family spoken in India and across South Asia. In 2016, Indic language speakers accounted for 60% of India's 409 million internet users. According to KPMG, 93% of the next 326 million new internet users in India will be local-language-first users.[97] These languages do not use the Latin alphabet. They use alphabets derived from Brahmic scripts, such as Devanagari, Bengali, Malayalam, Oriya, Sinhala, and Tamil. It is difficult for NLP software to understand these scripts. They also lack lexical resources, such as dictionary databases, terminology glossaries, and user-created content, which makes it challenging to build NLP algorithms. NLP algorithms' performance depends on the availability of a large amount of text containing all possible permutations and combinations of meaning. Documents such as legal contracts, news articles, and research reports are incorporated into an NLP algorithm to increase its performance. Only small quantities of such documents are available for most Indian languages.[98]

Likewise, a large population speaking African languages has not benefited from voice recognition innovation.[99] Over 2,000 languages are spoken

95. El Adl, "Debunking."

96. Purushottam Darshankar, "The unique challenge India presents to natural language processing," *Business Today*, October 8, 2018, https://www.businesstoday.in/opinion/columns/the-unique-challenge-india-presents-to-natural-language-processing/story/283650.html.

97. Jayanth Kolla, "How AI is helping firms tap users of Indian languages," *Live Mint*, August 27, 2018, https://www.livemint.com/AI/OHvuLFS2UCnItjk7CAcdpK/How-AI-is-helping-firms-tap-users-of-Indian-languages.html.

98. Darshankar, "The unique challenge."

99. Abdi Latif Dahir, "African languages are being left behind when it comes to voice recognition innovation," *Quartz Africa*, November 27, 2018. https://qz.com/africa/1475763/african-languages-are-lagging-behind-when-it-comes-to-voice-recognition-innovations/.

daily basis in Africa.[100] many of which are only spoken and not written.[101] As language technologies such as NLP advance and more sophisticated tools are built using AI, the divide between low-resource languages and others may even increase. This is because the advanced systems depend on the availability of a large amount of digital data. Many African languages that are low-resource languages are likely to be left behind.[102]

Low Degree of 4R Maturity

Among the major obstacles is also user-friendly applications.[103] Most AI models currently lack external validity and economic significance. AI can get results in well-defined application domains with no hidden information, such as the games go or chess. AI algorithms, however, perform poorly in navigating questions involving nuance.[104] A systematic review of articles published in the first nine months of 2020 that described ML models to detect and diagnose for COVID-19 from medical images using chest radiographs (CXR) and computed tomography (CT) images found that the ML models had major flaws and biases. The authors concluded that none of the 62 studies included in the review was of potential clinical use.[105]

A related challenge concerns opacity—the lack of interpretability or auditability of AI and ML methods. A growing concern has been expressed by activists and policymakers over the lack of explainability of AI solutions, a major hindrance in the wider acceptability and trust of such solutions. A survey conducted in early 2021 among 100 AI-focused leaders from the financial services sector from the U.S., Latin America, Europe, the Middle East, Africa, and Asia-Pacific that was published in Fair Isaac Corporation's

---

100. Mary-Ann Russon, "The push towards artificial intelligence in Africa," *BBC News*, May 28, 2019, https://www.bbc.com/news/business-48139212.

101. David M. Eberhard, Gary F. Simons, and Charles D. Fenning, *Ethnologue: Languages of the worlds*, 22nd ed. (SIL International, 2019).

102. Kathleen Siminyu, Jade Abbott, Sackey Freshia, and Vukosi Marivate, "AI4D—African language dataset challenge," *arxiv*, 2020, https://arxiv.org/pdf/2007.11865.pdf.

103. Joël Valenzuela, "Bitcoin remittances to Mexico see huge potential," Cointelegraph, December 21, 2015, https://cointelegraph.com/news/bitcoin-remittances-to-mexico-see-huge-potential.

104. Peter Judge, *What if AI is the problem, not the solution?*, Data Center Dynamics, June 11, 2021, https://www.datacenterdynamics.com/en/opinions/what-if-ai-is-the-problem-not-the-solution/.

105. University of Cambridge, *Machine learning models for diagnosing COVID-19 are not yet suitable for clinical use*, March 15, 2021, https://www.cam.ac.uk/research/news/machine-learning-models-for-diagnosing-covid-19-are-not-yet-suitable-for-clinical-use.

(FICO) State of Responsible AI report found that about 70% of respondents were unable to explain the way specific AI model decisions or predictions are made.[106] Only 35% of organizations made efforts to use AI in a transparent and accountable manner.[107]

The lower degree of technologies' maturity also has led to challenges related to interoperability and standardization. For instance, participants on a distributed ledger need to agree on common standards for an invoicing platform. For example, different banks need to agree on the number of data fields from an invoice to generate the hash value. They may also need common messaging standards. The banking industry, among the early adopters of blockchain, is characterized by a culture of competitiveness, which poses a challenge to working together.[108]

Poor Performance of Foreign Solutions

Some LMICs are using solutions developed by foreign companies. For instance, a Russian company was reported to be selling face recognition technology in South Africa and Kenya. These systems have low accuracy and precision in Africa and thus limited usability and functionality.[109] AI apps that rely on unstructured medical data need sophisticated NLP algorithms. Much of this technology is developed in the West or in Asia, and transferring it to African markets is challenging. It is important for a technology to be adaptable to the local language so that it can be modified to different languages, language structures, and accents. The medical situations may also be different, such as different diseases and health management systems.[110] Even simpler AI algorithms such as facial recognition have been trained to identify white males;[111] because of this, they have a higher tendency to mis-

106. FICO, *FICO Responsible AI*, https://www.fico.com/en/solutions/fico-responsible-ai.

107. Jonathan Greig, "Report finds startling disinterest in ethical, responsible use of AI among business leaders," *ZDNet*, March 25, 2021, https://www.zdnet.com/article/fico-report-finds-startling-disinterest-in-ethical-responsible-use-of-ai-among-business-leaders/.

108. Chanyaporn Chanjaroen and Darren Boey, "Fraud in $4 trillion trade finance has banks turning digital," *Bloomberg*, May 23, 2016, https://www.bloomberg.com/news/articles/2016-05-22/fraud-in-4-trillion-trade-finance-turns-banks-to-digital-ledger.

109. Jackie Snow, "How Africa is seizing an AI opportunity," *Fast Company*, March 10, 2019, https://www.fastcompany.com/90308114/how-africa-is-seizing-an-ai-opportunity.

110. Laura Sallstrom, Olive Morris, and Halak Mehta, *Artificial intelligence in Africa's healthcare: Ethical considerations*, Observer Research Foundation, September 9, 2019.

111. Sallstrom et al., "Artificial intelligence."

identify people of other races.[112] This means that they perform less well for most African faces than for Caucasian ones. For instance, facial recognition software is reported to identify the photo of a white man with a 99% accuracy. The error rates were as high as 35% for Black women.[113]

4R Technologies' Potential Malicious Use

The term *surveillance capitalism* is often used in connection with state-corporate surveillance, commercial exploitation of data, and internet governance. Attention has also been focused on issues of data monetization and algorithmic discrimination. Big data is a key component of surveillance capitalism. Corporations and states collect, store, process, and share large amounts of information, which enables them to infer traits about people (e.g., sexuality, religion, political affiliations, behavioral tendencies).[114]

For instance, social media companies can show contents that promote ideas that they want users to have. If a user's expressed views are approved by the owners of the social media, bots can "like" them.[115] A possible scenario is that ML algorithm observes social media users with a given psychological profile and delivers content to generate certain political views and behaviors. In general, content is served up based on users' profiles so that it is most effective. Over time, the algorithm can learn and generate effective content from scratch that is targeted to specific groups.[116] These risks are especially acute for LMICs that lack well-developed regulatory frameworks.

Some counties are also using 4R technologies, especially AI tools, to conduct surveillance activities on citizens. For instance, surveillance cameras in Chinese cities have AI-based facial recognition software that collects huge amounts of biometric information daily. In general, authoritarian regimes that are more interested in maintaining political control rather than maximizing economic performance and equity may use AI in a malicious manner to manipulate and control citizens' attitudes and behaviors.

---

112. Tom Simonite, "The best algorithms struggle to recognize black faces equally," *Wired* July 22, 2019.

113. Steve Lohr, "Facial recognition is accurate, if you're a white guy," *New York Times*, February 9, 2018, https://www.nytimes.com/2018/02/09/technology/facial-recognition-race-artificial-intelligence.html.

114. Michael Kwet, "Long read digital colonialism," *New Frame*, March 8, 2021, https://www.newframe.com/long-read-digital-colonialism/.

115. "AI social media could totally manipulate you," *Mind Matters News*, November 26, 2018.

116. François Chollet, "What worries me about AI," *Medium*, March 28, 2018, https://medium.com/@francois.chollet/what-worries-me-about-ai-ed9df072b704.

Stockholder Centric Bias in Technology Startups

Tech startups are under pressure to increase profits and shareholder value.[117] A result of this is that philanthropic activities are declining and the actions of technology startups that have launched products for low-income people seem to be more investor-centric. In the context of blockchain, a technology startup writer noted that "many of the highly publicized initial coin offerings (ICOs), which are a capital-raising method involving offers and sales of cryptotokens using blockchain, ICOs have yet to carry out much beyond upgrading the lifestyles of their founders and promoters."[118]

Many 4R technology systems are designed to benefit big companies rather than marginalized groups. For instance, BanQu developed blockchain systems in India, Uganda, and Zambia to track cassava and barley supplied to Anheuser-Busch. Anheuser-Busch can benefit tremendously from blockchain's use to promote supply chain transparency and traceability. Blockchain can help guarantee the quality of products with relevant data. Making digital payments to farmers may lower costs associated with payments. However, big firms have done very little to ensure that farmers can genuinely benefit from the integration of blockchain into supply chains.

Smallholder farmers who supply crops to Anheuser-Busch may theoretically enjoy additional benefits (e.g., low-cost loans from financial institutions). However, constraints related to information flows, transaction costs, and market access likely prevent them from realizing such benefits. For instance, farmers may not be able to present the information in a way that meets banks' requirements. They may lack persons in their social network who can understand the various available loan services. A lack of education may mean that many potential borrowers cannot fill out loan applications.[119] Poor people often need loans in small amounts. It is costly for financial institutions to deal with small transactions. In some cases, poor people may face prejudice and stereotypes. Some banks refuse their admission to bank branches.[120]

---

117. M. E. Porter and M. R. Kramer, "The competitive advantage of corporate philanthropy," *Harvard Business Review* 80, no. 12 (2002): 56–68.

118. Bill Wagner, "How this company's ICO quietly netted $600K," *Tech.co*, August 21, 2017, https://tech.co/news/ico-cryptocurrency-tokens-2017-08.

119. Nir Kshetri, "Blockchain-based financial technologies and cryptocurrencies for low-income people: Technical potential versus practical reality," *IEEE Computer* 53, no. 1 (2020): 18–29.

120. B. Thorsten, A. Demirgüç-Kunt, and P Honohan, "Access to financial services: measurement, impact, and policies," *World Bank Research Observation* 24 (2009): 119–

Chapter Summary and Conclusion

This chapter has provided a glimpse at the promising opportunities the 4R provides for nations, firms, and people. For instance, nations can enhance environmental sustainability, detect and prevent illegal activities, and track students' educational outcomes. 4R technologies can help increase firms' access to financing and markets.

In the past, costs to develop and deploy technology solutions were considerable. The drastic cost reductions of many 4R solutions open up new possibilities for deploying technologies to bring economic and social transformations to LMICs. Several examples have illustrated that LMICs are strengthening technological capability, with local firms playing key roles. Among other things, the emergence of local technology firms can help overcome concerns related to digital colonialism.

However, there are many obstacles that need to be overcome for the widespread adoption of 4R technologies. A large amount of high-quality data is needed to benefit from the 4R. If ML algorithms are trained with limited sets of features and small datasets, decisions made using those AI systems could be wrong, dangerous, and misleading. If ML algorithms are trained in developed countries, effectiveness likely cannot be achieved in developing countries. Likewise, solutions developed in foreign languages have limited usability for most of the population in LMICs. In general, available data in LMICs is of questionable quality. As a result, many ICT projects targeted at the poor fail to deliver the benefits promised by the initiators of such projects.

Concerns have also been raised about irresponsible and unethical acts using 4R technologies. In a discussion of AI's impacts, it is important to consider noneconomic costs such as loss of privacy. Such concerns exist at various stages from data gathering to develop AI applications to actual use of AI. When governments use AI for political gains and corporations use this technology for unethical behaviors and practices, there are often no repercussions for perpetrators.

145, https://academic.oup.com/wbro/article/24/1/119/1670507/Access-to-Financial-Services-Measurement-Impact.

CHAPTER 10

Economic Development Implications

4R technologies can help people with limited resources increase economic productivity and engage in economic exchange by reducing transaction costs. These mechanisms have the potential to help vulnerable individuals, households, and communities escape the poverty trap. An important question is how this potential can be better realized.

Since a large proportion of the B4B population lives in middle-income countries, 4R technologies' potential impact on the economic growth of these countries is a significant issue that cannot be ignored. In this regard, another question that needs to be answered is whether 4R technologies can help countries get out of the middle-income trap, which is a condition that prevents middle-income countries to achieve the levels of advanced countries.

4R technologies may also help countries make a jump in economic progress and technological development to pass over of stages in economic growth and modernization, facilitating what is referred to as leapfrogging.[1] Finally, valuable insights can be gained by looking at the 4R's impact on productivity growth in a country as compared to countries with similar levels of economic development.

This chapter seeks to address these issues by examining and interrogating key theories of economic development in the context of the 4R. Specifically, it draws on theories and concepts such as the poverty trap, middle-income trap, leapfrogging theory, and the flying geese paradigm (FGP) to present a general overview on the 4R's impact on the B4B.

1. W. E. Steinmueller, "ICTs and the possibilities for leapfrogging by developing countries," *International Labour Review* 140 (2001): 177–178.

4R Technologies and Economic Development

4R technologies are playing an increasing central role in the economic development of LMICs. There is some evidence that LMICs that make use of digital technologies can overtake their neighbors that rely on traditional resources such as oil (Figure 10.1).

I compare Angola and Kenya. Not long ago, oil-rich Angola achieved a long period of impressive growth. The country's oil sector accounts for one-third of GDP and over 90% of exports.[2] Following the end of its 27-year civil war in April 2002, Angola experienced double-digit annual GDP growth.[3] In 2020, Kenya overtook Angola to become the third-largest economy in Sub-Saharan Africa.[4] Over 2015–2019, Kenya's economic growth averaged 5.7%, making it one of the fastest growing economies in Sub-Saharan Africa (Figure 10.1).[5]

One observation about Kenya is that while the internet's contribution to GDP in Africa is 1.1%, the corresponding proportion for Kenya is 2.9%, higher than that of Canada, China, Brazil, and Russia. Senegal is the only country in Africa that outperforms Kenya in the internet's contribution to GDP, at 3.3%. Internet connectivity has been shown to stimulate economic growth.[6] Kenya is thus well positioned to be a leader among SSA economies.[7] Angola, however, is suffering from lower oil prices and decreasing production.

The availability of high-speed and cheap internet connections in Kenya has enabled even people living in poverty to take advantage of the 4R. For instance, the availability of digital infrastructure has helped Pipeline estate

---

2. World Bank, *Overview*, 2016, https://www.worldbank.org/en/country/angola/overview.
3. Arne Wiig and Ivar Kolstad, "Opinion: Is oil-rich Angola a development success?," *CNN*, August 30, 2012, https://www.cnn.com/2012/08/30/opinion/opinion-angola-development-elections/index.html.
4. Dominic Omondi, "Kenya is the third-largest economy in Sub-Saharan Africa Business," *The Standard*, June 6, 2020, https://www.standardmedia.co.ke/business/business/article/2001374157/kenya-economy-is-third-in-sub-saharan-africa.
5. World Bank, *Overview*, 2010, https://www.worldbank.org/en/country/kenya/overview.
6. McKinsey Global Institute, *Internet matters: The net's sweeping impact on growth, jobs and prosperity*, 2011.
7. Clare Akamanzi, Peter Deutscher, Bernhard Guerich, Amandine Lobelle, and Amandla Ooko-Ombaka, *Silicon savannah: The Kenya ICT services cluster*, April 24, 2016, https://www.isc.hbs.edu/Documents/resources/courses/moc-course-at-harvard/pdf/student-projects/Kenya%20ITC%20Services%202016.pdf.

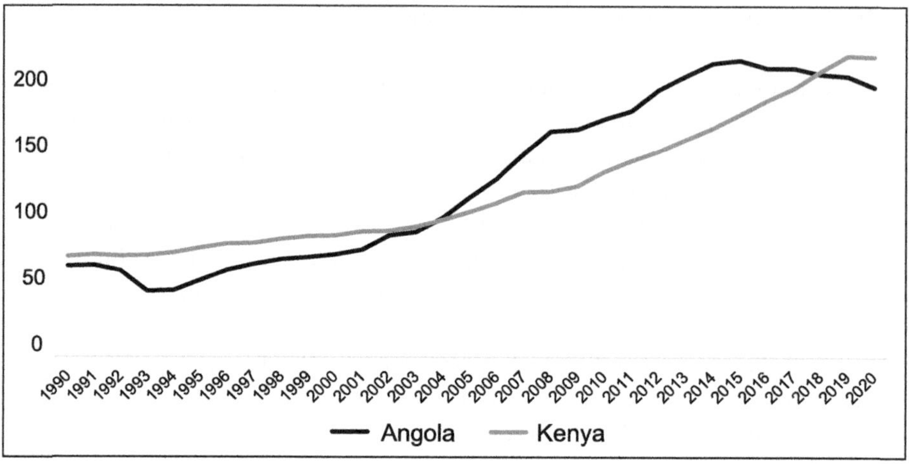

Figure 10.1. A comparison of the economies of Angola and Kenya (GDP, PPP, billion, constant 2017 international $). Data source: World Bank. *GDP, PPP (constant 2017 international $)*, https://data.worldbank.org/indicator/NY.GDP.MKTP.PP.KD.

in the Embakasi division of Nairobi attract high-technology 4R jobs such as AI data labeling. The lower-middle-class estate is home to 100,000 people. On March 31, 2022, Reuters published a story about Pipeline resident Daniel Nzoma, a data labeler for AI algorithms related to driverless cars and crop disease detection. Nzoma's job included "geotagging," which needs reliable and fast internet connectivity. Pipeline lacked such infrastructure before.[8]

Kenya has put significantly greater emphasis on modern technological infrastructures than other countries in Africa have. In 2021, the mobile network operator Safaricom launched a 5G network in Kenya, making it the second country in Africa to roll out the technology to customers.[9] As of March 2021, 5G infrastructures were available in Nairobi, Kisumu, Kisii, and Kakamega.[10]

Angola performs much lower than Kenya in the deployment and use of technological innovations. Angola's mobile or cellular coverage is lower than

8. Monica Njeri, "Fast internet brings tech jobs to Nairobi's poor neighbourhoods," *Reuters*, March 31, 2022, https://www.reuters.com/technology/fast-internet-brings-tech-jobs-nairobis-poor-neighbourhoods-2022-03-31/.

9. Carlos Mureithi, "Kenya becomes the second African country to launch 5G," *Quartz*, July 12, 2021, https://qz.com/africa/1990724/kenya-becomes-the-second-african-country-to-launch-5g/.

10. Juan Pedro Tomás, "Safaricom activates 5G network in Kenya," *RCR Wireless*, March 29, 2021, https://www.rcrwireless.com/20210329/5g/safaricom-activates-5g-kenya.

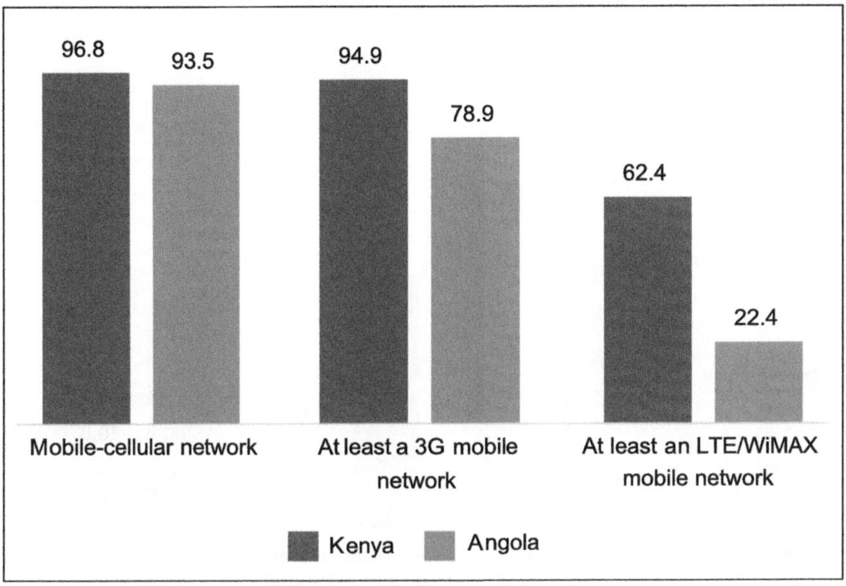

Figure 10.2. Percentage of population covered by a mobile-cellular network: Angola and Kenya (2020). Data source: The International Telecommunication Union (ITU).

Kenya's (Figure 10.2). The gap in coverage is especially large in broadband wireless systems such as LTE (Figure 10.2). As Figure 10.3 indicates, Kenya outperforms Angola in the penetration of various technologies and services. Figure 10.4 makes clear that a significantly higher proportion of the population in Kenya than in Angola uses the internet in key economic activities. In sum, digital technologies have driven economic growth in Kenya while Angola has fallen behind on many key indicators of digital transformation.

Poverty Trap

A poverty trap is a "self-reinforcing mechanism which causes poverty to persist."[11] Such a trap is more about staying poor for a long time rather than about being poor at any point in time. Poverty traps result from well-being "basins of attraction" in an economy.[12]

11. C. Azariadis and J. Stachurski, "Poverty traps," in *Handbook of economic growth* (Amsterdam: Elsevier, 2004), 295–384.

12. Christopher B. Barrett and Michael R. Carter, "The economics of poverty traps and

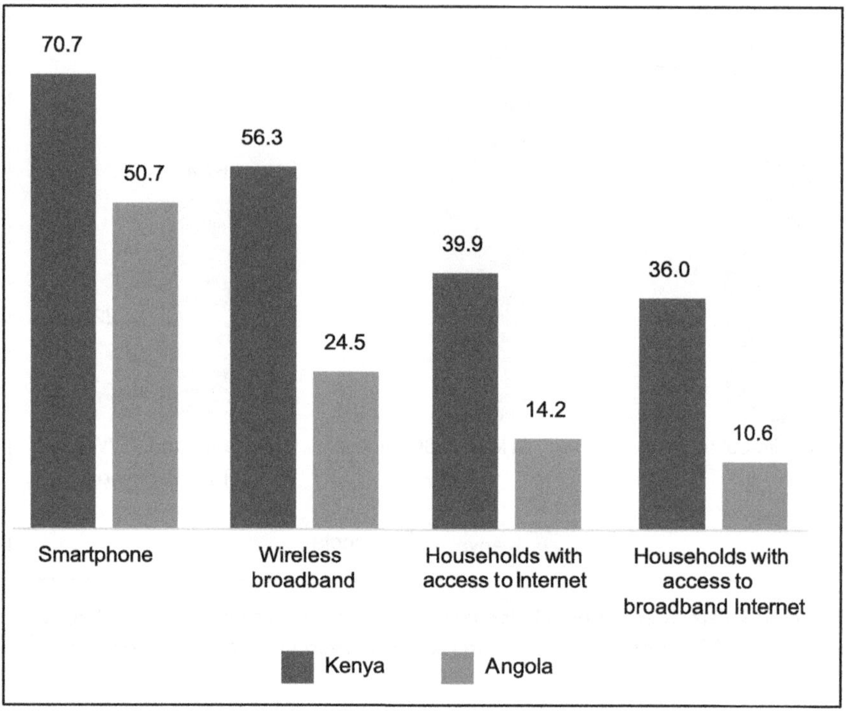

Figure 10.3. Percentage of population with access to various technologies and services: Angola and Kenya (2020). Data source: The International Telecommunication Union (ITU).

Various mechanisms associated with poverty traps operate at different levels: individual, household, community, regional, and national.[13] For instance, the World Bank has classified Colombia as an upper-middle-income country (https://data.worldbank.org/?locations=CO-XT). However, in 2019, monetary poverty in Colombia was experienced by 36% of the population.[14] People are at risk of monetary poverty when their equivalized disposable income—that is, the total household income available for spending or saving (after tax and other deductions) divided by the number of household members—is below the risk threshold. That threshold is 60%

persistent poverty: Empirical and policy implications," *Journal of Development Studies* 49, no. 7 (2013): 976–990.

13. Barrett and Carter, "The economics of poverty."

14. World Bank, *Poverty & equity brief Colombia Latin America & the Caribbean*, April 2021, https://databank.worldbank.org/data/download/poverty/987B9C90-CB9F-4D93-AE8C-750588BF00QA/AM2020/Global_POVEQ_COL.pdf.

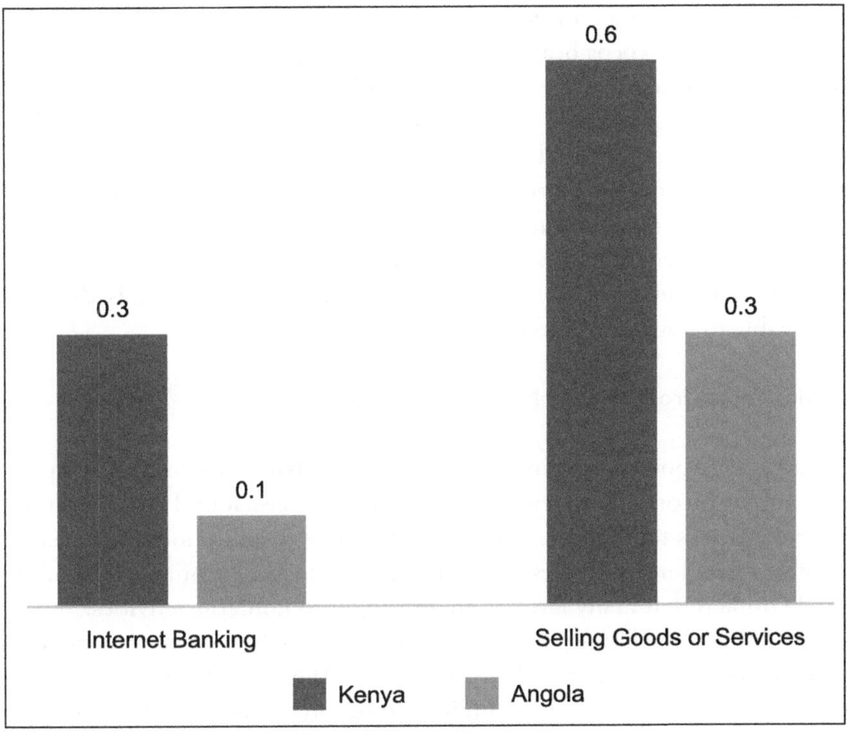

Figure 10.4. Internet use in key economic activities: Angola and Kenya (2020) (% of population). Data source: The International Telecommunication Union (ITU).

of the national median equivalized disposable income after social transfers, or financial help given by central, state, or local institutional units.[15]

Pulling People Out of the Poverty Trap

4R technologies can help lift poor people out of poverty and prevent them from falling into poverty (In Focus 10.1). Innovations can help farmers increase farm productivity, enhance access to formal financing, and earn more from their crops and livestock. For instance, more than half of households in rural cocoa-growing regions of Colombia live in poverty.[16] Farmers

15. Statistics Explained, *Glossary:At-risk-of-poverty rate*, April 13, 2021, https://ec.europa.eu/eurostat/statistics-explained/index.php?title=Glossary:At-risk-of-poverty_rate.

16. Antonie C. Fountain and Friedel Huetz-Adams, *Cocoa barometer 2020 index*, https://www.voicenetwork.eu/wp-content/uploads/2020/12/2020-Cocoa-Barometer.pdf.

in extreme poverty are often inclined to take fast cash by selling incompletely fermented cocoa beans rather than waiting for better cacao and making investment to ferment cacao beans, which can be sold at a higher price.[17] If solutions being developed by technology companies that utilize satellite data, IoT applications, and other apps (chapter 7) lead to increased productivity, many of these farmers can be lifted out of poverty. Likewise, by utilizing 4R technologies such as big data, AI, and blockchain, it is possible to make low-cost agricultural insurance available to farmers, which can help mitigate risk. Such a tool is especially important for households that are vulnerable to poverty due to natural disasters.

Moving People from Farm to Nonfarm Activities

A study of the poverty trap in rural China found that nonfarm activities produce higher income than farm activities. The study also found that education improves access to nonfarm activities and returns to education are higher in nonfarm than farm activities.[18] In this regard, a recent encouraging trend in rural China is that many farmers are moving into nonfarm activities.

In China some farmers have transformed themselves into internet entrepreneurs using e-commerce platforms such as Alibaba and JD.com. As early as in 2013, over 22% of the 7 million stores on Alibaba's Taobao Marketplace and Tmall.com originated from IP addresses in villages and towns. There were 16 Chinese villages that generated 5 billion yuan or more in online sales in 2013. A success story of a farmer from rural Qinghe County was reported by Bloomberg. The farmer made more than RMB 10 million (US$1.6 million) selling woolen yarns.[19] The number of Taobao villages—defined as villages that generate annual e-commerce transaction volume of RMB 10 million or more and have 100 or more active online shops—was 20 in 2013 and 3,202 in 2018.[20] By 2021 it was 5,425.[21]

17. "This company brings sustainability to Colombian fine cacao," *Chocolate Journalist*, March 30, 2017, https://www.thechocolatejournalist.com/blog/sustainability-colombian-fine-cacao.

18. John Knight, Li Shi, and Deng Quheng, "Education and the poverty trap in rural China: closing the trap," *Oxford Development Studies* 38, no. 1 (2010): 1–24, https://doi.org/10.1080/13600810903551595.

19. Lulu Yilun Chen, "China farmer turns yarn baron as villages embrace Alibaba," *Bloomberg*, November 6, 2013, http://www.bloomberg.com/news/articles/2013-11-06/china-farmer-turns-yarn-baron-as-villages-embrace-alibaba.

20. Xubei Luo, *In China's Taobao villages, e-commerce is one way to bring new jobs and business opportunities to rural areas*, World Bank, November 22, 2018, https://blogs.worldbank.org/eastasiapacific/china-s-taobao-villages-e-commerce-one-way-bring-new-jobs-and-business-opportunities-rural-areas.

21. Ali Research, *China Taobao Village report 2020*, February 8, 2021, http://www.aliresearch.com/en/Reports/Reportsdetails?articleCode=167153834769125376.

The government is also encouraging farmers to use online retailing. In 2014, Alibaba signed contracts to help farmers to do e-commerce in the western regions of Xinjiang and Gansu.[22] In 2015, Taobao Marketplace launched a shopping channel dedicated to farm products, where farmers could buy tractors and other machinery, pesticides, fertilizers, and seedlings. As of July 2015, about 100,000 merchants were offering farming products on Taobao's platform.[23]

It is also important to note that the economic impacts on low-income people may not be as attractive as the impression that Chinese e-commerce companies such as Alibaba are trying to create. For instance, rural Taobao partners reportedly receive a small commission or performance-based bonus from sales on the platform. Even the best-performing partners make less than what migrant workers earn. Rural sellers who want to open an official Tmall account to be a large-scale online business are required to pay about US$29,000. A Taobao partner suggested that the platform functions more to connect city sellers to rural buyers, not the other way around.[24]

Creation of New Economic Activities

Some new economic activities created by 4R technologies can help fight poverty. One such example is the evolving data-labeling industry (also known as content labeling or data annotation), which supports AI systems. Data labeling heavily relies on human-powered activities. Data labelers are referred to as the blue-collar workers of the AI age.[25]Data labeling is "a new type of blue-collar industry around curating the data that powers AI."[26]

Most data-labeling training can be completed easily and quickly, although some training may be time consuming. There are many jobs that computers cannot perform as well as humans. For instance, data labelers at the India- and U.S.-based data annotation company iMerit typically take a

22. "Undeveloped west China joins Singles' Day shopping spree," *Xinhua Insight*, 2015, http://news.xinhuanet.com/english/2015-11/11/c_134806559.html.

23. Greg Knowler, "Chinese farmers go internet shopping—for tractors," *Journal of Commerce*, July 15, 2015, http://www.joc.com/international-logistics/chinese-farmers-go-internet-shopping-%E2%80%93-tractors_20150715.html.

24. Yu Xueyi, "Why I quit Alibaba's big push toward countryside commerce," *Sixth Tone*, March 8, 2017, www.sixthtone.com/news/2031/why-i-quit-rural-taobao#jtss-twitter.

25. Maximilian Gahntz, "The invisible workers of the AI era," *Medium*, December 12, 2018, https://towardsdatascience.com/the-invisible-workers-of-the-ai-era-c83735481ba.

26. Kori Hale, "Google & Microsoft banking on Africa's AI labeling workforce," *Forbes*, May 28, 2019, https://www.forbes.com/sites/korihale/2019/05/28/google-microsoft-banking-on-africas-ai-labeling-workforce/#34b0fd0d541c.

seven-day online training course via video calls with U.S.-based trainers.[27] On the other hand, to develop an AI app to detect cancer on images from CT scan, experienced radiologists may have to train the algorithms.[28]

In China, which has already gained preeminence in AI, R&D activities are conducted in major cities such as Beijing, Shanghai, Hangzhou, and Shenzhen. Data labeling is performed in smaller towns and rural areas to take advantage of labor costs.[29] Shanxi's plan is to bring in more than 100 data-labeling companies and train more than 10,000 workers by 2022. It wants to have RMB 5 billion industry by 2025.[30]

Pushing Poor People into the Poverty Trap

Many technology companies are also using the 4R technologies in a way that will push poor people into further poverty. One such context is the P2P lending industry, which relies on big data and AI to come up with a credit score for people who lack credit history. In Kenya, digital lenders allegedly charge interest rates of more than 100% to borrowers. Users of such apps increased to 2 million in 2019 from 200,000 in 2016.[31]

Low financial literacy has been a key challenge in some LMICs.[32] P2P lenders mostly prey on consumers with low levels of financial literacy by charging extremely high interest rates.[33] Some borrowers lack an understanding of the consequences of not paying loans on time. For instance,

27. C. Metz, "A.I. is learning from humans, many humans," *New York Times*, August 16, 2019, https://www.nytimes.com/2019/08/16/technology/ai-humans.html.?auth=linked-google.

28. Gahntz, "The invisible workers."

29. Sarah Dai, "AI promises jobs revolution but first it needs old-fashioned manual labour—from China," *South China Morning Post*, October 8, 2018, https://www.scmp.com/tech/article/2166655/ai-promises-jobs-revolution-first-it-needs-old-fashioned-manual-labour-china.

30. Luna Lin, "Data labeling jobs are coming to underdeveloped regions in China, but can they stay?" *KrAsia*, October 16, 2019, https://kr-asia.com/data-labeling-jobs-are-coming-to-underdeveloped-regions-in-china-but-can-they-stay.

31. M. Ducan and M. Angus, "Kenyan parliamentary panel backs central bank to regulate digital lenders," *Reuters*, https://www.reuters.com/article/kenya-cenbank-idUSL8N2PG1TK.

32. Benno Ndulu and Tebello Qhotsokoane, "Harnessing fintech for a big leap in financial inclusion—lessons from East African success," *Pathways for Prosperity Commission*, 2020, https://pathwayscommission.bsg.ox.ac.uk/blog/harnessing-fintech-big-leap-financial-inclusion-lessons-east-african-success.

33. Nir Kshetri and H. Besada, "Big data's role in broadening financial inclusion in the Global South," *South-South Ideas*, UN Office for South-South Cooperation, https://www.ssc-globalthinkers.org/topic/south-south-ideas-big-datas-role-broadening-financial-inclusion-global-south.

during 2014–2017, 2.7 million Kenyans were blacklisted by the country's Credit Reference Bureaus.[34] About 400,000 of them had defaulted on loans of less than US$2.[35] It is argued that fintech companies have created credit bubbles among poor and vulnerable groups.[36] They extend credit to these groups with less stringent lending standards and without an expectation of a profitable use of the credit by the borrower.

According to Nabard's 2018 "all-India financial inclusion survey," about 30% of agricultural households in India took loans from high-interest private lenders. The RBI viewed this as a "cause of concern" because it suspected that a large proportion of such loans are unproductive, which means that they are not used in generating more farm income. Such loans may end up creating debt traps. Due primarily to poor returns from farming, farmers tend to use cheaper crop loans for their consumption needs. Farmers need to use their farm loans for capital investments or spending on farm assets, for which farmers need to pay higher interest.[37]

In Focus 10.1: Use of Blockchain and Smart Contracts to Lift Cambodian Farmers out of Poverty

According to official estimates, Cambodia's poverty rate was 13.5% in 2014. Of the country's 16.5 million people, about 4.5 million were precariously poor in 2021, which means that if economic and other external shocks hit them, they would be vulnerable to falling into poverty.[38] The agriculture sector employs 60% of the country's workforce, and many farmers lack contracts with their clients.[39] In April 2018, Oxfam launched the blockchain project Blockchain for Livelihoods

34. Zeituna Mustafa, Mercy Wachira, Vera Bersudskaya, William Nanjero, and Graham A. N. Wright, "Where credit is due customer experience of digital credit in Kenya," *MicroSave*, March, 2017.

35. Keren Weitzberg, "Mobile credit expands mass surveillance of ordinary Kenyans," *Coda Media*, September 17, 2019, https://codastory.com/authoritarian-tech/mobile-credit-kenya/.

36. Tom Collins, "Is fintech in Kenya too successful?," *African Business Magazine*, February 11, 2020, https://africanbusinessmagazine.com/african-banker/is-fintech-in-kenya-too-successful/.

37. "RBI shows concern over farmers pawning their gold to secure large loans," *Hindustan Times*, October 15, 2019, https://www.hindustantimes.com/india-news/farmers-pawning-gold-to-take-out-more-crop-loans-than-they-need-rbi/story-jijnVM2M1cAdew6P5GSSMN.html.

38. World Bank, *Overview*, 2017, https://www.worldbank.org/en/country/cambodia/overview.

39. "Oxfam launches project to connect rice supply chain in Cambodia," *Nation Thailand*, November 16, 2018, https://www.nationthailand.com/asean-plus/30358691.

from Organic Cambodian Rice (BlocRice). It uses smart contracts to improve Cambodian small-scale rice farmers' bargaining and negotiating power and digitally connect supply chain participants.[40] Agricultural cooperatives are parties to the contracts.[41] Other key include organic farmers and rice exporters in Cambodia and buyers in the Netherlands. The contract stipulates that the exporter will pay farmers the market price plus a premium.[42] Such a condition guarantees a market for the rice and reduces uncertainties for farmers.[43]

The Project

It started with organic rice farmers in Preah Vihear province in Cambodia, a province known for organic rice. Farmers own on average 1–2 hectares of land and produce 2.5–3 tons of rice per year.[44] BlocRice's first phase started with 50 farmers from the Reaksmei cooperative in one agricultural community in the province. Other parties were rice exporter AMRU Rice and rice-cake producer SanoRice. The consultancy Schuttelaar & Partners and the Dutch affiliate of the international Oxfam organization and Oxam Novib facilitated the process.[45] By selling directly to SanoRice, intermediaries were eliminated.

The 50 farmers who participated in the pilot agreed to provide 100 metric tons of rice. Since a late-season drought led to reduced production, they could deliver only 92.5 tons. The shipping took place in two containers to the Netherlands in March 2019.

Benefits to Farmers

Each farmer receives a digital identity that can be used to see details such as shipment weights and prices by logging into a website. The

40. Dean Pinkert, James Ton-that, and Ravi Soopramanien, "How blockchain can make supply chains more humane," *Stanford Social Innovation Review*, January 18, 2019, https://ssir.org/articles/entry/how_blockchain_can_make_supply_chains_more_humane#.

41. Sok Chan, "Oxfam's blockchain-based project gets official launch," *Khmer Times*, November 16, 2018, https://www.khmertimeskh.com/551240/oxfams-blockchain-based-project-gets-official-launch/.

42. Oxfam, *Can blockchain help rice farmers fight poverty? Oxfam in Cambodia*, https://cambodia.oxfam.org/can-blockchain-help-rice-farmers-fight-poverty.

43. Oxfam, *Can blockchain.*

44. Food Bev Asia, "Fair trade through blockchain technology," *Industry Sourcing*, July 1, 2019, http://www.industrysourcing.com/article/fair-trade-through-blockchain-technology.

45. Sok Chan, "Top rice exporter gets more hi-tech," *Khmer Times*, July 2, 2020, https://www.khmertimeskh.com/50747088/top-rice-exporter-gets-more-hi-tech/.

information is available in both Cambodian and English.[46] The BlocRice application allows farmers to see the database using their smartphones. In this way, they are informed about prices and other contract terms. Farmers enter their planting and harvest data, which helps AMRU Rice and SanoRice predict future yields.

The rice farmer and president of the cooperative Lyvoeung Chum noted that BlocRice provided more predictable prices than selling to traders. The farmers participating in BlocRice received US$0.24 per kilogram, and an additional US$.05. The premium amounted to about US$100 for most of the farmers that participated in the first phase of BlocRice.[47]

Plans to Expand

By July 2020, the project had expanded to 500 households in two communities in Preah Vihear province[48] The goal is to expand to 5,000 farmers by 2022.[49] Plans also include making more data available, such as weather forecasts, rice varieties, and farming practices, to help smallholders improve their yield and profitability forecasts.[50]

Current Challenges

Many challenges remain. In 2019, 10%–20% of the farmers had smartphones. Some farmers are illiterate or lack internet access and a mobile data plan. The BlocRice project manager Phay Cheth discussed the potential to develop an audio app to help illiterate farmers. Solutions are also being explored to develop apps to allow farmers to update data offline, which can be uploaded when devices are connected to the internet.

A benefit is that the system can improve farmers' bargaining position. Also, it is arguably a cheaper social certification mechanism than alternatives such as FairTrade.[51]

---

46. Chan, "Top rice exporter."
47. Oxfam, *Can blockchain.*
48. Chan, "Top rice exporter."
49. "Oxfam launches."
50. Ken Cottrill, "Blocrice makes a case for blockchain in smallholder farming: Sententia partners," *Chain Business Insights*, May 22, 2019, https://www.chainbusinessinsights.com/insights-blog/blocrice-makes-a-case-for-blockchain-in-smallholder-farming.
51. Joshua Hallwright and Elsa Carnaby, "Complexities of implementation: Oxfam Australia's experience in piloting blockchain," *Frontiers*, August 28, 2019, https://www.frontiersin.org/articles/10.3389/fbloc.2019.00010/full.

The solution overall provides only a small improvement. Farmers suffer other forms of uncertainties, such as weather. The availability of risk sharing and transfer mechanisms such as farm insurance is critical to improving the livelihood and development of farmers. As noted in chapter 8, Arbol has teamed up with Agribee Cambodia to offer blockchain-based crop insurance.

Leapfrogging Theories

A leapfrog involves making a jump in economic progress and technological development to pass over stages of economic growth and modernization.[52] Key mechanisms involved in leapfrogging include teaming up with more developed countries and copying applications from them.[53] Some argue that real-world examples of leapfrogging are rare.[54] Other have noted that the leapfrogging process requires cultural change.[55] A further observation is that positive effects of leapfrogging can be expected only in the presence of favorable institutional factors. These include strong political leadership, state intervention to create favorable conditions, and the open exchange of technologies and ideas among developed and developing countries.[56]

To illustrate the challenges associated with leapfrogging, I take the example of satellite technology in Nigeria. Prior research has emphasized the need for investment in human resources to make productive uses of the acquired technologies and to reduce reliance on Western countries.[57] Nigeria operates a communication satellite, NigComSat-1R, and an Earth observation satellite, NigeriaSat-2. However, Nigeria's space capabilities are underutilized. For example, data available from the Earth observation satellites is not used in town planning in most state and local governments.[58]

---

52. Steinmueller, "ICTs and the possibilities."

53. G. D. M. Wijers, "Determinants of the digital divide: a study on IT development in Cambodia," *Technology in Society* 32, no. 4 (2010): 336–341.

54. H. P. Howard, "Testing the leap-frog hypothesis: The impact of existing infrastructure and telecommunications on the global digital divide," *Communication and Society* 10 (2007): 133–157.

55. Steinmueller, "ICTs and the possibilities."

56. M. D. Chinn and R. W. Fairlie, "The determinants of the global digital divide: A cross country analysis of computer and internet penetration," *Oxford Economic Papers* 3 (2006): 1024–1068.

57. C. Alden, "Let them eat cyberspace: Africa, the G8 and the digital divide," *Millennium Journal of International Studies* 32 (2003): 468–469.

58. Samuel Oyewole, "One of Nigeria's satellites is on its last legs: Why this is worry-

Some have argued that China has leapfrogged some of the latest technologies. Former Google CEO and chairman Eric Schmidt estimated that China will surpass the U.S. in AI by 2025.[59] Other assessments see China further behind the U.S. on AI.[60] Some put China ahead already.[61] For instance, in perception AI, which involves using cameras, sensors, and other devices to capture faces, motion, people, and objects to derive economic value from them, China is reported to be slightly ahead of the U.S.[62] Some practical uses of perception AI include recognizing returning customers at a store and counting attendance in various settings such as an event or school. Kai-Fu Lee, chairman and CEO of China's Sinovation Ventures, noted that due to privacy concerns, the development and use of such applications is likely to be limited in the U.S. It is argued that Chinese citizens are willing to trade privacy for convenience, so China is likely to lead in this area.[63]

In his book *AI Superpowers*, Kai-Fu Lee estimated that, in internet AI, which capitalizes on data that users automatically label when they browse the internet, China and the U.S. were in a head-to-head race. Lee predicted that due to China's data advantage, Chinese tech companies will have a 60%–40% lead over U.S. companies by 2023. Their technology leverages data from recommendation engines, and algorithmic systems learn from personalized online content. Masses of data related to users' clicking behavior, time spent on a web page, and video-watching patterns are the key to building this type of AI. Some examples of uses include Amazon's product recommendation systems, YouTube's video recommendation, and Facebook's ad

---

ing," *Conversation*, August 4, 2021, https://theconversation.com/one-of-nigerias-satellites-is-on-its-last-legs-why-this-is-worrying-165068.

59. Joshua P. Meltzer and Cameron F. Kerry, "Strengthening international cooperation on artificial intelligence," *Brookings*, February 17, 2021, https://www.brookings.edu/research/strengthening-international-cooperation-on-artificial-intelligence/.

60. Jeffrey Ding, "China's current capabilities, policies, and industrial ecosystem in AI," *Testimony before the U.S.-China Economic and Security Review Commission, Hearing on Technology, Trade, and Military-Civil Fusion: China's pursuit of artificial intelligence, new material and new energy*, June 7, 2019.

61. Amy Webb, "China is leading in artificial intelligence—and American businesses should take note," *Inc. Magazine*, https://www.inc.com/magazine/201809/amy-webb/china-arttifical-intelligence.htm.

62. Martin Wolf, "China battles the US in the artificial intelligence arms race," *Financial Times*, April 16, 2019, https://www.ft.com/content/2f295a9e-5f96-11e9-b285-3acd5d43599e.

63. McKinsey Global Institute, *Kai-Fu Lee's perspectives on two global leaders in artificial intelligence: China and the United States*, June 14, 2018, https://www.mckinsey.com/featured-insights/artificial-intelligence/kai-fu-lees-perspectives-on-two-global-leaders-in-artificial-intelligence-china-and-the-united-states.

delivery. By building a detailed picture of a user's personalities, habits, and desires, labeled data can help tailor contents to individual users.[64]

However, China has not leapfrogged the U.S. in the overall AI industry. In business AI, which involves exploit data by businesses, China is far behind the U.S.[65] In general, researchers outside of China more frequently cite U.S. patents and papers than Chinese ones. This indicates that Chinese patents may be of inferior quality.

Regarding the AI workforce, according to the 2018 *China AI Development Report*, by the end of 2017, China had 18,200 AI scientists and engineers, which made it the world's second-largest pool of such people; the U.S. had about 29,000.[66] However, it is reported that some workers in China who claim AI expertise may only have associate or technical certificates.[67]

The open exchange of technologies and ideas among developed and developing countries is needed for leapfrogging and to maximize its positive effects.[68] The tensions between China and the West have hindered such exchanges. Among other factors, the U.S. government's use of trade policy instruments such as export bans against Chinese technology companies has made it difficult for China to leapfrog and achieve its technological ambition.

Middle-Income Trap

A middle-income trap is referred to a situation in which countries that have got out of the poverty trap and grew to middle-income levels subsequently fail to grow further to achieve the levels of advanced countries.[69] Slowdowns in economic growth are reported to occur in two modes, one in the \$10,000–11,000 range and another at \$15,000–\$16,0000 in PPP dollars, which is adjusted to take into account the cost of living.[70]

64. Wolf, "China battles."

65. Wolf, "*China battles.*"

66. China Institute for Science and Technology Policy at Tsinghua University, *China AI Development Report 2018*, July 2018, https://indianstrategicknowledgeonline.com/web/China_AI_development_report_2018.pdf.

67. Jonathan Vanian, "What people get wrong about China and artificial intelligence: eye on AI," *Fortune*, July 9, 2019, https://fortune.com/2019/07/09/china-data-artificial-intelligence/.

68. Chinn and Fairlie "The determinants of."

69. H. Kharas and H. Kohli, "What is the middle income trap, why do countries fall into it, and how can it be avoided?," *Global Journal of Emerging Market Economies* 3, no. 3 (2011): 281–289.

70. Barry Eichengreen, Donghyun Park, and Kwanho Shin, *Growth slowdowns redux:*

Countries are likely to avoid slowdowns if the population has a high level of secondary and tertiary education and if high-technology products account for a large share of exports. It is important to move up the technology ladder to avoid the middle-income trap.[71]

Countries in the middle-income trap face two major institutional and political challenges. First, the policies that are required to upgrade productivity—human capital and innovation—require huge investments in institutional capacity. Second, these institutional challenges arise when political capacity for building these institutions is weak due to the fragmentation of potential support coalitions. Fractured social groups, especially business and labor and inequality, are key challenges.[72]

To better understand the challenges faced by middle-income countries in achieving the levels of advanced countries, I consider the case of China. China's per capita GDP in 2020 was US$10,500.40;[73] it was US$17,310 in PPP terms.[74] While the Chinese Communist Party (CCP) says it has achieved its poverty alleviation target, the bar has been set very low. In 2021, of China's 1.4 billion people, 600 million had a per-person income of about US$150 a month or less.

Despite a significant rural poverty problem, there is a high degree of satisfaction in rural low-income areas for the CCP. A large number of intellectuals, entrepreneurs, migrant workers, and activists, however, have shown their discontent and dissatisfaction with the direction of the CCP. The country's so-called creative class has emphasized the importance of broader reforms to realize the full potential of the population.[75]

China is nowhere close to fully utilizing modern technologies in achieving high economic growth. In 2021, China was about 50% as efficient as the U.S. in its combination of labor and capital, which, according to Bloomberg

---

New evidence on the middle-income trap, National Bureau of Economic Research Working Paper 18673, January 2013, https://www.nber.org/papers/w18673.

71. Eichengreen et al., "Growth slowdowns redux."

72. R. F. Doner and B. R. Schneider, "The middle-income trap," *World Politics* 68, no. 4 (2016): 608–644.

73. World Bank, *GDP per capita (current US$)—China, World, Data*, https://data.worldbank.org/indicator/NY.GDP.PCAP.CD?locations=CN-1W&most_recent_value_desc=true.

74. World Bank, *GDP per capita.*

75. Ann Scott Tyson, "Vilified abroad, popular at home: China's Communist Party at 100," *Christian Science Monitor*, February 18, 2021, https://www.csmonitor.com/World/Asia-Pacific/2021/0218/Vilified-abroad-popular-at-home-China-s-Communist-Party-at-100.

Economics, will reach 70% by 2050.[76] The country lags the U.S. in semiconductor development, hardware, research and development, and dynamic commercial AI.[77]

China's annual imports of semiconductors were expected to exceed US$300 billion in 2020,[78] mostly from U.S. semiconductor companies.[79] China's chip-manufacturing capability is estimated to be two generations (seven to ten years) behind global leaders such as South Korea, Taiwan, and the U.S.[80] SMIC and other Chinese chipmakers produce chips for lower-end products such as IoT devices. China hoped that SMIC's foundry services would become a source of export revenue and help the domestic AI industry.[81]

The Fallacy of China's Large Data Pool

On the bright side, Chinese firms have access to huge amount of data, which is China's often-touted advantage on the AI front. In a note provided to clients, analysts of the Swiss multinational investment bank UBS noted that China is overtaking the U.S. in access to huge amounts of user data needed to build robust AI systems. Restrictive data collection laws in the U.S. and Europe hinder such efforts. The UBS analysts cited San Francisco's ban on the use of facial recognition.[82]

---

76. Eric Zhu and Tom Orlik, "When will China rule the world? Maybe never. The Communist Party wants the world to see China's continued rise as inevitable. In reality, it's anything but," *Bloomberg*, July 5, 2021, https://www.bloomberg.com/news/features/2021-07-05/when-will-china-s-economy-beat-the-u-s-to-become-no-1-why-it-may-never-happen?srnd=premium.

77. "America and China's great AI competition: What is driving it," *National Interest*, March 24, 2019, https://nationalinterest.org/blog/buzz/america-and-chinas-great-ai-competition-what-driving-it-48677.

78. Anniek Bao and Liu Peilin, "China's stumbling sprint to semiconductor self-sufficiency," *Caixin Global*, November 20, 2020, https://www.caixinglobal.com/2020-11-20/chinas-stumbling-sprint-to-semiconductor-self-sufficiency-101630701.html.

79. Christopher A. Thomas, "Lagging but motivated: The state of China's semiconductor industry," *Brookings*, January 7, 2021, https://www.brookings.edu/techstream/lagging-but-motivated-the-state-of-chinas-semiconductor-industry.

80. Yvette To, "China chases semiconductor self-sufficiency," *East Asia Forum*, February 22, 2021, https://www.eastasiaforum.org/2021/02/22/china-chases-semiconductor-self-sufficiency.

81. Simon Sharwood, "Chinese chipmaker SMIC says US sanctions mean it will struggle to develop 10nm products," *The Register*, December 21, 2020, https://www.theregister.com/2020/12/21/smic_says_us_bans_bite.

82. Benjamin Pimentel, "China is 'mostly world class' in artificial intelligence software,

Access to data offers advantages in some areas. In the area of cybersecurity, for instance, not all types of data can be obtained to train AI algorithms. Due to privacy concerns and regulations such as the General Data Protection Regulation (GDPR), good and representative training datasets are not always available to test and validate hypotheses related to how user interactions across applications are linked to successful cyberattacks against them.[83]

However, the quality and relevance of data are important. Despite China's huge amount of data, the domain-specific nature of data means that data collected for one purpose is not of much use for other purposes. This means that a large amount of surveillance data may not give the country an advantage in developing AI algorithms for drug discovery or self-driving cars.[84]

Institutional, Governance, and Regulatory Barriers

To avoid the middle-income trap, the approach to policy formulation has to be "pragmatic rather than doctrinaire."[85] The CCP's doctrine and ideology have been the most powerful and potent force in China. Following the 1989 Tiananmen Square events, the CCP "has firmly adhered to one core principle: uphold the rule of the CCP at all cost."[86]

A lesson from some successful countries such as Japan and South Korea is that the decentralization of power and effective institutions is the key to move up the value chain.[87] China clearly lacks such structure. Policymakers need to address local issues of opportunity and distribution, which require a high degree of decentralization in decision-making—not easy in a centralized system.[88] For instance, 4R technologies currently offer limited opportunities for the rural poor in China. It is also critical to reform the *hukou* system that blocks rural migrants from accessing the same benefits as urban residents.[89]

---

but still relies on America for the processors to run it, say Wall Street analysts," *Business Insider*, July 18, 2019, https://www.businessinsider.com/china-world-class-ai-depends-us-chips-2019-7.

83. Raffael Marty, "Cybersecurity is the next frontier for AI and ML," *Venturebeat*, June 18, 2021.

84. Carl Benedikt Frey and Michael Osborne, "China won't win the race for AI dominance," *Foreign Affairs*, June 19, 2020, www.foreignaffairs.com. https://www.foreignaffairs.com/articles/united-states/2020-06-19/china-wont-win-race-ai-dominance.

85. Kharas and Kohli, "What is the middle income trap."

86. C. Minzner, *End of an era: How China's authoritarian revival is undermining its rise* (New York: Oxford University Press, 2018), 18.

87. Kharas and Kohli, "What is the middle income trap."

88. Kharas and Kohli, "What is the middle income trap."

89. Alice Su, "China is purging celebrities and tech billionaires. But the problem is big-

Improved governance and regulatory agencies are needed to avoid the middle-income trap.[90] In many middle-income countries, firms have to deal with institutional barriers such as obsolete laws and regulations that often favor powerful players.[91] For instance, bankruptcy laws mostly favor creditors and penalize failure instead of providing as a learning experience. Small and new firms have few options for financing. Large and well-connected firms often dominate, which limits growth opportunities for SMEs.[92]

In China's case, perhaps the greatest barrier is its highly centralized governance structure, which requires the country's firms to help the government to maintain social and political control, for instance, by facilitating the censorship measures. Unlike in the U.S., where IT companies such as Google and Facebook have shown some willingness to fight for consumers' privacy, this is not the case in China.[93]

The CCP's propaganda agencies provide censorship pressures on major Chinese companies. In 2011, then propaganda chief Li Changchun reportedly met Robin Li, chief of Baidu, and other senior propaganda officials had meetings with the chiefs of Youku (a YouTube equivalent) and Sina (which owns microblogging website Sina Weibo). The goal was to harmonize these firms' commercial policies with the government's political goals.[94]

Chinese technology companies are expected to work with the CCP's propaganda apparatus to censor and filter user-generated content.[95] It is also a common practice among big internet firms to seek government input before launching a product or service. In this way, they involve regulators in product development. When Sina was designing Weibo, it worked closely with regulators.[96]

---

ger than 'sissy men,'" *Los Angeles Times*, 2021, https://www.latimes.com/world-nation/story/2021-09-14/common-prosperity-china-crackdown-analysis.

90. Kharas and Kohli, "What is the middle income trap."

91. Kharas and Kohli, "What is the middle income trap."

92. Kharas and Kohli, "What is the middle income trap."

93. M. Persson, M. Vlaskamp, and F. Obbema, "China rates its own citizens—including online behaviour," *de Volkskrant*, 2015, http://www.volkskrant.nl/buitenland/china-rates-its-own-citizens-including-online-behaviour~a3979668.

94. Reporters without Borders, *Internet leaders cooperate with government efforts to reinforce online controls*, September 11, 2011, http://en.rsf.org/chine-respect-for-free-speech-continues-27-10-2011,41303.html.

95. Li Hui and Megha Rajagopalan, "At Sina Weibo's censorship hub, China's Little Brothers cleanse online chatter," *Yahoo News*, September 11, 2013, http://finance.yahoo.com/news/sina-weibos-censorship-hub-chinas-210105714.html.

96. "An internet with Chinese characteristics," *The Economist*, July 30, 2011, http://www.economist.com/node/21524821.

The algorithms and data developed and collected by private companies are also reported to be used in China's Social Credit System. Ten private companies have permission to launch internet-based credit-rating and credit-ranking systems, which include Alibaba's Ant Group and Tencent's paywall.[97] Alibaba's Sesame Credit is arguably the system that is most admired by government authorities.[98]

China's state media have acknowledged the costs of censorship to the national economy. They argue, though, that the benefits exceed the costs.[99] While these strategies help the CCP maintain control over society, there are significant costs to the economy.

The requirement to comply with the government's censorship policies forms a major component of product development costs for China's internet startups. Nonetheless, while Chinese technology firms are expected to use their resources to help the CCP achieve its political goals and objectives, these firms also gain political favor. It is argued that the rise of some of China's big companies can be attributed to government policies that have made difficult for foreign companies to enter.[100] The industries in which these companies operate have been insulated from foreign competitors. The success of China's biggest internet giants—Baidu, Alibaba, and Tencent (the so-called BAT) could be credited to China's web filters (the Great Firewall). Blocking of sites such as YouTube, Twitter, and Google provided Chinese internet companies with the space they needed to grow.[101] The *Global Times* noted that without the firewall, "China would become the realm of Google China, Yahoo China and Facebook China."[102] This also means that major internet markets in China are dominated by the monopoly of firms such as Alibaba and Tencent.

---

97. Zheping Huang, "All Chinese citizens now have a score based on how well we live, and mine sucks," *Quartz*, October 9, 2015, http://qz.com/519737/all-chinese-citizens-now-have-a-score-based-on-how-well-we-live-and-mine-sucks/.

98. "Discipline and Punish: The Birth of China's Social-Credit System," *The Nation*, 2019, https://www.thenation.com/article/china-social-credit-system/.

99. "Xinhua Insight: Changing China transforms Internet landscape," *Xinhua*, 2014, http://news.xinhuanet.com/english/indepth/2014-04/20/c_133276467.htm.

100. B. Einhorn, "How China's government set up Alibaba's success," *Bloomberg*, 2014, http://www.bloomberg.com/news/articles/2014-05-07/how-chinas-government-set-up-alibabas-success.

101. T. P. Chen, "China owns 'Great Firewall,' credits censorship with tech success," *Wall Street Journal*, January 28, 2015, http://blogs.wsj.com/chinarealtime/2015/01/28/china-owns-great-firewall-credits-censorship-with-tech-success/.

102. Chen, "China owns."

Investment and Capital Accumulation

To avoid the MT, innovations must accompany investment and capital accumulation.[103] Chinese technology companies have faced barriers raising capital from Western capital markets. In March 2020, the U.S. Securities and Exchange Commission (SEC) adopted the Holding Foreign Companies Accountable Act, which required auditing by a U.S. watchdog of certain companies identified by the SEC. These companies are required to submit documents to prove they are not owned or controlled by a foreign governmental entity. Chinese companies are required to name each board member who is a Chinese Communist Party official. The U.S. regulator can stop the trading of securities of companies that do not satisfy the requirement. Following the adoption of this regulation, shares of U.S.-listed Chinese tech such as Alibaba, Baidu, JD.com, and NetEase dropped sharply.[104]

Many have questioned the morality and ethics of Western capital markets that have funded Chinese companies such as Alibaba and Sina Weibo, which have made the censorship system work. Moreover, Western multinationals have been accused of tailoring operations in China or providing the systems and technologies needed for censorship.[105]

In addition to financial capital, Chinese technology firms have also faced obstacles to accumulating physical capital, such as equipment. In December 2020, the U.S. Department of Commerce's Bureau of Industry and Security added the SMIC semiconductor company and ten of its related entities to the U.S. blacklist called the Entity List due to its "relationships of concern with the military-industrial complex, China's aggressive application of military-civil fusion mandates and state-directed subsidies."[106] This means that U.S. firms cannot do business with Huawei without obtaining a government license.[107] The U.S. concluded that there was an "unacceptable risk" that equipment supplied to SMIC could be used for military purposes.[108]

103. Kharas and Kohli, "What is the middle income trap."

104. A. Kharpal, "Chinese tech stocks hammered as U.S. law threatens to delist firms from American exchanges," *CNBC*, 2021, https://www.cnbc.com/2021/03/25/chinese-tech-stocks-fall-as-us-sec-begins-law-aimed-at-delisting.html.

105. M. Dickie, "China traps online dissent," *Financial Times*, 2007, http://www.ft.com/intl/cms/s/0/ef0e7d64-9138-11dc-9590-0000779fd2ac.html.#axzz3Vzs5a5VR.

106. Grant Leach and Cortney O'Toole Morgan, "BIS adds over 70 new entities to the entity list, including SMIC," *Global Trade Magazine*, January 8, 2021, https://www.globaltrademag.com/bis-adds-over-70-new-entities-to-the-entity-list-including-smic.

107. Jenny Leonard and Ian King, "Five months after Huawei export ban, U.S. companies are confused," *Los Angeles Times*, October 24, 2019, https://www.latimes.com/business/story/2019-10-24/huawei-export-ban-us-companies-confusion.

108. "U.S. tightens exports to China's chipmaker SMIC, citing risk of military use,"

For one thing, China's technological ambition cannot be achieved without a developed chip and semiconductor industry. China also lags the U.S. in AI hardware because most of the world's biggest AI-enabled semiconductor chips companies are U.S. based (e.g., Nvidia, Intel, Apple, Google, Advanced Micro Devices).[109] AI algorithms need to analyze huge datasets extremely fast, which would require tens of trillions of calculations. Sophisticated semiconductors are needed to achieve such capabilities.

Capital and Skill Intensiveness

To avoid the middle-income trap, growth needs to be more capital intensive and skill intensive in manufacturing, which is also referred to as "moving up the value chain." Growth also needs to be more heavily oriented toward services.[110]

On the skills front, an encouraging trend is that China has been successful in developing domestic AI talent. According to a 2019 study by the Chicago-based think tank MacroPolo, which analyzed authorship of papers accepted to prestigious international AI conferences NeurIPS during 2009–2018, the number of authors who completed undergraduate studies in China had increased by about tenfold. In 2009, there were about 100 Chinese researchers, or 14% of authors, which increased to about 1,000 in 2018, or about 25%. The largest increase was between 2017 and 2018 after the 2017 AI national strategy was released. Many universities had started AI specialization and degree programs.[111] By the end of 2019, the total number of AI authors on arXiv was 57,654, of which 26,818 (48%) were from the U.S. and 6,401 (11%) from China.[112]

China has faced problems in retaining the talent in the country. MacroPolo found that about three-quarters of Chinese authors were working outside China; 85% of those were working in the U.S. at technology companies and universities.[113] Since professional talent is increasingly mobile interna-

Reuters, September 26, 2020, https://www.reuters.com/article/usa-china-smic/u-s-tightens-exports-to-chinas-chipmaker-smic-citing-risk-of-military-use-idUSKBN26H0LO.

109. Sarah O'Meara, "Will China lead the world in AI by 2030?," *Nature*, August 21, 2019, https://www.nature.com/articles/d41586-019-02360-7.

110. Kharas and Kohli, "What is the middle income trap."

111. MacroPolo, *China's AI talent base is growing, and then leaving*, July 30, 2019, https://macropolo.org/chinas-ai-talent-base-is-growing-and-then-leaving/?rp=e.

112. J. Desrosiers, "Global AI talent report 2020," *jfgagne*, 2020, https://jfgagne.ai/global-ai-talent-report-2020/.

113. "China's path to AI domination has a problem: Brain drain," *MIT Technology Review*, August 7, https://www.technologyreview.com/f/614092/china-ai-domination-losing-talent-to-us/.

tionally, middle-income countries, especially small ones, need to be proactive in providing attractive opportunities and lifestyles.[114] Such conditions are lacking in China. In 2012, over 150,000 Chinese, mainly rich and educated elites, obtained overseas citizenship. The main destination countries were the U.S. (87,000), Canada and Australia (30,000 each), and New Zealand (6,000). Many reportedly left China to search for a more democratic society, a cleaner environment, and better educational opportunities.[115]

Finding New Markets

To grow exports, firms need to introduce new processes and find new markets instead of relying on sales of the same product to existing markets.[116] It is worth noting that Chinese technology firms have demonstrated a high level of success in finding new markets, especially in LMICs. China has sold AI and facial recognition software in several foreign countries. For instance, the Philippines' Bonifacio Global City has been equipped with mass-surveillance systems developed by China's Huawei, which works with police to link cameras to data-collection tools such as a plate recognition system. The system gathers evidence and identifies suspects using facial recognition.[117] Privacy advocates have been concerned about the potential misuse of data.[118] Serbia, Turkey, Russia, Ukraine, Azerbaijan, Angola, Laos, Kazakhstan, Kenya, Uganda,[119] Ecuador, Bolivia, and Peru[120]—all these countries are also using China-developed facial recognition software.

---

114. Kharas and Kohli, "What is the middle income trap."

115. "Brain drain should send message to China's leaders," *South China Morning Post*, January 6, 2013, https://www.scmp.com/comment/insight-opinion/article/1120829/brain-drain-should-send-message-chinas-leaders.

116. Kharas and Kohli, "What is the middle income trap."

117. N. Mandhana, "Huawei's video surveillance business hits snag in Philippines," *Wall Street Journal*, 2019, https://www.wsj.com/articles/huaweis-videosurveillance-business-hits-snag-in-philippines-11550683135.

118. B. O'Rourke and G. Choy, "Big brother Huawei kitted out this Philippine city: Is China watching?," *South China Morning Post*, 2019, https://www.scmp.com/week-asia/economics/article/2183540/big-brother-huaweiwatches-philippine-city-does-china-too 35.

119. "Chinese facial recognition tech installed in nations vulnerable to abuse," *CBS News*, October 16, 2019, https://www.cbsnews.com/news/china-huawei-face-recognition-cameras-serbia-other-countries-questionable-human-rights-2019-10-16/.

120. C. Rollet, "Ecuador's all-seeing eye is made in China," *Foreign Policy*, August 9, 2018, https://foreignpolicy.com/2018/08/09/ecuadors-all-seeing-eye-is-made-in-china/; Paul Mozur, Jonah M. Kessel, and Melissa Chan, "Made in China, exported to the world: The surveillance state," *New York Times*, April 24, 2019, https://www.nytimes.com/2019/04/24/technology/ecuador-surveillance-cameras-police-government.html.

To find new markets, it is important for exporters to understand the quality, price, and consumer preference in each market they want to enter. Most firms develop such understanding in domestic markets before developing into global brands.[121] A main challenge some Chinese technology companies are facing is that, due to the CCP's censorship activities, it has been difficult for Chinese technology companies to understand true consumer preference in the domestic market, which has been among the key mechanisms affecting Chinese multinationals' performance in foreign countries. Consider the Chinese search engine Baidu, which is also one of the biggest AI companies in China. Baidu's Portuguese version, Baidu Busca, went live in Brazil on July 7, 2014.[122] The Chinese president Xi Jinping and Brazil's then president Dilma Rousseff had jointly pressed a button initiating the Baidu Busca service. A blogger noted that Baidu had applied Chinese political censorship on certain searches on Baidu Busca.[123] Baidu subsequently apologized for the error.

Fei Chang Dao, who tracks censorship on Baidu and Weibo, provided a comparison of screenshots;[124] this showed that search results for "Hu Jintao" in Chinese on Baidu's Brazilian Busca and its China-based search engine were almost identical. He noted that Baidu's China-based search engine told users: "In accordance with relevant laws, regulations, and policies, some search results have not been displayed." Fei Chang Dao provided a screenshot showing the results of a search for "Falun Gong" on br.baidu.com, which showed only results from the *People's Daily*. Likewise, a search for "Xi Jinping" returned results from websites operated by China's central government and the CCP, while the same search on the Brazilian version of Bing returned tens of thousands of results.[125]

Baidu is also allegedly engaged in censorship activities in other markets. In 2011, eight New York residents filed a lawsuit against Baidu. The plaintiffs alleged that their articles or videos about the Chinese democratic movement had been censored by Baidu. An example cited in the lawsuit involved the 1989 military action against protesters in Tiananmen Square.

---

121. Kharas and Kohli, "What is the middle income trap."

122. Felipe Torres, "Baidu takes on Brazil! Will Brazil take to Baidu?" *Webcertain*, July 29, 2014, http://blog.webcertain.com/baidu-takes-on-brazil-will-brazil-take-to-baidu/29/07/2014/.

123. Fei Chang Dao, "Baidu launches Brazilian search engine—apparently With Chinese political censorship," July 19, 2014, http://blog.feichangdao.com/2014/07/baidu-launches-brazilian-search-engine.html.

124. Fei Chang Dao, "Baidu launches."

125. Fei Chang Dao, "Baidu launches."

They blamed Baidu as an "agent and enforcer of the anti-democracy policies" of China.[126]

With the expansion of international service tradability, new ICT technologies to digitize and store services, and easy and cheap transportability through modern telecommunications networks, services have become a powerful engine of growth for many middle-income countries.[127] Some level of success has been achieved by Chinese firms in exporting services based on 4R technologies in LMICs.

China's 4R-related products and services, however, are facing political and other barriers to tradability in international service markets, especially in high-income countries in the West. Many high-profile Chinese technology firms have allegedly violated human rights, among the huge obstacles to entering these markets. China's use of AI and other technologies for surveillance and control has been especially unacceptable in the U.S. and other democratic countries.[128] Concerns have also been raised regarding AI's implementation in large-scale social schemes such as the use of AI in the administration of social credit programs.[129]

Chinese technology firms' alleged cooperation with the government to crack down and suppress dissidents has prevented their export to other foreign markets. European Commission president Ursula von der Leyen emphasized the importance of strengthening the transatlantic partnership as a response to "an illiberal China."[130]

Among a high-profile Chinese technology company sanctioned by the U.S. is Huawei. In May 2019, the Bureau of Industry and Security put the company on its Entity List.[131] For instance, since Huawei is in the Entity List, Google cannot supply Android services, updates, or apps to Huawei 5G handsets.[9] Huawei decided to delay sales of its 5G-enabled Mate 30 smartphone series in Europe, its biggest mar-

---

126. Owen Fletcher, "Baidu accused of aiding Chinese censorship in U.S. suit," *Wall Street Journal*, May 19, 2011, http://www.wsj.com/articles/SB10001424052748703482104576332073063272688.

127. Kharas and Kohli, "What is the middle income trap."

128. White House, *The United States approach to the Peoples Republic of China*, May 2020, https://www.whitehouse.gov/wp-content/uploads/2020/05/U.S.-Strategic-Approach-to-The- Peoples-Republic-of-China-Report-5.20.20.pdf.

129. R. Andersen, "Chinese AI Is Creating an Axis of Autocracy," *The Atlantic*, 2020, https://www.theatlantic.com/magazine/archive/2020/09/china-ai-surveillance/614197.

130. Joshua P. Meltzer and Cameron F. Kerry, "Strengthening international cooperation on artificial intelligence," *Brookings*, February 17, 2021, https://www.brookings.edu/research/strengthening-international-cooperation-on-artificial-intelligence/.

131. Bureau of Industry and Security, Department of Commerce, *Entity list*, https://www.bis.doc.gov/index.php/policy-guidance/lists-of-parties-of-concern/entity-list.

ket outside China. The company noted that the value and usability of these 5G smartphones would be reduced without access to Google apps[2]. Other countries such as Australia and New Zealand also banned Huawei's 5G equipment. In December 2018, the Japanese government banned Huawei and another Chinese technology company, ZTE, from network hardware procurement.[132]

In an attempt to place more restrictions on Huawei and ZTE, in October 2019, the U.S. Federal Communications Commission (FCC) put forward a proposal that bans companies receiving government money from purchasing equipment or services from these firms. The FCC argued that if Huawei's 5G network equipment operates in sensitive locations such as near a U.S. military base, China could ask the company to install a secret backdoor or malware without U.S. officials knowing it.[133]

Many other high-profile Chinese technology companies are on the Bureau of Industry and Security's Entity List. In October 2019, the U.S. Commerce Department put SenseTime,[134] valued at US$7.5 billion (the highest valuation in the world for an AI startup), and seven others on the Entity List.[135] Hefei Bitland Information Technology was added to the July 2020 Entity List for alleged human rights violations. Hefei Bitland makes graphics cards and liquid crystal monitor (LCM) modules; its main clients include Lenovo and Hewlett-Packard (HP).[136] Likewise, Nanchang O-Film Tech supplies parts to Lenovo, Apple, HP, and other computer manufacturers.[137] Its products include cameras, touchscreens, and fingerprint sensors.[138]

---

132. "Japan bans Huawei and ZTE 5G networking hardware, will Canada be next?," *VentureBeat*, December 10, 2018, https://venturebeat.com/2018/12/10/japan-bans-huawei-and-zte-5g-networking-hardware-will-canada-be-next/.

133. Sherisse Pham, "Huawei and ZTE could lose what little business they have in the United States," *CNN Business*, October 30, 2019, https://www.cnn.com/2019/10/29/tech/fcc-huawei-5g-ajit-pai/index.html.

134. "Chinese companies are scrambling to survive Trump's blacklist," *Business Times*, October 29, 2019, https://www.businesstimes.com.sg/technology/chinese-companies%C2%A0are-scrambling-to-survive-trump%E2%80%99s-blacklist.

135. "How China's AI tech founders plan to survive Trump's blacklist," *South China Morning Post*, October 29, 2019, https://www.scmp.com/tech/start-ups/article/3035279/china-ai-champions-sensetime-are-scrambling-survive-trumps-blacklist.

136. Kenneth Rapoza, "The 11 sanctioned Chinese companies: what they sell, and to whom they sell it," *Forbes*, July 22, 2020, https://www.forbes.com/sites/kenrapoza/2020/07/22/the-11-sanctioned-chinese-companies-what-they-sell-and-to-whom-they-sell-it/#20eb022f47eb.

137. T. Grubb, "Alleged human-rights abuses in China delay laptop shipments to NC students," *News and Observer*, 2020, https://www.newsobserver.com/news/local/education/article244852172.html.

138. Amber Neely, "Apple supplier O-film Tech accused of human rights violations in

Flying Geese Paradigm

Japanese economists developed the flying geese paradigm (FGP) to study technological development in Southeast Asia. The model viewed Japan as a leading power.[139] The developing nations could be viewed as "aligned successively behind the advanced industrial nations in the order of their different stages of growth in a wild-geese-flying pattern."[140] A key point of this paradigm is that Asia's hierarchical regional order, in which the production of commoditized manufacturing goods moves from more advanced to less advanced countries, makes it possible for Asian nations to catch up with the West.

It is argued that 4R technologies have challenged conventional wisdom on the path of development. In the past, low-cost jobs such as those in apparel and assembly lines provided the path to industrialization. These are being replaced by robots and AI systems. Low-income countries are no longer in a position to take advantage of their lower wages and export labor-intensive goods. They lack foreign exchange to import infrastructure capital to generate domestic savings to invest in human capital. The "flying geese pathway to development is blocked."[141]

Many low-income countries' lack of ability to invest in infrastructure and human capital means that these economies are not well positioned to be among geese on the front. At the same time, while 4R creates new types of jobs requiring new skill sets, the education systems are mostly unprepared.

Chapter Summary and Conclusion

This chapter discussed the 4R with respect to key theories of economic development. The discussion indicates how 4R technologies have various mechanisms that can help poor people get out of the poverty trap. That is, by taking appropriate measures, the 4R can take the B4B out of the "basin of poverty." Some success stories on 4R technologies pulling poor people out

---

China," *AppleInsider*, July 20, 2020, https://appleinsider.com/articles/20/07/20/apple-supplier-o-film-tech-accused-of-human-rights-violations-in-china.

139. K. Akamatsu, "A historical pattern of economic growth in developing countries," *Journal of Developing Economies* 1, no. 1 (1962): 3–25.

140. Akamatsu "A historical."

141. Jeffrey D. Sachs, "Some brief reflections on digital technologies and economic development," *Ethics & International Affairs, Roundtable: AI and the future of Global Affairs* 33 (2019): 159–167.

of the poverty trap have been reported, including the evolving data-labeling industry.

However, the rural population lacks opportunity to benefit from the 4R. Moreover, due to the lack of regulatory oversights, some technologies companies have been using 4R technologies in a way that has forced poor people to fall deeper into the poverty trap. The millions of Kenyans who have been blacklisted by the country's Credit Reference Bureaus indicates that low-income people have been pushed into deeper poverty by lenders that claim to use AI and big data to assess creditworthiness. To prevent the poverty traps, regulations are needed to prevent companies from engaging in exploitative behavior and to promote productive entrepreneurial activities.

Whether, and the extent to which, poor people can escape the poverty trap is a function of how stocks of assets and resulting flows of income evolve. Among the most relevant is that 4R technologies have tremendous potential to revolutionize the agriculture and livestock sector. For instance, these technologies can help provide affordable agricultural and health insurance products to poor households to help them increase assets and income. In this way, the 4R can improve poor people's welfare and prevent vulnerable groups from falling into poverty.

To analyze the context and mechanisms by which 4R technologies may help economies avoid the middle-income trap, this chapter looked at the situation facing China. China has encountered obstacles in meeting the conditions required to avoid the trap due primarily to the lack of appropriate institutional support. China's case also illustrates how different institutional conditions may favor the development of different technologies. While some middle-income countries can use 4R technologies such as AI to leapfrog in some areas, they may still fall in the middle-income trap.

This chapter has illustrated this dynamic by comparing the economic growth rates of Kenya and Angola in the past few years. The integration of digital technologies in economic activities could be crucial as enablers for the economic performance of LMICs and may help explain why some economies grow much faster than other economies.

CHAPTER 11

Political, Social, and Ethical Implications

4R technologies have attracted growing political, social, and ethical attention. First, the importance of policy frameworks related to 4R technologies in economic and social development is being recognized. In 2019, a specialized technical committee on communication and information technologies of the African Union (AU), held in Sharm El Sheikh, Egypt, expressed the view that Africa's digital transformation depends on political commitment, appropriate policies and regulations, and the scaling up of investment and dedication of resources. The 2019 Sharm El Sheikh Declaration also emphasized the importance of the harmonization of legal and regulatory frameworks to create a single common digital market.[1]

A related point is that LMICs lag in terms of adequate legislation and enforcement to ensure responsible uses of 4R technologies such as AI and big data. In this regard, it is worth noting that in Europe, 96% of countries had data protection and privacy legislation in place as of October 2021 compared to 52% in Africa.[2] Of the world's 47 LDCs, only 21 (45%) had such legislation. LMICs that lack data protection laws are especially vulnerable to predatory and unethical practices of technology companies.

A related point is that some high-income countries are developing regulations to ensure that the uses of AI systems cause no harm to consumers. For instance, the European Commission released its proposed AI regulation on April 21, 2021.[3] The legislation included restrictions on facial recognition

1. W. Gravett, "Digital neo-colonialism: The Chinese model of internet sovereignty in Africa," *African Human Rights Law Journal* 20, no. 1 (2020).

2. UNCTAD, *Data protection and privacy legislation worldwide*, 2021, https://unctad.org/page/data-protection-and-privacy-legislation-worldwide.

3. European Commission, *Proposal for a regulation on a European approach for artificial*

technologies.[4] Specifically, it prohibits the use of AI systems that cause or may cause "physical or psychological" harm through the use of "subliminal techniques" or by exploiting vulnerabilities of a "specific group of persons due to their age, physical or mental disability." It also bans AI systems from providing social scoring for use by public authorities, and it precludes the use of "real-time" remote biometric identification systems, such as facial recognition, in publicly accessible spaces for law enforcement purposes.[5] Most LMICs currently lack such initiatives.

Second, there are a number of ethical issues associated with the 4R. A key ethical concern posed by the 4R, for instance, is related to collecting, storing, sharing, and accessing personal data. Technology startups desperately need more data. When companies gather an enormous amount of data, among the most pressing ethical issue is protecting people's privacy and empowering marginalized communities instead of exploiting them.[6] The hunger for data compounded by underdeveloped regulations and the lack of understanding of existing regulations to access personal information can lead to unethical data use.

It has been recognized that there is no easy short cut to developing and implementing ethical behavior in the AI industry in LMICs. At the First African Region Data Protection and Privacy International Conference organized by the UN Global Pulse, the Ministry of Communications for Ghana, and the Data Protection Commission for Ghana, participants noted that responsible and ethical AI practices require more than just laws. The key ingredients for such practices include increasing investment in AI education, educating technology-savvy lawyers and policymakers, and promoting gender equality in AI education.[7]

---

intelligence: Shaping Europe's digital future, August 24, 2022, https://digital-strategy.ec.europa.eu/en/library/proposal-regulation-european-approach-artificial-intelligence.

4. Madhumita Murgia, "Europe attempts to take leading role in regulating uses of AI," *Financial Times*, 2021, https://www.ft.com/content/360faa3e-4110-4f38-b618-dd695deece90.

5. Mark MacCarthy and Kenneth Propp, "Machines learn that Brussels writes the rules: The EU's new AI regulation," *Brookings*, May 4, 2021, https://www.brookings.edu/blog/techtank/2021/05/04/machines-learn-that-brussels-writes-the-rules-the-eus-new-ai-regulation/.

6. Karen Hao, "In 2020, let's stop AI ethics-washing and actually do something," *MIT Technology Review*, December 27, 2019, https://www.technologyreview.com/2019/12/27/57/ai-ethics-washing-time-to-act/.

7. Mila Romanoff and P. Hidalgo-Sanchis, "Building ethical AI approaches in the African context," *UN Global Pulse*, August 28, 2019, https://www.unglobalpulse.org/2019/08/ethical-ai-approaches-in-the-african-context/.

Third, while many positive impacts of 4R technologies have been discussed in this book, these technologies also have important negative social impacts. For instance, issues such as AI bias affect some groups more than others. Some also argue that it is "overly simplistic" to describe this phenomenon as algorithm biased. They may be working as intended and predictions they make are accurate. The biases in AI could thus be the result of broader biases that exist in society.[8]

This chapter looks at the current stage of policies and regulations related to 4R technologies from the B4B populations' perspective. Social and ethical implications are also discussed.

National Policies, Regulations, and Enforcement

Broad Policy and Institutional Processes

Some countries have turned their attention to broad policy and institutional processes to benefit from the 4R. African economies such as Kenya, Tunisia, South Africa, Ghana, and Uganda are working to develop data protection and ethics strategies as a component of broader AI strategies and policies.[9] Many Latin American economies are also reported to be developing national AI plans[10]

In some economies, institutional capacity is being strengthened to benefit from the 4R. Rwanda's Centre for the Fourth Industrial Revolution brings together key stakeholders such as government agencies, the private sector, civil society, and academia. These actors work together to design, test, and refine policy frameworks and governance protocols to maximize the benefits of 4R technologies.[11] The center's main focus areas are AI and data policy.

---

8. Will Knight, "AI programs are learning to exclude some African-American voices," *MIT Technology Review*, August 16, 2017, https://www.technologyreview.com/s/608619/ai-programs-are-learning-to-exclude-some-african-american-voices/.

9. Romanoff and Hidalgo-Sanchis, "Building ethical AI approaches."

10. Angelica Mari, "Artificial intelligence gathers pace in Latin America," *ZD Net*, June 11, 2020, https://www.zdnet.com/article/artificial-intelligence-gathers-pace-in-latin-america/.

11. Julius Bizimungu, "Inside Rwanda's Centre for Fourth Industrial Revolution," *New Times*, November 25, 2020, https://www.newtimes.co.rw/news/inside-rwandas-centre-fourth-industrial-revolution.

Regulations Enabling the Use of 4R Technologies

Some LMICs have created regulations enabling the use of 4R technologies to bring economic and social transformation. For instance, in September 2019, a crop-spraying drone passed the South African Civil Aviation Authority (SACAA) regulations. Drones can spray in lands with challenging terrain that traditional aircraft cannot reach. They can also get closer to the crops, one to three meters above them, and programmed to follow mapped routes via GPS.[12]

Similarly, Zipline worked closely with policymakers in Ghana and Rwanda. Zipline reported that policymakers in the two countries were more willing to accommodate delivery drones in their airspace than the U.S., the EU, and other nations.[13]

The Center for the Fourth Industrial Revolution in San Francisco worked together with the Rwandan government to develop a regulatory framework for drone operations.[14] In 2016, the government of Rwanda approved regulations on drones, which made commercial drone delivery services for medical supplies possible.[15]

The government of Rwanda's policy frameworks around the use of drones is viewed as a model for other countries that want to use this technology. It is argued that developing countries, whose regulatory environments tend to be less robust than their developed country counterparts, can also be ideal places to test emerging technologies uses.[16]

---

12. "South Africa's first crop-spraying drones can legally fly from today—saving farmers millions," *Farming Portal*, October 1, 2019, http://www.farmingportal.co.za/index.php/agri-index/74-tegnology/3155-south-africa-s-first-crop-spraying-drones-can-legally-fly-from-today-saving-farmers-millions.

13. Faine Greenwood, "Assessing the impact of drones in the global COVID response," *Brookings*, July 30, 2021, https://www.brookings.edu/techstream/assessing-the-impact-of-drones-in-the-global-covid-response.

14. Shusuke Murai, "New Tokyo research center aims to boost Japan's "Fourth Industrial Revolution," *Japan Times*, July 5, 2018, https://www.japantimes.co.jp/news/2018/07/05/business/tech/new-tokyo-research-center-aims-boost-japans-fourth-industrial-revolution/#.XaARbOMzaUk.

15. Amar Toor, "This startup is using drones to deliver medicine in Rwanda," *Verge*, April 5, 2016, https://www.theverge.com/2016/4/5/11367274/zipline-drone-delivery-rwanda-medicine-blood.

16. Catherine Cheney, "Rwanda could become a model for drone regulation," *Devex*, January 23, 2018, https://www.devex.com/news/rwanda-could-become-a-model-for-drone-regulation-91868.

Legislation on Responsible Use of 4R Technologies

Legislative measures are being put in place to control irresponsible and unethical business practices involving 4R technologies. One area has been the P2P lending industry. Digital P2P lenders allegedly charge high interest rates.[17] P2P lenders also harvest data from phones and used the data to embarrass debtors by calling family members.[18] The negative consequences of these practices disproportionately affect low-income people living in poverty.

Kenya is reported to have dozens of mobile phone microlenders such as the Silicon Valley–backed Tala that are not covered by the existing regulatory laws.[19] In August 2021, a parliamentary committee in Kenya backed a law to regulate digital lenders.[20] Likewise, Indonesia's Financial Services Authority (OJK) introduced draft proposals in 2020 to toughen existing rules by increasing paid-up capital requirements and mandating more frequent board meetings.[21]

Transforming the Government Sector

The 4R technologies can help governments in LMICs make intelligent decisions to benefit the poor and vulnerable groups. As discussed in chapter 6, remote-sensing technologies helped deliver vaccines, medical supplies, and equipment to remote areas in Ghana and Rwanda. The ultra-cold-chain technology needs a drop site of about two standard parking spaces. Big data can be used to improve existing services. For instance, in the education sector, data from devices, exams, and other sources can help design innovative ways to monitor student performance and improve teaching practices. Such tools can also be used in crime fighting and law enforcement. In Colombia's capital Bogotá, researchers are using big data to understand how crimes are related to public infrastructure such as bus stations, hospitals, schools, and drugstores.[22] Technologies such as blockchain can also help fight corruption in public procurement (In Focus 11.1)

---

17. M. Ducan and M. Angus, "Kenyan parliamentary panel backs central bank to regulate digital lenders," *Reuters*, August 9, 2021, https://www.reuters.com/article/kenya-cenbank-idUSL8N2PG1TK.

18. M. Ruehl, "Asian authorities clamp down on digital lenders," *Financial Times*, 2021, https://www.ft.com/content/b72c33a4-b6af-4a8d-8475-256fb7075546.

19. Ducan and Angus, "Kenyan parliamentary."

20. Ducan and Angus, "Kenyan parliamentary."

21. Ruehl, "Asian authorities."

22. World Bank, *Big data in action for government: Big data innovation in public services, policy and engagement*, 2017, https://documents1.worldbank.org/curated/en/176511491287380986/pdf/114011-BRI-3-4-2017-11-49-44-WGSBigDataGovernmentFinal.pdf.

In Focus 11.1: Blockchain in Colombia's School Meal Procurement

In Colombia, many corrupt practices have been reported in school meal procurement, and various scandals have arisen in recent years. The Colombian newspaper *El Tiempo* reported that that chicken breasts were sold to schools prices four times higher than local supermarkets.[23] In some cases, purchased goods are not delivered. The former mayor of the Caribbean port city Cartagena was charged for illegally contracting a deal of over COL$23 million (about US$7,000). Of 2.6 million purchased loaves of bread, 1 million were never delivered to schools. Public figures and officials and a small number of food contractors have been involved in procurement fraud.[24]

In an attempt to address corrupt practices in public procurement, the World Economic Forum teamed up with the Inter-American Development Bank (IDB) and the Office of the Inspector General of Colombia (Procuraduría General de Colombia) to investigate, design, and trial the use of blockchain for public procurement activities. A software proof-of-concept focused on a school meals program (Programa de Alimentación Escolar) for young people from low-income households was developed. A public blockchain procurement system was used to track the process of supplier selection in this program in the city of Medellín. Ethereum blockchain was used.[25]

The WEF's initial blockchain project in Colombia focused on contractor selection. The goal is to improve transparency, fairness, and competitiveness in a bidding process. A tenderer publicly commits to contract terms and selection criteria prior to eliciting bids. In this way, risks of favoring specific contractors are eliminated. For competing vendors, a blockchain-based solution's permanent and tamper-proof bid records can ensure that a firm cannot alter submitted bids after learning new information about competing bids. An additional benefit is that by increasing the

23. "Una pechuga de pollo a $ 40.000 y huevo a $900, en sobrecostos del PAE," *El Tiempo*, 2017, https://www.eltiempo.com/justicia/investigacion/sobrecostos-en-programa-de-alimentacion-escolar-en-colombia-153590.

24. M. Sarralde Duque, "En un año se robaron 32,8 millones de raciones de comida del PAE," *El Tiempo*, 2017, https://www.eltiempo.com/justicia/investigacion/desfalco-de-32-8-millones-en-raciones-de-comida-escolar-de-los-ninos-153928.

25. I. Hall, "Colombian blockchain trial cause for 'cautious optimism,' says WEF," *Global Government Forum*, 2020, https://www.globalgovernmentforum.com/colombian-blockchain-trial-cause-for-cautious-optimism-says-wef/.

perception of fairness, blockchain-led transparency can attract more vendors to the procurement process. A set of clearly defined selection criteria would increase the possibility that an outsider can win. During the auction and vendor evaluation processes, actions and decisions are automatically recorded. These records are permanent and publicly viewable, which increases auditability. It is also possible to include a user interface, which can be used by the public to monitor actions and decisions so that risks can be flagged in real time. These enable monitoring authorities such as the inspector general to investigate potential corrupt activity even before an auction concludes.

Blockchain-based solutions can also be used to monitor the chosen contractor's performance. For instance, information regarding actual deliveries can be made available to key stakeholders such as parents, teachers, enforcement officials, and the press. Their participants can be used to report meal deliveries and quality in real time. By improving observation in the delivery process and allowing stakeholders to monitor and engage, contractor accountability can be improved.[26] By allowing the participation of diverse groups, such systems can promote informal accountability, which, as prior research has shown,[27] can improve public procurement.

Efforts of Multilateral Organizations and Western Governments

Multilateral organizations, such as the African Development Bank and Western governments, have also provided support to undertake measures to strengthen national policies and strategies for utilizing 4R technologies and improving regulatory systems and enforcement measures. In March 2021, the African Development Bank approved a grant of US$1.024 million for AI to develop systems to process customer complaints on behalf of the national banks of Ghana and Rwanda and also Zambia's Competition and Consumer Protection Commission. The project aims to develop multilingual chatbots and artificial intelligence for financial regulators to handle complaints. It will incorporate key local languages for ease of use,

26. C. Barrera, S. Hurder, and A. Lannquist, "Here's how blockchain could stop corrupt officials from stealing school lunches," *World Economic Forum*, 2019, https://www.weforum.org/agenda/2019/05/heres-how-blockchain-stopped-corrupt-officials-stealing-school-dinners/.

27. B. Romzek, K. LeRoux, and J. Blackmar, "A preliminary theory of informal accountability among network organizational actors," *Public Administration Review* 72, no. 3 (2012): 442–453.

record customer complaints, and track their resolution. Audio complaints will be used for those unable to read and write. The solution's deployment is planned in Kinyarwanda, Swahili, French, and English in Rwanda; English and Nyanja/Chewa in Zambia, and English and Twi in Ghana.[28] If successfully developed and deployed, the solution is likely to benefit linguistic minorities in these countries that are among the poorest groups.

International cooperation in law enforcement has also accelerated. The U.S. law enforcement authorities trained over 50 Nigerian investigators and prosecutors to handle cryptocurrency-enabled organized crimes. In August 2021, the International Computer Hacking & IP Attorney Advisers (ICHIP) in Addis Ababa, Ethiopia, and Abuja, Nigeria, and the Federal Bureau of Investigations (FBI) had organized a webinar focusing on cryptocurrency basics for investigators and prosecutors in Nigeria.[29]

Areas Needing Regulatory Attention

While recent developments on the policy and regulatory fronts are encouraging, some areas need regulatory attention. This section focuses on two key areas in which such attention is particularly needed.

Data Protection and Data Ownership Laws

Many people hold ill-informed or misguided views on the importance of data privacy for low-income people. Some argue that only wealthier and more privileged people need to worry about their privacy and have the ability to afford privacy- and security-enhancing measures such as encryption. It is argued that the digital divide has extended to become a "privacy divide."[30] For instance, one commentator noted that about 90% of the discussion at the 2013 internet Governance Forum (IGF) referred to big data as a surveillance tool. At the same time, the debate focusing on developing countries treated big data as a means to "observe" people to fight poverty. The argument provided by IGF participants was that

28. Godfrey Ivudria, "$1 million for financial inclusion-African Development Bank," *EABW News*, March 10, 2021, https://www.busiweek.com/afdb-provides-1million-for-ai-ghana-rwanda-and-zambia/.

29. Akon, "Akon: Why crypto could transform Africa's future," *Fortune*, April 14, 2021, https://fortune.com/2021/04/14/akon-city-senegal-africa-akoin-cryptocurrency.

30. Jason Koebler, "Why mass surveillance is worse for poor people," *Motherboard*, February 24, 2015, http://motherboard.vice.com/read/why-mass-surveillance-is-worse-for-poor-people.

data can help provide access to clean drinking water, healthcare, and other necessities.[31]

It might be argued that poor people need more—not less—privacy for their data. In countries where interethnic or tribal tensions and violence exist or where people live in the aftermath of a civil war, privacy is a genuine concern. Moreover, in countries characterized by conflict, crisis, and weak law enforcement, privacy breaches may lead to physical security risk.[32]

Of special concern is the data of 500 million smallholder farms worldwide, which employ more than 2 billion people. Such farms produce about 80% of food products consumed in Asia and SSA economies.[33] Farmers' personal data, such as demographic characteristics and household size, and crop and livestock farming data, such as land size, land ownership, average productivity and production value, number and value of inputs (e.g., seeds, fertilizers), livestock holding, and irrigation capability are highly valuable to businesses involved in the farm supply chain and financial institutions. Unsurprisingly, these data are highly sought after by the providers or technological solutions to farmers. For instance, the EcoProMIS project in Colombia (chapter 7) involves selling farming data to external partners, such as the service industry, government agencies, NGOs, and food processors.[34] Likewise, the Nigerian precision farming technology company Zenvus encourages farmers to share their farm data with banks to get loans. However, adequate protection of farmers' data is far from guaranteed.

To provide further insights into farmers' perceptions and responses to possible misuse of their farming data by various actors, it is important to contrast the situations in LMICs with those in high-income countries. In developed countries such as the U.S., farmers' awareness and activism is pushing the privacy agenda. Some farmers are concerned that big agricultural firms such as Monsanto might influence them to buy specific seeds, sprays, and equipment and be likely to profit from the costs of their services

31. Linnet Taylor, *Surveil the rich, observe the poor: Big data at the Internet Governance Forum 2013*, October 25, 2013, http://linnettaylor.wordpress.com/2013/10/25/surveil-the-rich-observe-the-poor-big-data-at-the-internet-governance-forum-2013.

32. E. Letouzé, "Big data for development: What may determine success or failure?" *OECD Technology Foresight*, Paris, 2012, https://www.oecd.org/sti/ieconomy/Session_5_Letouz%C3%A9.pdf.

33. International Fund for Agricultural Development, *Enabling poor rural people to overcome poverty: Smallholders can feed the world*, 2011, https://www.ifad.org/documents/38714170/40706188/Smallholders+can+feed+the+world_e.pdf./460ca6c2-7621-40d8-9f79-a56f6f8fa75e.

34. Agri-Innovation Den, *Meet our 2019 Agri-innovation Den Finalists*, 2019, https://www.aginnovationden.com/roelof-kramer-meet-our-2019-agri-innovation-den-finalists/.

and higher seed sales.[35] Another key concern that farmers have expressed is that their data and information could be used by competitors. For example, other farmers' access to crop-yield information may create direct and unwanted competition to rent farmland, which may cause a new spike in land values and seed prices.[36] The issue regarding who owns farmers' crop data is of equal concern.[37]

Another fear is that Wall Street traders could use the data to make bets that hurt farmers. For instance, if conditions early in the growing season lead to lower futures contract prices, it may reduce the profits farmers could have made from crops.[38] Likewise, farmers are concerned that hedge funds or big companies might use real-time data at harvest time from many combines to speculate in commodities markets long before official crop-production estimates are available.[39] Such fears indeed rely on facts, as technology makes it possible to do so. For instance, a group at the MIT Media Lab used location data from mobile phones to estimate the number of people in Macy's parking lots on Black Friday. The model they developed made it possible to estimate the retailer's sales on that day even before Macy's had recorded sales. Insights like this are expected to provide competitive advantage to Wall Street analysts and managers.[40]

U.S. farmers also have raised additional concerns. For instance, if agriculture-related companies misuse or accidentally release their information, government auditors may scrutinize their finances or environmental activists may protest fertilizer and pesticide use.[41] However, such concerns have not yet been substantiated in LMICs.

Data security breaches in the U.S. have been eye-opening events for U.S.

---

35. T. J. Seppala, "Monsanto pushes big data-driven planting but farmers are skeptical," *Engadget*, February 26, 2014, https://www.engadget.com/2014-02-26-monsanto-prescriptive-farming.html.

36. Jacob Bunge, "Big data comes to the farm, sowing mistrust," *Wall Street Journal*, February 25, 2014, https://www.wsj.com/articles/SB10001424052702304450904579369283869192124.

37. Seppala, "Monsanto pushes."

38. Bunge, "Big data."

39. "American farmers confront big data revolution," *Fox News*, December 2, 2015, http://www.foxnews.com/us/2014/03/29/american-farmers-confront-big-data-revolution/.

40. Andrew McAfee and Erik Brynjolfsson, "Big data: The management revolution," *Harvard Business Review*, October 2012, http://hbr.org/2012/10/big-data-the-management-revolution/ar/1.

41. Shruti Singh and Jack Kaskey, "Farmers press agribusiness giants for data security," January 23, 2014, *Bloomberg*, https://www.bloomberg.com/news/articles/2014-01-23/farmers-press-agribusiness-giants-for-data-security.

farmers. In 2013, the Environmental Protection Agency, which gathers farm and livestock information to monitor air and water quality, inadvertently released to environmental groups personal information of 80,000 farmers.[42] Such events have provided the U.S. farmers with an environment to learn and become more familiar with data security and privacy issues. Such mechanisms are yet to evolve in many developing countries. Farmers in LMICs, who are often poor and less educated, often lack awareness and understanding of potential privacy and security risks.

Another challenge is related to data ownership. First, it is worth noting that in some advanced countries, insurers provide a high degree of privacy. For instance, in June 2019, the insurance firm Wakam (formerly La Parisienne Assurances) and Sigfox France announced a first usage-based insurance solution using connected objects, which focused initially on the mobility market (e.g., cars, motorcycles, electric scooters). It minimized data collection for General Data Protection Regulation (GDPR) compliance.[43]

Data misuse by service providers, third-party intermediaries, and other players is a concern due to underdeveloped data privacy regulations in LMICs. For instance, data created and made available through IoT makes it possible for insurers to better understand risk. The question is, Does the data belong to the insurance company or the customer? Customers may argue they have rights over their personal data. They might need access to historical data to switch insurers at renewal.[44] However, insurers may not give such data to consumers.

Regulations of Cryptocurrency Transactions

There has been a rapid rise in cryptocurrency transactions to facilitate business activities in LMICs. For instance, cryptocurrency transactions to and from Africa are primarily related to remittances and other legitimate business dealings.[45] As discussed in chapter 3, crypto-denominated international commerce has become increasingly common in LMICs.

---

42. Singh and Kaskey, "Farmers press."

43. Jorge Rudes, "La Parisienne Assurances automates contract management via the Quorum blockchain," *BitcoinDynamic*, April 6, 2020, https://bitcoindynamic.com/news/la-parisienne-assurances-automates-contract-management-via-the-quorum-blockchain/.

44. SAS, *5 Challenges for IoT in the insurance industry*, 2019, https://www.sas.com/en_us/insights/articles/big-data/5-challenges-for-iot-in-insurance-industry.html.

45. Richard Chelin, "Africa: new playground for crypto scams and money laundering—analysis," *Eurasia Review*, August 11, 2021, https://www.eurasiareview.com/11082021-africa-new-playground-for-crypto-scams-and-money-laundering-analysis/.

Some have developed innovative approaches based on blockchain that help employees receive faster payment from their employers. The Kenya-based blockchain startup BitPesa is helping speed the flow of cash from businesses in China to their African employees, who send money to their families. Launched in 2013, BitPesa uses bitcoin to facilitate low-cost instant payments.[46] As of the end of 2016, BitPesa operated in Kenya, Nigeria, Uganda, and Tanzania.[47]

It is important for governments in LMICs to ensure that crypto service providers have sufficient mechanisms in place to tackle money-laundering risks. Crypto businesses also should work with law enforcement to flag suspicious activities and help in the investigation.[48]

Digital Colonialism: Political and Social Consequences

It is argued that the operations of developed world technology companies competing for economic advantage in the cyberspace is akin to European colonial powers' competition over claiming the resources and control over colonies.[49] In this regard, it is helpful to begin by pointing to important political and social consequences of colonialism. For instance, colonialism involves political domination by colonialists, which leads to a significant reduction of the level of political sovereignty of the colonized. Likewise, "colonially-induced social transformation through immigration, proselytization and partition" has been reported in SSA economies.[50]

These long-standing concerns have been exacerbated by the activities of Chinese and Western technology firms in Africa and other LMICs. This phenomenon is often referred to as *digital colonialism*, defined as the "use of digital technology for political, economic and social domination of another nation or territory."[51] A primary goal of digital colonialism is data extraction.

46. Malaka Gharib, "Blockchain could be a force for good. but first you have to understand it," *NPR*, January 11, 2017, http://www.npr.org/sections/goatsandsoda/2017/01/11/503159694/blockchain-could-be-a-force-for-good-but-first-you-have-to-understand-it.

47. Zama Dyani, "Op ed: Africa needs more bitcoin and blockchain education," *Bitcoin Magazine*, December 20, 2016, https://bitcoinmagazine.com/articles/op-ed-africa-needs-more-bitcoin-and-blockchain-education-1482249305/.

48. Chelin, "Africa."

49. H. McClure, "Winter, the wild, wild web: The mythic American west and the electronic frontier," *Western Historical Quarterly* 31 (2000): 457–476.

50. Patrick Ziltener and Daniel Künzler, "Impacts of colonialism: A research survey," *Journal of World-Systems Research* 19, no. 2 (2013): 290–311.

51. Michael Kwet, "Long read digital colonialism," *New Frame*, March 8, 2021, https://www.newframe.com/long-read-digital-colonialism/.

To take an example, Berlin-based crypto startup Worldcoin uses metallic orbs to scan people's irises and collect other biometric data on faces and bodies. Worldcoin recruits country-level operators to collect biometric data. The operators receive commission in the stablecoin Tether for each person's biometric data that they collect. It conducted field testing mostly in LMICs such as Indonesia, Sudan, and Kenya. A civil engineer in Kenya's Nakuru city reported that he recruited between 150 and 200 people, at KES 50 (US$0.44) per scan. Likewise, in Indonesia's West Java province, an elementary school madrasa was used as a Worldcoin registration site to scan irises and other biometric data. The principal of the school was paid 2,000 Indonesian rupiah (IDR) ($0.14 US) for each person successfully scanned.[52]

In recent years, concerns have been mainly directed at Chinese and U.S. technology companies. In Africa, for instance, these technology companies frame their digital infrastructure projects as a humanitarian endeavor. Their ulterior motive, however, is arguably to extract data.[53] Their motivations are clear: the bigger and more diverse the dataset is, the more valuable the AI algorithm will be.

The Chinese Approach

Chinese AI corporations are leveraging global infrastructure projects initiated by the government, such as the Belt and Road Initiative, to secure contracts with African governments. In these deals, AI and other technology solutions are provided for free or at a low cost in exchange for access to local citizens' data. Chinese AI firms are relatively transparent about their goals. There is virtually no check or accountability as to how the information extracted from Africa will be used.[54]

To take an example, the Chinese company CloudWalk works with the Zimbabwean government to develop a facial recognition program. China described this as a "win-win" deal. Chinese AI companies can train ML algorithms on Africans to diversify their datasets, and Zimbabwe gets access to use the latest technology to monitor its population.[55]

---

52. Eileen Guo and Adi Renaldi, "Deception, exploited workers, and free cash: How Worldcoin recruited its first half a million test users," *MIT Technology Review*, April 6, 2022, https://www.technologyreview.com/2022/04/06/1048981/worldcoin-cryptocurrency-biometrics-web3/.

53. Nima Elmi, "Is big tech setting Africa back?," *Foreign Policy*, November 11, 2020, https://foreignpolicy.com/2020/11/11/is-big-tech-setting-africa-back/.

54. Elmi, "Is big tech setting."

55. "Exporting repression? China's artificial intelligence push into Africa," *Council on*

Regarding the motivation of Chinese technology companies' operations in Africa, the mass surveillance program implemented in Xinjiang has allowed Chinese companies to achieve some diversity in the data pool since the population of the province is mainly composed of Muslim Uyghurs, who are of Turkic descent.[56] But the datasets are still far from diverse enough. The CloudWalk deal in Zimbabwe provided access to a data pool with a much higher degree of diversity than China's ethnic composition can offer. It is the first Chinese AI project in Africa.

In terms of the upfront payment, it is possible that the Zimbabwean government paid for the technology and tools at a lower cost than it would have on the open market. However, CloudWalk received much more than monetary payment: data on the Zimbabwean people, which will provide the company an opportunity to improve its facial recognition systems. The Zimbabwean government failed to consider—or ignored—that high-dimensional biometric data such as faces are highly valuable. Natasha Msonza, cofounder of the Digital Society of Zimbabwe, noted: "It feels like [CloudWalk] is looking for guinea pigs. I don't believe that the Zimbabwe government gave this proposition much thought before volunteering its citizens to be subjected to racial facial recognition experiments."[57] Eric Olander, founder of the China Africa Project, noted that many Western companies "aren't willing to make that step that the Chinese are willing to do. . . . [The Chinese] are willing to make an investment in a market as volatile as Zimbabwe, where companies from other countries are not."[58]

In such a situation, what Gregory Unruh refers to as the techno-institutional complex (TIC) arises.[59] The idea is that radical systems such as those facilitated by the 4R should be viewed as a complex system of technologies embedded in a conditioning social context of public and private institutions. They should not be seen as a set of discrete technological artifacts.[60]

Foreign Relations, December 17, 2018, https://www.cfr.org/blog/exporting-repression-chinas-artificial-intelligence-push-africa.

56. Amy Hawkins, "Beijing's big brother tech needs African faces. Zimbabwe is signing up for China's surveillance state, but its citizens will pay the price," *Foreign Policy*, July 24, 2018, https://foreignpolicy.com/2018/07/24/beijings-big-brother-tech-needs-african-faces/.

57. Hawkins, "Beijing's big brother."

58. Hawkins, "Beijing's big brother."

59. Gregory C. Unruh, "Understanding carbon lock-in," *Energy Policy* 28 (2000): 817–830.

60. Totti Konnola, Gregory C. Unruh, and Javier Carrillo-Hermosilla, "Prospective voluntary agreements for escaping techno-institutional lock-in," *Ecological Economics* 57, no. 2 (2006): 239–252.

The development of TIC takes place through a path-dependent, coevolutionary process. Such a process entails positive feedback among technological infrastructure and the organizations and institutions (e.g., regulatory agencies, and domestic and foreign creators) involved in creating and using them. Once countries and organizations are locked in, it becomes difficult to displace the existing system, which can lock out alternative technologies for a long period.[61]

To put things in context, in the long run, existing data projects increase Africa's dependence on China (and the West) and their technologies for policy formulation. For example, as Chinese firms accumulate data related to the movement of people and vehicles, African policymakers are more likely to rely on this private market information to develop solutions to infrastructure and transport problems.[62] African researchers referred to this as a form of digital neocolonization, similar to the exploitation of Africa by Western countries. China's "digital neocolonialism" in Africa arguably consists of three principal elements: promoting the Chinese model of internet sovereignty in African nations, exporting surveillance technologies that support authoritarian regimes, and deploying AI and data-mining techniques.

The Western (U.S.) Approach

Western corporations' approach is arguably more covert and embedded in seeming corporate social responsibility (CSR) activities. U.S. technology companies, for instance, are establishing digital hubs and providing free internet access. Some examples include Google's Internet Balloons and Facebook's Undersea Cable. It is argued that these initiatives "encourage more Africans to get online, use their services, and, by convenient coincidence, relinquish their data in the process."[63]

Some argue that through the domination of digital technology, the U.S. has reinvented colonialism in the Global South.[64] Commenting on U.S. technology companies' educational technologies in Brazil, Giselle Ferreira and her coauthors state that "when GAFA [Google, Amazon, Facebook, Apple] companies generously offer technologies to disadvantaged students,

61. Konnola et al., "Prospective voluntary."

62. Khwezi Nkwanyana, "China's AI deployment in Africa poses risks to security and sovereignty," *Strategist*, May 5, 2021, https://www.aspistrategist.org.au/chinas-ai-deployment-in-africa-poses-risks-to-security-and-sovereignty/.

63. Elmi, "Is big tech setting."

64. M. Kwet, "Digital colonialism: US empire and the new imperialism in the Global South," *Race & Class* 60, no. 4 (2019): 3–26.

data is unimpededly extracted and subsequently treated in a manner that renders local specificities devoid of importance."[65]

To illustrate this, it is helpful to consider Google's approach. It is important to stress that Google's facial recognition project was alleged to be engaged in deceptive tactics targeting racial and ethnic minorities. Some of the contract workers, referred to by the company as temporary, vendor, and contractor (TVCs), which make up more than half of its total staff, were assigned to collect biometric data from minorities. They offered US$5 gift cards as incentive to collect face scans from minorities, which included homeless people in Atlanta, students in various universities, and attendees at the BET Awards event in Los Angeles.[66]

Tech companies such as Google lack access to data of diverse populations to realize the full benefits of AI-based facial recognition technology. One way to tackle this problem is to use the populations, such as those in Africa for data extraction. In 2019, Google opened an AI lab in Accra, Ghana.[67]

Some critics have questioned the motivation of Google's parent firm, Alphabet's Project Loon, which utilized a fleet of balloons to beam high-speed internet to some remote parts of the world. The project was shut down in early 2021. The defense company Raven Aerostar had built Google's balloons and flight-control systems. Raven Aerostar also manufactures balloons for the U.S. government's Columbia Scientific Balloon Facility, run by NASA. Raven Aerostar stated that its super-pressure balloons' "mission possibilities" include "scientific data collection, remote communications, GPS augmentation, intelligence gathering, persistent surveillance, reconnaissance, radar calibration, satellite simulation, incremental testing, and research and development of sensor."[68] The editor and wireless technology analyst Brad Reed at "Boy Genius Report" was quoted as saying: "If Google's claims about the Loon balloons' navigability are true, it is in fact an 'unmanned aircraft,' sometimes more pejoratively referred to as a drone. And what's worrisome is not so much Google's stated goal, but that with unprecedented proprietary technology,

65. G. M. dos S. Ferreira, L. A. da S. Rosado, M. S. Lemgruber, and J. de S. Carvalho, "Metaphors we're colonised by? The case of data-driven educational technologies in Brazil," *Learning, Media and Technology* 45, no. 1 (2019): 46–60.

66. Ginger Adams Otis and Nancy Dillon, "Google using dubious tactics to target people with 'darker skin' in facial recognition project: sources," *Daily News*, October 2, 2019, https://www.nydailynews.com/news/national/ny-google-darker-skin-tones-facial-recognition-pixel-20191002-5vxpgowknffnvbmy5eg7epsf34-story.html.

67. Kwasi Gyamfi Asiedu, "Google is throwing its weight behind artificial intelligence for Africa," *Quartz Africa*, June 14, 2018.

68. Will K. Butler, "Can we trust Google with the stratosphere?," *The Atlantic*, August 20, 2013.

scant law on the books, and a few key government connections, Project Loon may only be a harbinger of a new era in our relationship to the skies overhead, one that our laws are dramatically unprepared for."[69] The editor and wireless technology analyst Brad Reed at "Boy Genius Report" was quoted as saying: "It's easy to understand why Google is working to bring mobile broadband connectivity to Africa: The company wants to get an early start on securing its next generation of customers."[70]

Another complaint is that Western technology companies are not providing their services to poor people who need them urgently. For instance, Loon had secured the government of Kenya's approval to launch balloons to provide commercial internet connectivity services. Some critics suggested that the balloons would have been better in another African country because Kenya already has an estimated 39 million out its 48 million people online.[71] Many technology companies want to extract as much valuable data as possible from their operations in LMICs.

U.S. technology companies have also been criticized for providing inferior products and services in LMICs. For instance, Facebook in India allegedly lacks key mechanisms to fight misinformation and hate speech that the company has deployed in the U.S. and other mostly-English-speaking countries for many years. For instance, Facebook had not developed algorithms to detect hate speech in Hindi and Bengali, the fourth and seventh most spoken languages in the world. It was also reported that some nefarious political actors had used multiple Facebook accounts to spread anti-Muslim messages across people's news feeds that clearly violated Facebook's rules.[72]

Government Surveillance and Espionage

Several counties are using 4R technologies to conduct surveillance activities on citizens. For instance, China is reported to have about 630 million

69. "Google's plans to expand Internet access in Africa is about the data," *Deutsche Welle*, June 27, 2013, https://www.dw.com/en/googles-plans-to-expand-internet-access-in-africa-is-about-the-data/a-16903897.

70. "Google's plans."

71. "4G internet balloons take off over Kenya," *BBC News*, July 7, 2020, https://www.bbc.com/news/technology-53321007.

72. Cat Zakrzewski, Gerrit De Vynck, Niha Masih, and Shibani Mahtani, "How Facebook neglected the rest of the world, fueling hate speech and violence in India," *Washington Post*, October 24, 2021, https://www.washingtonpost.com/technology/2021/10/24/india-facebook-misinformation-hate-speech/.

AI-based facial recognition cameras.[73] Chapter 9 discussed the Chinese government's, extensive surveillance mainly targeting the Uyghur community living in Xinjiang.

In the global debate over technology, the proliferation of AI-based surveillance systems and other aspects of exportable techno-authoritarianism have been among the major issues.[74] In recent years, China has exported AI-based surveillance to authoritarian markets and liberal democracies.[75]

Data and information services provider IHS Markit estimated that China accounted for nearly half the global facial recognition business in 2018.[76] Freedom House documented that 18 countries purchased AI surveillance tools from China in 2018:[77] These include Serbia,[78] Turkey, Russia, Ukraine, Azerbaijan, Angola, Laos, Kazakhstan, Kenya, Uganda,[79] Ecuador, Bolivia, and Peru.[80] Others suggest that only Huawei provides AI surveillance technology to at least 50 countries worldwide (In Focus: 11.2). No other company comes close.[81]

---

73. Anna Fifield, "Orwell's nightmare? Facial recognition for animals promises a farmyard revolution," *Washington Post*, August 24, 2020, https://www.washingtonpost.com/world/asia_pacific/facial-recognition-china-animals-farms-agriculture/2020/08/23/9808c710-d6fb-11ea-b9b2-1ea733b97910_story.html.

74. Andres Ortega, *The U.S.-China race and the fate of transatlantic relations, part 1: Tech, values, and competition*, Center for Strategic and International Studies, January 2020, http://csis-website-prod.s3.amazonaws.com/s3fs-public/publication/200113_USChinaTranstlanticRelations.pdf.

75. Steven Feldstein, "The global expansion of AI surveillance," *Carnegie Endowment for International Peace* September 2019, https://zhizhi88.com/wp-content/uploads/2019/10/2019%E5%B9%B4%E5%85%A8%E7%90%83%E4%BA%BA%E5%B7%A5%E6%99%BA%E8%83%BD%E7%9B%91%E6%B5%8B%EF%BC%88AIGS%EF%BC%89%E6%8C%87%E6%95%B0%E6%8A%A5%E5%91%8A.pdf.

76. Y. Yang, "Facial recognition: how China cornered the surveillance market," *Financial Times*, 2019, https://www.ft.com/content/6f1a8f48-1813-11ea-9ee4-11f260415385.

77. T. Mai, S. Adrian, W. Jessica, and F. Allie, *The rise of digital authoritarianism* (Freedom House, 2018), https://freedomhouse.org/sites/default/files/10192018_FINAL_FOTN_2018.pdf.

78. "Huawei's facial recognition technology causes anxiety in Serbia," *South China Morning Post*, October 17, 2019, https://www.scmp.com/news/world/europe/article/3033267/huaweis-facial-recognition-technology-causes-anxiety-serbia.

79. "Chinese facial recognition tech installed in nations vulnerable to abuse," *CBS News*, October 16, 2019, https://www.cbsnews.com/news/china-huawei-face-recognition-cameras-serbia-other-countries-questionable-human-rights-2019-10-16/.

80. Charles Rollet, "Ecuador's all-seeing eye is made in China," *Foreign Policy*, August 9, 2018, https://foreignpolicy.com/2018/08/09/ecuadors-all-seeing-eye-is-made-in-china/.

81. Feldstein, "The global."

In Focus 11.2: Huawei's AI Surveillance

Privacy advocates are concerned about the potential misuse of data collected by AI-based surveillance systems provided by Chinese companies such as Huawei.[82] Deployment of China-developed solutions in countries with poor records on human rights is a concern.

In 2020, Ugandan police used Huawei's facial recognition systems to identify and arrest 836 suspected supporters of opposition leader Bobi Wine.[83] There is a lack of any judicial oversight and concerns of backdoor access to the system for illegal facial recognition surveillance. The system can be used to stifle anti-regime comments and peaceful civil action.[84] Through its Safe City projects, Huawei has supplied surveillance technologies to 16 African countries, often funded by China Exim Bank loans.[85]

The Philippine's Bonifacio Global City has been equipped with mass-surveillance systems developed by Huawei. The cameras are linked to data collection tools such as plate recognition. Huawei works with the police to fight crimes. It gathers evidence and identifies suspects using facial recognition technology.[86]

Nur Sultan, Kazakhstan, had over 2,000 cameras installed by Huawei working with Kazakhtelecom, Kcell, Beeline, and Tele2. The Tajikistan government implemented Huawei's "Safe Cities" system in Dushanbe in 2013. Over 800 cameras were reported to be watching monuments, parks, and other public spaces.[87]

---

82. Ben O'Rourke and Gigi Choy, "Big brother Huawei kitted out this Philippine city: Is China watching?," *South China Morning Post*, January 28, 2019, https://www.scmp.com/week-asia/economics/article/2183540/big-brother-huawei-watches-philippine-city-does-china-too.

83. "Three bodies of last week protests not claimed," *Monitor*, November 23, 2020, https://www.monitor.co.ug/uganda/news/national/three-bodies-of-last-week-protests-not-claimed-3207394.

84. Stephen Kafeero, "Uganda is using Huawei's facial recognition tech to crack down on dissent after protests," *Quartz*, November 27, 2020, https://qz.com/africa/1938976/uganda-uses-chinas-huawei-facial-recognition-to-snare-protesters/.

85. Khwezi Nkwanyana, "China's AI deployment in Africa poses risks to security and sovereignty," *Australian Strategic Policy Institute*, May 5, 2021, https://www.aspistrategist.org.au/chinas-ai-deployment-in-africa-poses-risks-to-security-and-sovereignty/.

86. Nihrika Mandhana, "Huawei's video surveillance business hits snag in Philippines," *Wall Street Journal*, February 20, 2019, https://www.wsj.com/articles/huaweis-video-surveillance-business-hits-snag-in-philippines-11550683135.

87. Yau Tsz Yan, "Smart cities or surveillance? Huawei in Central Asia," *Diplomat*, August 7, 2019, https://thediplomat.com/2019/08/smart-cities-or-surveillance-huawei-in-central-asia/.

In December 2020, Myanmar rolled out the first phase of its "Safe City" initiative using a system of 335 surveillance cameras in eight townships in the capital, Naypyidaw.[88] In Myanmar, the launch of facial and license plate recognition was approved without public consultation or transparency. It is unclear how authorities plan to mitigate the technology's potential impact on human rights and the right to privacy.[89]

In Russia, rights groups say that migrants, particularly from Central Asia, are often subjected to racial profiling, arbitrary detention, and violence. The Russian IP video surveillance software company AxxonSoft's cameras have ethnicity analytics. The company said this feature was included in its product "inadvertently" when integrated with third-party software.[90]

Brazil is also allegedly sliding toward techno-authoritarianism.[91] In October 2019, then President Jair Bolsonaro signed a decree that requires all federal bodies to share their data they hold on Brazilian citizens, which will be consolidated in a centralized database. They include huge amounts of data, such as employment and health records, and biometric information such as face and voiceprint. While the government's stated goal is to use the data to improve public services and fight crimes, due to Bolsonaro's government's authoritarian path, critics have been concerned that the data may be used to spy on political dissidents[92]

88. Thura Myat, "Nay Pyi Taw authorities activate 335 cameras able to detect faces," *Myanmar Times*, December 23, 2020, https://www.mmtimes.com/news/nay-pyi-taw-authorities-activate-335-cameras-able-detect-faces.html.

89. "Myanmar: Facial recognition system threatens rights," *Human Rights Watch*, March 12, 2021, https://www.hrw.org/news/2021/03/12/myanmar-facial-recognition-system-threatens-rights.

90. "'Racist' facial recognition sparks ethical concerns in Russia, analysts say," *Reuters*, July 5, 2021, https://www.voanews.com/europe/racist-facial-recognition-sparks-ethical-concerns-russia-analysts-say.

91. Richard Kemeny, "Brazil is sliding into techno-authoritarianism," *MIT Technology Review*, August 19, 2020, https://www.technologyreview.com/2020/08/19/1007094/brazil-bolsonaro-data-privacy-cadastro-base/.

92. "Podcast: COVID-19 is helping turn Brazil into a surveillance state," *MIT Technology Review*, September 16, 2020, https://www.technologyreview.com/2020/09/16/1008495/podcast-covid-19-brazil-surveillance-state/; Presidência da República Secretaria-Geral Subchefia para Assuntos Jurídicos Decreto Nº 10.046, October 9, 2019, http://www.planalto.gov.br/ccivil_03/_Ato2019-2022/2019/Decreto/D10046.htm.

Ethics of Companies' Questionable Practices

Many companies are carrying out highly questionable practices from an ethical perspective. In this section, I illustrate some such practices.

Global Sweatshops

The emergence of global sweatshop factories of the 4R era has been a concern. One area that is especially important to consider is the data-labeling industry. The so-called ghost workers who do data annotation or labeling are distributed across the world.[93] They use online annotation platforms or work in annotation companies. In some cases, prisoners[94] and other economically vulnerable groups[95] do the labeling in jurisdictions that have limited labor laws. Some data-labeling firms have been accused of paying low wages. In this way, organizations engaged in this industry are allegedly facilitating a "new kind of slavery in the digital era."[96]

Although some data-labeling firms have claimed to produce positive social effects, such claims cannot be easily verified. There are virtually no regulations that govern the working conditions of data labelers. Industry standards related to ethical sourcing are weak. There is also a lack of third-party validation. We cannot thus really take self-reported information provided by these firms as proof that they are creating positive economic and social impacts in the developing world. There is little hard evidence to counter critics' concern that the firms claim that they engage impact sourcing is nothing more than marketing gimmicks or "impact washing."[97]

Another daunting challenge in getting data-labeling firms to engage in ethical practices is that the firms provide their services to businesses so face little public pressure. By contrast, fair-trade goods are directly sold to consumers.[98]

93. M. L. Gray and S. Suri, *Ghost work: How to stop Silicon Valley from building a new global underclass* (Eamon Dolan Books, 2019).

94. Karen Hao, "An AI startup has found a new source of cheap labor for training algorithms: prisoners," *MIT Technology Review*, March 29, 2019, https://www.technologyreview.com/f/613246/an-ai-startup-has-found-a-new-source-of-cheap-labor-for-training-algorithms/.

95. L. Yuan, "How cheap labor drives China's A.I. ambitions," *New York Times*, November 25, 2018.

96. Nicola Croce, "The new assembly lines: Why AI needs low-skilled workers too," *World Economic Forum*, August 12, 2019.

97. Kate Kaye, "These companies claim to provide 'fair-trade' data work. Do they?," *MIT Technology Review*, August 7, 2019.

98. Kaye, "These companies."

Some data-labeling firms, such as CloudFactory, DDD, iMerit, and Samasource, are members of the Global Impact Sourcing Coalition (GISC), founded in 2016. The GISC established an "impact sourcing standard" that defines minimum requirements and voluntary best practices for employment. The GISC requires member performance on criteria such as nondiscrimination and equal pay to be assessed every two years. Some well-known data-labeling firms such as Alegion are not GISC members.[99]

When it comes to promoting ethical and socially responsible behaviors, however, the GISC is at best of questionable effectiveness and value. For instance, violators face no penalty or sanction; they do not lose GISC membership. GISC members also differ significantly in the information they publish.

Samasource's impact audits report includes indicators such as workforce demographics and number of people lifted from poverty. DDD's reports contain information about employees' earnings and increase in lifetime income. CloudFactory had not published a social impact report since 2015. As of 2019, iMerit had not published any such reports[3]. The U.S.-based provider Alegion, which is not a GISC member, has outlined broad targets that it seeks to achieve but lacks specific metrics[3].

Moral Distance

Researchers have come up with the concept of moral distance, which refers to the use of data by organizations far from where it was collected.[100] For instance, human genome studies often ask participants to sign a form that gives them little direct control over the use of their data.[101]

There are concerns that international research-funding agencies and researchers from high-income economies set research priorities and establish data-sharing rules in Africa. Research participants in Africa are vulnerable to exploitation due to illiteracy and disease that are common in Africa. Also, many people have poor access to medical care. People participate in research to access doctors and other health professionals, when they are unwell.[102]

The argument is that if data is collected from an African village, the

99. Kaye, "These companies."

100. R. Abebe, K. Aruleba, A. Birhane, S. Kingsley, G. Obaido, S. L. Remy, and S. Sadagopan, "Narratives and counternarratives on data sharing in Africa," *Proceedings of the 2021 ACM Conference on Fairness, Accountability, and Transparency*, 2021.

101. Linda Nordling, "Give African research participants more say in genomic data, say scientists," *Nature*, February 15, 2021, https://www.nature.com/articles/d41586-021-00400-9.

102. Nordling, "Give African research."

primary beneficiaries of that project should be the community in the village that serves as the data source. In most cases, the way data is collected and shared reflects the values and interests of organizations that are neither connected to nor have a vested interest in creating value for the communities providing the data. Many data initiatives in Africa are driven by well-intentioned efforts to fight against poverty and inequality. These initiatives, however, are driven by "deficit narratives," which means they focus on the continent's negative perception and ignore its positive contributions and assets. Data sharing is often inherently extractive. That is, data is collected from African communities without considering how communities where data is extracted from should be paid back. Data-sharing initiatives are mostly driven by non-African stakeholders. Data subjects themselves are not even viewed as stakeholders in the data-sharing process. There is also "data colonialism" where data of African communities is accessed and shared by heterogeneous geographies of people due to non-African organizations' unfair data-sharing practices. Local communities realize little benefits and are mostly harmed by these practices. Data sharing should be considered just like the farm-to-table movement, which involves growing and harvesting local ingredients and using them to serve meals to the local community. Data should contribute to fight poverty and help local communities rather than extract resources from them.[103]

Some initiatives are being taken to address the situation. In February 2021, the report Recommendations for Data and Biospecimen Governance in Africa was published by a committee of 13 African scientists.[104] The committee was formed by the African Academy of Sciences, based in Nairobi, and the African Union Development Agency, based in Addis Ababa, in June 2019. The report argues that the current way of data use can fuel distrust between researchers and participants, which needs to change.[105]

Workplace Surveillance

There has been a significant increase in workplace surveillance in recent years. It is argued that China's technology company workers are facing an environment similar to sweatshops. They work long hours to meet objectives

---

103. Abebe et al., "Narratives and counternarratives."

104. *Recommendations for data and biospecimen governance in Africa*, Alliance for Accelerating Excellence in Science in Africa, 2020, https://www.aasciences.africa/sites/default/files/Publications/Recommendations%20for%20Data%20and%20Biospecimen%20Governance%20in%20Africa.pdf.

105. Nordling, "Give African research."

set by big data analytics. They are constantly monitored, which, due to poor labor regulations, has created the potential for abuse.[106]

Adverse Social and Welfare Effects

Regarding the differential welfare effects on sophisticated and unsophisticated consumers, the median voter theory, developed and refined by Hotelling and others is of interest.[107,108] These authors specified the conditions and mechanisms under which competition between political parties would lead to an outcome that favors the median voter. Extending a median voter model in the context of big data, Strahilevitz predicted that the U.S. laws will "systematically favor the interests of sophisticated consumers, which are congruent with those of data miners, since sophisticated consumers are on the whole more politically engaged people who pay attention to legislative policy proposals and vote their interests."[109] Note that sophisticated consumers tend to be wealthier and better educated and have a higher tendency to vote. These consumers arguably think they will lose nothing from policies that allow firms to access their data,[110] and they are likely to make the necessary efforts to fight businesses' informational advantage. Some argue that the general public outside this group may not necessarily be a "winner" in economic or other terms in corporations' big data initiatives that rely on "data accessibility and manipulation."[111] Others maintain that gains associated with data-driven personalization primarily accrue to businesses that are resourceful and can see clear benefits of big data.[112]

Key aspects of the median voter theory can be extended to apply at the global level. A deduction from the median voter theory is that B4B populations face more adverse conditions for privacy protections. That is, they are more susceptible to privacy violations and face other social costs. 4R technologies are being used to exploit vulnerable populations in LMICs.

---

106. Nikki Sun, "China's tech workers pushed to their limits by surveillance software," *Financial Times*, June 15, 2021, https://www.ft.com/content/b74b6ad6-3b8d-4cd8-9dd6-3b49754aa1c7.

107. H. Hotelling, "Stability in competition," *Economic Journal* 39 (1929): 41–57.

108. B. A. Abrams and A. L. Kenneth, "A median-voter model of economic regulation," *Public Choice* 52 (1987): 125–142.

109. Strahilevitz, "Toward a positive theory."

110. Strahilevitz, "Toward a positive theory."

111. A. L. Allen, "Privacy law: Positive theory and normative practice," *Howard Law Journal* 56, no. 3 (2013): 241–251.

112. Aspen Institute, *The promise and peril of big data* (Queenstown, MD, 2010).

P2P Lenders and Fintech's Questionable Practices

P2P lenders and other fintech companies have engaged in unethical and questionable activities. In China, for instance, Ant Group uses its vast data and AI to analyze credit risk, which enabled the company to offer cheaper loans.[113] Ant Group and other Chinese fintech companies that use AI and big data analytics in extending loans are being accused of abusing personal data to collect debts. One such example is Ant Group's Ant Check Later, which allows users to delay payments and pay in installments. An online user reported being contacted by Ant Check Later for information about a friend who owed money to the payment service. Ant Group reportedly said that the practice of contacting a borrower's friends or relatives to help with collecting debts is common in the financial sector. TMTpost cited a China Youth Daily poll, which showed that 76% of respondents believed there was data abuse.[114] Users of JD.com reported similar problems. The JD.com financial unit operates JD Baitiao, similar to Ant Check Later. In a question posted on the online legal advice site 110.com, a JD.com user asked if it was legal for JD.com to give a third-party service his personal information for the purpose of debt collection. Fintech companies in Kenya are also reported to use "societal shaming" for debt collection.[115]

Customers have reported that friends and family were harassed after they were late in repaying loans offered by Silicon Valley–based personal credit startups in Kenya. Debt shaming has led to at least one suicide in Kenya.[116] A January 2021 *Wired* article reported that at least eight people in India who had been the victims of harassment by quick loan apps had committed suicide in the previous two months.[117] In China, it has been reported that

113. Gillian Tett, "Artificial intelligence is reshaping finance," *Financial Times*, November 19, 2020, https://www.ft.com/content/c7d9a81c-e6a3-4f37-bbfd-71dcefda3739.

114. "Online credit services in China accused of abusing personal data," *Want China Times*, 2015, www.wantchinatimes.com/news/content?id=20150927000087&cid=1203.

115. Olumuyiwa Olowogboyega, "Countdown: The 10 most important African tech companies of the decade," *Techcabal*, December 26, 2019, https://techcabal.com/2019/12/26/countdown-the-10-most-important-african-tech-companies-of-the-decade/.

116. Antoaneta Roussi, "Kenyan borrowers shamed by debt collectors chasing Silicon Valley loans: Customers say friends and family were harassed after they were late repaying US startup branch," *Financial Times*, September 9, 2020, https://www.ft.com/content/16c86479-e88d-4a28-8fa4-cd72bace5104.

117. Varsha Bansal, "Shame, suicide and the dodgy loan apps plaguing Google's Play Store," *Wired*, January 20, 2021, https://www.wired.co.uk/article/google-loan-apps-india-deaths.

many women were required to share their nude pictures as collateral. In 2016, a student from Jiangsu Province reported that she agreed to provide nude photos to private lenders in exchange for a loan of 120,000 RMB (US$18,781) to start a small business.[118]

Global and Regional Norms

It is important to look at the role and importance of global norms in influencing the use of 4R technologies. For instance, global and national money-laundering regulations including customer due diligence, reporting of suspicious transactions, and record keeping are needed to increase the adoption of payment systems such as cryptocurrencies.[119]

A complaint that is often heard is that developed countries are disproportionately benefiting from global norms, which are being shaped to benefit them. LMICs arguably have not been given an opportunity to participate in discussions involving global AI governance.[120] LMICs are highly underrepresented in international AI advisory boards, expert panels, and councils appointed by international organizations, which has resulted in the lack of regional diversity. This underrepresentation in AI governance can be attributed to geopolitical power imbalances of the colonial era.[121] For instance, in the UNICEF's AI for Children project's expert advisory group, there are no representatives from regions that have the highest concentration of children and young adults, including the Middle East, Africa, and Asia.[122]

Some have refused to participate in international data agreements.[123] In the 2019 G20 Summit held in Japan, India did not sign the Osaka Track

118. Stuart Leavenworth, "China's 'naked loans' force female students to bare all in return for more cash," *The Guardian*, June 15, 2016, https://www.theguardian.com/world/2016/jun/15/chinas-naked-loans-force-female-students-to-bare-all-in-return-for-more-cash.

119. Richard Chelin, "Africa: New playground for crypto scams and money laundering—analysis," *Eurasia Review*, August 11, 2021, https://www.eurasiareview.com/11082021-africa-new-playground-for-crypto-scams-and-money-laundering-analysis/.

120. Hao, "The problems AI."

121. Hao, "The problems AI."

122. Abhishek Gupta and Victoria Heath, "AI ethics groups are repeating one of society's classic mistakes," *MIT Technology Review*, September 14, 2020, https://www.technologyreview.com/2020/09/14/1008323/ai-ethics-representation-artificial-intelligence-opinion/.

123. Hao, "The problems AI."

declaration, launched by Japan's then prime minister Shinzo Abe to promote cross-border data flow.[124]

The dominance of the EU and other developed economies in the development of global AI governance and ethical frameworks has been a concern for LMICs that lack institutional capacity. For instance, it is argued that such frameworks are harder to adhere to for Latin American companies. Since the EU has stricter regulations, it could hamper the development of the Latin American AI industry. It is argued that while countries such as Japan, despite having different regulatory regimes, have negotiated agreements with the EU, Latin America lacks such negotiating strength.[125]

The global AI ecosystem can also benefit from regional collaboration. For instance, it is argued that due to the disjointed regulatory and policy environment, countries in Latin America lack a unified voice on this issue.[126]

Chapter Summary and Conclusion

The discussion in this chapter makes it clear that the 4R has important political, social, and ethical dimensions that policymakers, businesses, consumers, and other actors should consider. It is encouraging that new regulations are being enacted in some LMICs to benefit from the 4R. Some countries are also putting in place legislative measures to control irresponsible and unethical business acts utilizing 4R technologies.

Nonetheless, there is a divergence between LMICs and high-income economies in some key regulatory areas related to the 4R. Weak and underdeveloped regulatory environments in many LMICs have facilitated the misuse and abuse of personal data. Poor people's lack of capability and resources to defend themselves would make data privacy more—not less—important for them. Due to such environments, in many cases, big technology companies are enriching themselves using data obtained from low-income people and small businesses. In some cases, these companies' activities bring little or

124. Ankur Taliyan, "G20 Summit: Why India refused to sign Osaka declaration on global data flow," *Times Now News*, July 4, 2019, https://www.timesnownews.com/india/article/g20-summit-why-india-refused-to-sign-osaka-declaration-on-global-data-flow/446887.

125. "The global AI agenda: Latin America," *MIT Technology Review*, 2020, https://mittrinsights.s3.amazonaws.com/AIagenda2020/LatAmAIagenda.pdf.

126. Angelica Mari, "Artificial intelligence gathers pace in Latin America," *ZD Net*, June 11, 2020, https://www.zdnet.com/article/artificial-intelligence-gathers-pace-in-latin-america/.

no benefits to these vulnerable groups. A related point is that some uses of 4R technologies have adverse welfare effects on unsophisticated consumers. A core concern is that there is a relative lack of voice and representation of the B4B in the global AI industry.

Some governments have identified an opportunity to use data to achieve their political agenda. Surveillance and espionage have also increased because 4R technologies have greatly increased the ease with which these activities can be performed. They are misusing 4R technology to conduct espionage on citizens and engage in activities that violate their privacy and human rights. These concerns have been especially strongly voiced against countries that have used China-developed solutions. This chapter also illustrated businesses' certain questionable practices related to the uses of 4R technologies from an ethical perspective.

This chapter has provided an account of how LMICs are playing key roles in the global AI industry: However, due to global technology companies' practices of ethics dumping and ethics shirking, minimal benefits have been observed in LMICs. Many companies are going for a data grab under cover of helping poor and marginalized groups. Currently, the nascent technology industries of most developing countries cannot compete globally with those from China or the U.S. However, companies from China and the U.S. can maintain control over local policies.

CHAPTER 12

Conclusion and Recommendations

Low-income populations in LMICs have already started experiencing the economic and social impacts of 4R technologies. These technologies are facilitating low-income consumers' access to diverse products and services, ranging from LED solar home systems and clean-burning LPG to bank loans to buy agricultural inputs and low-cost crop insurance. But what is coming is even bigger.

There are many new and promising developments on the horizon that are aimed at fighting poverty by enabling the B4B population (also known as the poverty market or submerged market) to participate in the market economy. These technologies can also help LMICs provide the evidence required to demonstrate product quality and comply with various sustainability standards to improve their firms' access to foreign markets.

LMICs' use of 4R technologies to provide evidence of product quality has become necessary since the adverse-selection problem often put them in disadvantage due to greater information asymmetry. In the absence of sufficient information to judge a product, buyers rely on indirect indicators and cues to assess a product.[1] These include brand name and country of origin as extrinsic cues, which work against firms based in LMICs. 4R technologies can help provide direct and objectively verifiable indicators quality as well as means of verification.

Among the encouraging developments are several examples of leapfrogging facilitated by 4R technologies in a number of industries and economic sectors in the LMICs. These include solar power's roles in leapfrogging the

1. J. C. Olsen, *Cue utilisation in the quality perception process: A cognitive model and an empirical test*, unpublished doctoral thesis, 1972, Purdue University, Lafayette, IN.

lack of a fully deployed power grid in India, Kenya and other African countries, and drone delivery's roles in leapfrogging the lack of well-developed roads and refrigerators for storing vaccines and temperature sensitive pharmaceutical products in Ghana and Rwanda. Regarding the last point, Zipline's cofounder and head of product and engineering Keenan Wyrobek put the issue this way: "We leapfrog broken refrigerators, we leapfrog the lack of roads."[2] These examples indicate that even LMICs may leapfrog in a specific industry. Firms' entrepreneurial activities and political leadership may help countries to leapfrog in one or more economic sectors.

In rich countries, the 4R has raised some fears that machines will displace humans. For instance, among the obstacles facing Gulf Cooperation Council (GCC) countries' acceptance of the development of AI include the fear that robots will take people's jobs and the fear of Western dominance in the technology market.[3] Such concerns are less relevant in most LMICs. First, in Africa, most industries rely on low-paid workers, so machines are less likely to replace jobs. It is not economically feasible to replace or augment low-paid labor until the cost of robotic labor reduces dramatically. Second, the impacts of AI on the informal sector, which employees a large proportion of the population in these countries, are likely to be minimal. Third, there is relatively little R&D in AI in Africa.[4]

As mentioned, the B4B population is likely to differ from richer consumers in the ways the 4R technologies are used. Due to the lack of huge investments, AI-enabled robots and intelligent machines may not be as relevant and important in most LMICs. The most valuable uses of AI would be in performing tasks that humans are not currently capable of doing, such as detecting dangerous gas leaks in mines, dealing with traffic congestion, and providing real-time actionable recommendations to farmers. In the financial sector, for instance, AI-driven customer service is focused more on enabling inclusion in Africa.[5] In the GCC economies, however, AI is being used to

2. Ryan Nakashima, "Drone company demos how blood air-drops will work in Rwanda," *HuffPost*, April 4, 2016, https://www.huffpost.com/archive/ca/entry/drone-company-demos-how-blood-air-drops-will-work-in-rwanda_n_9610378.

3. Tarek Abd El-Galil, "Arab narratives about artificial intelligence are explored in new report," *Al-Fanar Media*, July 15, 2021, https://www.al-fanarmedia.org/2021/07/artificial-intelligence/.

4. Oxford Insights and the International Development Research Centre, *Government artificial intelligence readiness index 2019*, 2019, https://www.oxfordinsights.com/ai-readiness2019.

5. "The global AI agenda: The Middle East and Africa," *MIT Technology Review Insights*, June 19, 2020, https://www.technologyreview.com/2020/06/19/1004121/the-global-ai-agenda-the-middle-east-and-africa.

provide financial services such as wealth management, which is primarily targeted at high-net-worth individuals.

Having said that, it is apparent that AI-based systems in LMICs are also being used to deliver high-quality and personalized financial services. In Nigeria, Diamond Bank's chatbot Ada uses AI to provide personalized experience to customers. It is deployed on Facebook Messenger.[6] The chatbot can be used to conduct transactions such as airtime purchase, bill payment, and stock trading. Ada interacts with customers in English and Pidgin. Likewise, Kenya's Absa Bank announced a plan to launch a digital platform that will use AI, ML, and analytics to gauge and anticipate customers' needs and requests. It offers multiple solutions through core systems and delivery platforms. The customers get access to a 24/7 "digital personal banker" that can help perform transactions. The customers receive instructions on WhatsApp chat.[7]

4R technologies can also provide effective enforcement mechanisms. As illustrated in the case of South African Breweries, as a self-enforcement tool, IoT devices can monitor and measure machine performance in real time. Examples such as M-KOPA solar panels have suggested the value of the IoT as a low-cost mechanism for second-party enforcement.

However, much more needs to be done to stimulate the adoption of 4R technologies to bring positive economic and social change in LMICs. While many bold claims have been made about the effectiveness of solutions based on 4R technologies such as AI in addressing many socioeconomic challenges, supporting evidence has not been convincing and conclusive. In the healthcare sector, for instance, a systematic literature review of research articles published in PubMed, which comprises biomedical literature from Medline, life science journals, and online books, found that only 66 articles included the terms *cost effectiveness*, *economic impact*, or *cost saving*, and only six provided detailed assessments. The authors found that even the six articles met the criteria for detailed assessment had provided only fragmented cost or cost-saving aspects but none provided a complete cost-benefit analysis.[8]

---

6. Yinka Ogunlami, "Diamond Bank launches artificial intelligence-powered chatbot," *Pulse*, March 15, 2018, https://www.pulse.ng/news/chat-with-ada-diamond-bank-launches-artificial-intelligence-powered-chatbot/merhtn1.

7. D. Simmonds, "Kenya's Absa Bank to launch banking via WhatsApp," *Umaizi*, 2021, https://umaizi.com/kenyas-absa-bank-to-launch-banking-via-whatsapp/.

8. J. Wolff, J. Pauling, A. Keck, and J. Baumbach, "The economic impact of artificial intelligence in health care: Systematic review," *Journal of Medical Internet Research* 22, no. 2 (2020): e16866.

Many technology companies' business models currently often work against marginalized groups such as small businesses, low-income people, and rural households. Many solutions offered by big technology companies to these groups are often seemingly promising and useful, but in reality, they are mostly useless. In some cases, poorly performing apps developed in high-income countries are being exported to LMICs. Such apps are likely to exhibit even worse performance in serving poor people in LMICs.

Some unethical and corrupt government administrations and large companies make use of these technologies to take advantage of vulnerable groups. For instance, some authoritarian governments are using these technologies to intimidate and suppress oppositions and to protect their regimes.

Data is one of the most important ingredients in achieving competitive advantage at the organizational and national levels for various reasons. For instance, AI algorithms need to be trained and tested on distinct datasets. Commenting on the role of data in the growth of the Chinese AI industry, a *New York Times* article put the issue this way: "If China is the Saudi Arabia of data, its data factories are the refineries, turning raw data into the fuel that can power China's goal of A.I. supremacy."[9] The 4R's potential to transform LMICs hinges on the availability of useful data.

A related problem is that low-income populations are often delivered poor-quality products and services thanks to the lack of high-quality data or limited availability of local data. For instance, ACRE's weather-monitoring stations (chapter 8) are primarily installed in heavily farmed areas in Kenya and Rwanda. This means that areas that are not heavily farmed lack required data.

The 4R's Effects on Industries and Markets

As discussed throughout this book, the various technologies that comprise the 4R are having dramatic impacts on diverse industries and markets in LMICs. A sample of examples discussed in this book are presented in Table 12.1. In addition to the three industries and economic sectors covered in detail in chapters 6, 7, and 8, the 4R is likely to have impacts on a wide range of economic and social outcomes. These technologies have created many other positive effects, some of which are described in this section.

9. Li Yuan, "How cheap labor drives China's A.I. ambition," *New York Times*, November 25, 2018, https://www.nytimes.com/2018/11/25/business/china-artificial-intelligence-labeling.html.

Table 12.1. The 4R's effects on diverse industries and markets: some examples

	Healthcare and pandemic preparedness	The agricultural sector	Finance, banking, and insurance
AI	Aravind Eye Hospital's diagnosis of diabetic retinopathy (chapter 2)	Demetria's sensors to assess biochemical markers of coffee taste (chapter 7)	MyBucks' banking and lending software (Haraka) in Malawi (chapter 5)
Blockchain	Uthabiti's system to enhance traceability and trackability of pharmaceutical supply chain in Kenya (chapter 6)	Sarafu used by farmers in Kenya (chapter 8)	Insurance products offered by ACRE in Kenya and Arbol in Cambodia (chapter 8)
Remote sensing and satellite imagery	China's use to track patients' position, monitor cargo transports and for large-scale disinfection missions during COVID-19 (chapter 4)	EcoProMIS in Colombia and Acquahmeyer in Ghana (chapter 7)	OKO's agricultural insurance in Mali and Uganda (chapter 8)
IoT	Sanku's Project Healthy Children (chapter 6)	Kenya's smart irrigation system Illuminum Greenhouses (chapter 5)	ACRE's plan to install soil moisture sensors in the weather stations to complement remote-sensing data in its crop insurance program (chapter 8)

Detecting and Preventing Illegal Activities

4R technologies are effective in detecting and preventing illegal activities. For instance, the prevention of unregulated gold mining makes a promising use case for DE Africa data, with an economic impact of US$900 million or more thanks to the potential reduction in environmental damage and fiscal evasion. According to a 2016 study, tax losses associated with illegal mining were US$550 million in Ghana and US$2.2 billion in South Africa.[10] As

10. Nicolò Andreula and Anne-Marie Engtoft Larsen, "Climate data presents a $2 billion opportunity in Africa alone: Here's why," *World Economic Forum*, January 22, 2021, https://www.weforum.org/agenda/2021/01/climate-data-2-billion-opportunity-in-africa-davos-agenda/.

to the environmental impact, illegal gold mines are estimated to account for the loss of over 1% of primary forest in Ghana. In 2016, the country's Western Region spent US$250 million to recover lands and water bodies damaged by illegal gold mines.[11]

Satellite data have also been used to fight illegal mining activities. In Ghana, data from a population census, aerial technology such as drones, and satellite images are combined to fight illegal mining. To identify *galamsey* activities,[12] government agencies utilize data from satellite images of concession maps and vegetation areas in high-risk areas.[13]

Protection of Environment

The previous section discussed how the prevention of illegal gold mining can contribute to the protection of environment in countries such as Ghana and South Africa. Likewise, in Brazil, 20% of areas officially designated as protected have been already destroyed to use for agriculture and cattle farming. Brazil's newest satellite, Amazonia 1, which orbits 752 kilometers above Earth to monitor the rain forest, is expected to reduce such activities. Brazil's Amazon monitoring program relied on flyovers from the U.S. satellite Landsat, which provides high-definition image data every 16 days. Two satellites codeveloped by Brazil and China—CBERS-4 and CBERS-4A—provide more timely alerts, every three to four days. Amazonia-1, whose cameras cover an 850-kilometer swath at 65-meter resolution, provides timely and better-quality data. It generates new images every day or two. That frequency gives authorities faster alerts about deforestation. A near-real-time monitoring can help enforcement teams to go to the right place at the right time.

There was also a "data blackout" in 2012 due to a problem with Landsat. The National Institute for Space Research or Instituto Nacional de Pesquisas Espaciais (INPE) was forced to buy expensive satellite data from the UK government that had poorer image quality.[14] The new satellite is expected to help Brazil overcome these challenges.

---

11. Mikaela Weisse and Elizabeth Dow Goldman, "The world lost a Belgium-sized area of primary rainforests last year," *World Resources Institute*, April 25, 2019, https://www.wri.org/blog/2019/04/world-lost-belgium-sized-area-primary-rainforests-last-year.

12. *Galamsey* is a Ghanaian term, which means illegal small-scale gold mining.

13. "Ghana is using satellite imagery to combat illegal mining," *Space in Africa*, June 17, 2019.

14. Sofia Moutinho, "Brazil's first homemade satellite will put an extra eye on dwindling Amazon forests," *Science*, February 26, 2021, https://www.science.org/news/2021/02/brazil-s-first-homemade-satellite-will-put-extra-eye-dwindling-amazon-forests.

Tracking Educational Outcomes

4R technologies are also being used in tracking educational outcomes. In April 2021, Ethiopia's minister of education reported that the country started working with IOHK to develop a blockchain-based system to track student performance in the country's schools.[15] IOHK is research and development (R&D) company behind the Cardano blockchain. The plan is to use IOHK's Atala PRISM identity solution to create records of educational performance across 3,500 schools consisting of 5 million students, and 750,000 teachers. The goal is to identify the causes of underachievement and allocate resources effectively. The country also launched Digital Ethiopia 2025. A component of this digital transformation strategy is to develop a blockchain-based national identity system. The government has issued a national identity standard. The Atala PRISM blockchain ID is expected to be the first system to issue IDs based on this standard.[16]

Taking Advantage of the 4R

The potential benefits of 4R technologies have not been fully realized because of several barriers discussed in chapter 9. Economic barriers are among the key factors that prevent a large proportion of populations from using solutions based on 4R technologies. For instance, by adopting solar power, developing countries such as India should be able to leapfrog the lack of a fully deployed power grid.[17] While solar power offers leapfrogging opportunities, economic barriers prevent low-income people from taking advantage of such opportunities. High costs are the main reasons many households do not use solar systems. An Indian farmer who earns about US$75 per month for a family of six was reported to pay US$10 to rent the solar unit from Simpa. To cover that expense, he rents out his tractor for a nominal fee to neighbors, who are too poor to afford solar technology.[18]

---

15. Marc Hochstein and Anna Baydakova, "Ethiopian education minister confirms Cardano blockchain partnership," *CoinDesk*, April 30, 2021, https://www.coindesk.com/ethiopian-education-minister-confirms-cardano-blockchain-partnership.

16. "IOHK partners with Ethiopian government for student IDs on Cardano blockchain" *CryptoNinjas*, April 27, 2021, https://www.cryptoninjas.net/2021/04/27/iohk-partners-with-ethiopian-government-for-student-ids-on-cardano-blockchain/.

17. Varun Sivaram, "The global warming wild card," *Scientific American*, May 1, 2017, https://www.scientificamerican.com/article/can-india-save-the-warming-planet.

18. Michael Edison Hayden, "To know the true meaning of solar energy, ask a power-

A similar point can be made about the adoption of LPG. For instance, in the Tanzanian capital Dar es Salaam, KopaGas's US$150 gas stove and canister is half the average monthly wage.[19]

Underdeveloped financial infrastructure such as payment systems is also a concern. While some developing countries such as Kenya have well-developed mobile payment systems, many LMICs still have a low penetration of electronic payment methods, which would prevent the development of solutions such as a fully automated crop insurance scheme. For instance, most farmers in Sri Lanka managed transactions with cash or check only, which inhibited the process of automated payouts.[20]

Policymakers and key decision makers in some countries lack motivation and initiative to optimally utilize 4R technologies despite their capabilities. For instance, India has its own satellite systems and data-processing capacity, which has the potential to address many poverty-related problems facing the country. The challenge has been to convince health, water and environmental agencies to accept and use the data.[21]

Even worse, some authoritarian regimes are using 4R technologies for nefarious purposes. They lack commitment to the general welfare of the broader population. For instance, China has been exporting its techno-authoritarian models to other countries, especially with regimes that are sympathetic to Chinese policy. These regimes use the same combination of solutions, equipment, and ideology to monitor and control citizens at home.[22]

While entrepreneurial activities related to 4R technologies in LMICs are encouraging, it is very difficult, if not impossible, for firms in these coun-

---

less rice farmer," *NPR*, December 8, 2015, https://www.npr.org/sections/goatsandsoda/2015/12/08/458931828/to-know-the-true-meaning-of-solar-energy-ask-a-powerless-rice-farmer.

19. Megan Rowling, "Smart gas cooking seeks to break African cities' dirty charcoal habit," *Reuters*, August 6, 2019, https://www.reuters.com/article/us-africa-energy-cooking-feature-idUSKCN1UW161.

20. Etherisc, "Oxfam, Etherisc, and Aon Deliver Pay-Outs with First Blockchain-Based Agricultural Insurance *Policies for Smallholder* Farmers in Sri Lanka," *PR Newswire*, November 4, 2019, https://www.prnewswire.com/news-releases/oxfam-etherisc-and-aon-deliver-pay-outs-with-first-blockchain-based-agricultural-insurance-policies-for-smallholder-farmers-in-sri-lanka-300949728.html.

21. T. Anderson, "Launching your own satellite—The pros and cons," *SciDev.Net*, November 11, 2009 https://www.scidev.net/global/features/launching-your-own-satellite-the-pros-and-cons/.

22. C. Bartholomew, "China and 5G," *Issues in Science and Technology* 36, no. 2 (2020): 50–57.

tries to compete with established companies such as Facebook, Google, and Amazon. Three challenges have been identified. First, the lack of network effects creates considerable barriers to entry for Global South firms. The more users online platforms have, the bigger the network effect is. A second barrier concerns the economies of scale. It is costly to operate centralized social networks because of costly cloud infrastructure, the lack of teams of skilled programmers to develop quality products, and the inability to curate data effectively and monetize the service to cover costs. In addition, global technology giants that already dominate the market have accumulated brand equity and trade secrets, and they possess the power to acquire smaller companies.[23]

Many 4R applications developed in labs suffer from low external validity. A key challenge to address real-world problems and increase the external validity of lab experiments is to bring different types of data together, which are controlled by different entities.[24] However, there is an unwillingness to share data. For instance, while universities and research centers constitute a key source of data and knowledge, scientists working in these institutes can be against making relevant data accessible due to security, privacy, and other concerns. They often use reasoning against data sharing such as "I don't want to share it," "It's mine," or "It's government property."[25]

A further problem is the nascent development of solutions based on 4R technologies. Despite digital tools' potential to support more people's healthcare needs affordably and efficiently, critics have raised concerns that claims about digital tools such as AI can be overhyped and unproven. Much of the analyses of such systems are funded by the companies themselves. They are not published in peer-reviewed journals. The studies are conducted under tight restrictions on the medical conditions that are examined. The analyses that are publicly available are limited in scope, often providing no comparison with competing products or not discussing the final outcomes for patients.[26] For instance, as of 2021, the Babylon system was still a more "rudimentary" form of telemedicine. It had plans to test AI within a few months.[27]

23. Michael Kwet, "Long read digital colonialism," *New Frame*, March 8, 2021, https://www.newframe.com/long-read-digital-colonialism/.

24. David Pilling, "AI in Africa healthcare falls short of potential," *Financial Times*, May 18, 2020, https://www.ft.com/content/90fa8f44-6847-11ea-a6ac-9122541af204.

25. Prachi Patel, "Feeding the world with big data," IEEE Spectrum, May 14, 2013, http://spectrum.ieee.org/computing/networks/feeding-the-world-with-big-data.

26. A. Jack, "Rwanda venture tests digital health potential in developing world," *Financial Times*, 2021, https://www.ft.com/content/4fe33c92-cbd5-459a-8df6-20d0d1f57ec8.

27. Jack, "Rwanda."

There is also no level playing field for researchers from LMICs. For instance, researchers in these countries have reported that they have experienced barriers in knowledge exchange due to Western governments' actions, such as visa denial to attend prestigious conferences. This is in contrast to Western multinational companies' access to a wide range of resources in the Global South. As noted in chapter 9, Google's Loon balloons and flight-control systems involved extensive data collection in Africa. Likewise, Google researchers trained their deep-learning system using a deidentified dataset of 248,445 patients from India's multicity hospital network Apollo Hospitals.[28] The model's generalizability to distinguish normal from abnormal chest X-rays across datasets and settings was assessed using six independent datasets from India, China, and the U.S.[29]

In addition to disparities in skills from LMICs' economic situations and institutional arrangements, there is also a racial disparity. It was reported that at a 2016 conference in Barcelona, which had more than 5,000 participants, there were fewer than 10 black people.[30]

Policy Implications

The 4R has no how-to manual for countries and policymakers to copy from.[31] A well-developed national strategy is the key to benefiting from the 4R. However, each country's case is unique, and its 4R strategy should be based on careful analysis of economic, social, environmental, and political situations. As discussed in chapter 11, whereas some LMICs have developed national digital strategies, most lack such initiatives. LMICs that lack such strategies can use other economies' strategies as a model to develop national 4R strategies.

Due to the lack of appropriate regulatory frameworks, many LMICs are not currently in a position to take advantage of the 4R's potential in bank-

---

28. Debolina Biswas, "Google uses Apollo Hospitals' X-ray data to identify chest abnormalities," *Analytics India Magazine*, September 6, 2021, https://analyticsindiamag.com/google-uses-apollo-hospitals-x-ray-data-to-identify-chest-abnormalities/.

29. Z. Nabulsi et al., "Deep learning for distinguishing normal versus abnormal chest radiographs and generalization to two unseen diseases tuberculosis and COVID-19," *Scientific Reports*, 2021, https://www.nature.com/articles/s41598-021-93967-2#Sec9.

30. Moustapha Cisse, "Look to Africa to advance artificial intelligence," *Nature* 562, no. 7728 (2018): 461.

31. Murat Sönmez, "How Japan could soon offer lessons for the Fourth Industrial Revolution," *ITU News*, July 4, 2018.

ing, finance, and insurance. For instance, as of mid-2021, no African country had clear legislation for open banking.[32] In early 2021, the Central Bank of Nigeria (CBN) released the Regulatory Framework for Open Banking. Among other things, the framework discusses risk-based data access levels and exchange of data and services, and standards for data exchange among financial services sector participants.[33]

Governments in LMICs can take measures that have a great impact on their digital landscapes. For instance, in 2014, the Bank Verification Number, a centralized biometric banking identification number, was introduced in Nigeria. It allowed fintech companies to remotely recruit and perform regulatory checks on customers. It is far more common to make payments via mobile banking app in Nigeria than it is in most Western countries. The country's demographics help. About 70% of Nigerians are younger than 30 and tech-savvy.[34] Policymakers need to promote 4R technologies in a way that benefits the B4B by providing economic incentives (e.g., the Government of Togo's 50% subsidy to buy IoT-based solar irrigation systems) as well as using of legal instruments (e.g., Kenya's proposed a law to regulate digital lenders). Reform measures should be introduced to stimulate competition among service providers to ensure more affordable energy sources such as solar units to low-income households.

Especially since P2P lenders' unethical business practices have led to painful experiences such as suicides among people in poverty, appropriate regulations to require some degree of explainability in AI algorithms are needed. For instance, P2P lenders claim that their algorithms capture the aspects of social capital that are relevant to the ability and willingness to pay. The lack of explainability means that the algorithms instead may tell the lenders whether they can make the borrower repay the loan by engaging in harassment or by applying public pressure by shaming the borrower. For instance, lenders can impose higher social costs on the borrowers with higher number of contacts by shaming them in the eyes of social media friends. The social costs are likely to be even higher for borrowers with more reputable

32. Victor Oluwole, "Open banking in Africa," *Business Insider Africa*, February 10, 2021, https://africa.businessinsider.com/local/markets/open-banking-in-africa/0rtzd5g.

33. Pius Okwuanya, "Open banking in Nigeria: what next for financial services sector," *This Day Live*, February 25, 2021, https://www.thisdaylive.com/index.php/2021/02/25/open-banking-in-nigeria-what-next-for-financial-servicessector/#:~:text=The%20Central%20Bank%20of%20Nigeria's,services%20in%20Nigeria%2C%20define%20risk%2D.

34. Neil Munshi, "Explosion in electronic payments powers start-up boom in Nigeria," *Financial Times*, August 19, 2021, https://www.ft.com/content/5fa49678-9eed-45e8-9c3d-6e19a2237b81.

contacts or those that interact with their networks more frequently. To make enforcement through public shaming more effective, borrowers are also asked to provide sensitive content. The serious adverse economic and social consequences of quick loan apps are associated and facilitated by regulators' indifference toward misuse and abuse of personal data. Fintech companies in many Global South economies face little regulatory scrutiny.[35]

Stricter data privacy and data protection regulations may discourage such practices. Countries should also enact regulations that prohibit or impose restrictions on lending money to borrowers who lack proven entrepreneurial skills. Policy reforms are needed to ensure that low-income people can benefit from these solutions. Financial institutions using AI to make lending decisions should be required to develop algorithms that reliably and accurately detect borrowers' capability to profitably utilize the loans. It is important to ensure that individual borrowers are responsible and that microentrepreneurs borrowing money have sufficient knowledge and skills to start and manage a business.

Jimson Olufuye, the former president of African Information Communication Technology Alliance (AfICTA), noted that embracing 4R technologies requires a critical mass, which is lacking in countries such as Nigeria.[36] One important way forward is to increase international collaboration in the 4R technologies. A number of areas of collaboration can be proposed. As noted, due to their low research output, most Global South economies rely on technology systems developed by countries in the Global North. Such technologies often perform poorly in the Global South. Global South economies can pool together research resources and teams to develop solutions based on latest technologies to address various poverty-related issues facing the B4B. Multilateral agencies and global North economies also need to help Global South economies in such initiatives.

Many innovative digital solutions developed in LMICs—such as Safaricom's Jubilee Insurance's, and United Bank's chatbots—are being used mainly in their home countries. It is likely that these solutions with, or without, slight modifications can be used in other LMICs to address the diverse challenges facing these economies. These solutions have not been internationalized mainly because the companies that developed them often lack resources to internationalize them. Governments and multilateral agen-

35. Tom Collins, "Is fintech in Kenya too successful?," *African Business Magazine*, February 11, 2020, https://africanbusinessmagazine.com/african-banker/is-fintech-in-kenya-too-successful/.

36. "Nigeria not ripe for Artificial Intelligence—Expert," *Pulse*, April 15, 2019, https://www.pulse.ng/news/local/nigeria-not-ripe-for-artificial-intelligence-expert/wt358my.

cies in the Global South can help these companies internationalize their solutions to other Global South economies.

Regarding the question of how LMICs can take advantage of advancement in satellite technologies, two competing views have been expressed: they should launch their own satellite systems to achieve technological autonomy and balance consumption and production of satellite data, and they should only focus on building capacity to use and interpret freely available as well as paid for satellite data. The latter argument may have had some validity several years ago when the costs of developing and launching satellites were prohibitively high, but the data presented in this book should largely render this either-or argument obsolete. LMICs must do both if their economies are to prosper.

China's efforts to transform its economy with digital technologies offers useful lessons for policies in other middle-income countries, which can use 4R technologies such as AI to leapfrog in some areas, even though they may still fall in the middle-income trap. The discussion in chapter 10 makes clear that institutional reforms should be a priority to avoid being stuck in the middle-income trap. Institutional reforms can also help fulfill other conditions required to achieve the status of a high-income economy. For instance, meeting longer-term AI goals would require China to increase contributions to fundamental theories and develop enabling technologies such as semiconductor electronic devices and integrated circuits. In 2018, China ranked sixth in terms of number of top AI researchers based on an index that measures research productivity and citations.[37] Reform measures such as more democratic political culture and better educational opportunities will play a role in improving this situation by luring educated elites and scientists back home.[38] Some measures have been initiated to achieve this goal. For instance, China's highly publicized government program Thousand Talents Plan, launched in 2008 and expanded in 2011 to include younger and foreign researchers,[39] attracted an estimated 7,000 Chinese scientists back home by 2018.[40] Nonetheless, China's current intervention measures are far

37. Sarah O'Meara, "Will China lead the world in AI by 2030?," *Nature*, August 21, 2019, https://www.nature.com/articles/d41586-019-02360-7.

38. "Brain drain should send message to China's leaders," *South China Morning Post*, January 6, 2013, https://www.scmp.com/comment/insight-opinion/article/1120829/brain-drain-should-send-message-chinas-leaders.

39. Smriti Mallapaty, "China hides identities of top scientific recruits amidst growing US scrutiny," *Nature*, October 26, 2018, https://www.nature.com/articles/d41586-018-07167-6.

40. Erica Pandey, "The scientists caught between the U.S. and China," *Axios*, November

from satisfactory given that hundreds of thousands of Chinese are obtaining overseas citizenship annually in the Global North economies such as the U.S., Canada, Australia, and New Zealand.[41]

It is important to ensure that low-income people have access to the highest levels of connectivity, bandwidth, and internet access. New generations of telecommunications and technologies such as 5G have diffused slowly in LMICs compared to high-income countries (Figure 12.1). Policy measures are also needed to improve connectivity by increasing investments.

Finally, benefits of the 4R cannot be effectively realized with an overreliance on foreign companies for technological solutions. For instance, there are concerns regarding whether efforts by China and the West to facilitate Africa's leapfrogging are genuine or just data colonialism. Many critics have argued that foreign technology companies' main objectives in Africa include extraction and monetization of data and the creation of a technology monopoly.[42] The AI applications developed in other regions lack contextual relevance, particularly due to different cultural and infrastructural factors. For example, a self-driving lorry developed in the U.S. cannot effectively function in African countries for a variety of reasons and is unlikely to be successful on the roads without substantial adaptation.[43] Local capacity building should thus be another key policy area.

4R technologies can also help overcome some of limitations of the current data-collection techniques. For instance, Somalia's official COVID-19 death count as of October 2021 was about 1,100. But other stakeholders and movement actors such as media and NGOs have argued that actual numbers are much higher. Thus, the official data may not capture the accurate and full picture of the impact of COVID-19. The London School of Hygiene & Tropical Medicine used data from satellite imagery to identify deaths associated with the pandemic. High-resolution satellite data from the space technology company Maxar was used to count the number of burials in the capital Mogadishu's Banadir district for the period January 2017–September 2020. A comparison of burials during the COVID-19 with those before COVID-19 indicated that COVID-19 deaths were highly

10, 2018, https://www.axios.com/scientists-china-thousand-talents-program-economic-espionage-b5f0542e-3368-40cc-8df5-bf9184192dd6.html.

41. "Brain drain."

42. Nima Elmi, "Is big tech setting Africa back?," *Foreign Policy*, November 11, 2020, https://foreignpolicy.com/2020/11/11/is-big-tech-setting-africa-back.

43. Oxford Insights and the International Development Research Centre, *Government artificial intelligence.*

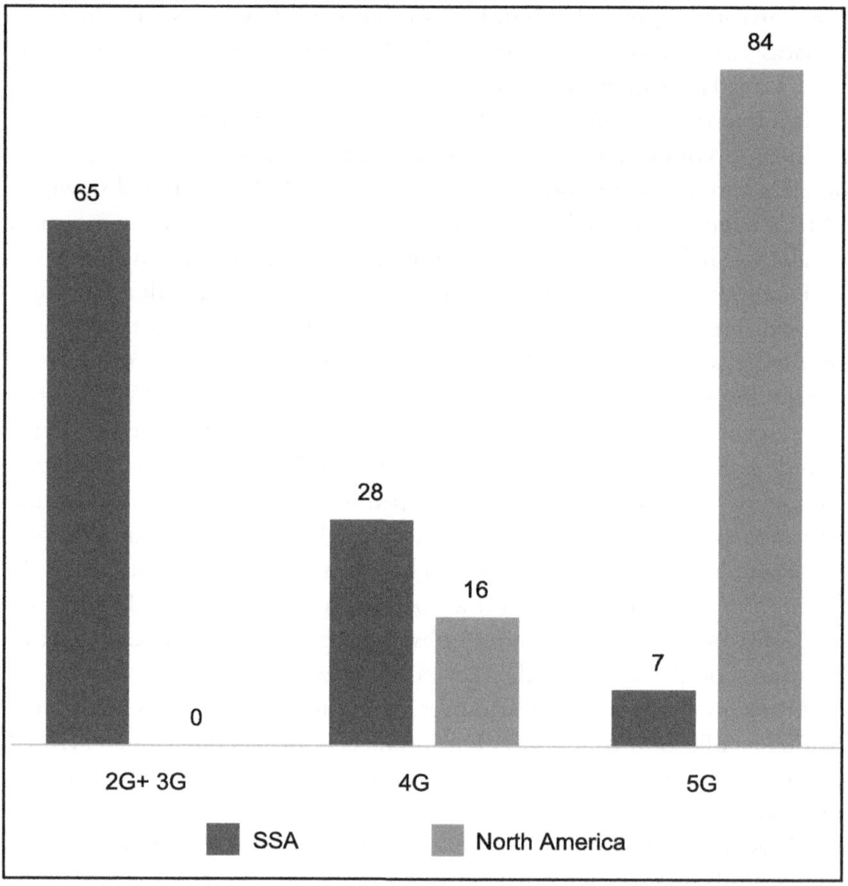

Figure 12.1. Mobile subscriptions by region and technology (%, estimates for 2026). Data source: Ericsson, *Ericsson mobility report, 2021*, https://www.ericsson.com/49f7c7/assets/local/mobility-report/documents/2021/june-2021-ericsson-mobility-report.pdf.

underreported in the official source.[44] Earth observation data such as satellite imagery data thus can be used for monitoring impacts of public health issues and other societal problems.

Studies have found that people who are able to escape from extreme poverty may again fall back into it. Some key mechanisms to help people escape

44. Abdihamid Warsame, "Somalia and COVID-19: How we used satellite data to track the toll of the pandemic," *Conversation*, October 26, 2021, https://theconversation.com/somalia-and-covid-19-how-we-used-satellite-data-to-track-the-toll-of-the-pandemic-167497.

permanently from poverty include increasing investments in education and associated infrastructure, creating opportunities for employment and self-employment, and ensuring that farmers make productive use of their land.[45] 4R technologies can help achieve these goals in a number of ways. Technologies such as drones and AI can help smallholder farmers better use their land and increase agricultural productivity (chapter 7). AI and other 4R technologies can also ensure access to finance for educational needs (chapter 8).

Facilitating Inclusive Democratic Participation

A study of 317 presidential elections in 40 African countries from 1960 to 2016 found that voter participation is low during times of slow economic growth.[46] People suffering from adverse economic conditions feel they need to deal with more urgent "bread-and-butter" issues. Concerns related to politics become a much lower priority for them.[47] When conditions such as broadband access and digital user skills improve and 4R technologies such as AI and blockchain develop further, voting access in poor and marginalized communities can be increased. For instance, eligible voters can cast a ballot anonymously using a computer or a smartphone. Blockchain uses an encrypted key and tamper-proof personal IDs.[48] Facial recognition can help verify a voter's identity. In this way, 4R technologies can also facilitate inclusive democratic participation.

Implications for Businesses

I now turn to the implications of the discussion of this book to businesses from LMICs and from rich countries. Regarding technology firms in LMICs, such firms are latecomers, characterized by relatively lower technological capabilities. A strategic-asset-seeking strategy is more likely to be

45. Katie Ellis and Angela Hawk, "The chronic poverty report 2014–2015: the road to zero extreme poverty," Overseas Development Institute, 2014, https://cdn.odi.org/media/documents/8834.pdf.

46. Halfdan Lynge and Ferran Martinez i Coma "The effect of economic downturns on voter turnout in Africa," *Electoral Studies* 76 (April 2022): 102456.

47. S. J. Rosenstone, "Economic adversity and voter turnout," *American Journal of Political Science* 26, no. 1 (1982): 25–46.

48. Nir Kshetri and J. Voas, "Blockchain-enabled e-voting," *IEEE Software* 35, no. 4 (2018): 95–99.

pursued by such firms,[49] as their access to strategic assets is critical in gaining competitive advantage in the marketplace. Strategic assets are resources and capabilities that can help a firm gain competitive advantages.[50] Such a strategy can help these firms reduce gaps with competitors by acquiring needed resources.[51]

Firms based in LMICs firms may need to utilize different strategies in the 4R era than did firms in earlier technological revolutions. For instance, to get superior resources and skills not available at home, Asian firms tend to seek such assets in advanced host countries.[52] In new industries such as AI, the gaps between firms in LMICs and rich countries are relatively low compared to in established industries.

Data are a key strategic asset in 4R technology industries such as AI. Increasing access to high-quality and relevant data thus should be one of the key focus areas to succeed in the 4R era. Organizations and people should be open to changing business processes and adopting new ways of thinking. According to a survey conducted by the MIT Technology Review Insights for its report *The global AI agenda: The Middle East and Africa*, for instance, changing business processes around AI was the greatest challenge worldwide. Globally 51% of the respondents considered this a challenge, compared to 58% the Middle East and Africa.

Among the major obstacles that technology companies from LMICs face are lack of education and information about 4R technologies.[53] Technology companies need to develop effective education and awareness programs to help key stakeholders understand the economic and social transformation potential of 4R technology. For instance, there has been a lack of awareness of blockchain among key stakeholders. The company Saldo needed to educate underbanked communities. The company also worked with the Mexican foreign ministry to hold financial literacy events. Saldo found that it is too complex to talk about blockchain. It started educating financial institutions first, which are more familiar with blockchain.[54]

For multinational firms from advanced economies, 4R technologies can

49. P. Deng, "Why do Chinese firms tend to acquire strategic assets in international expansion?," *Journal of World Business* 44, no. 1 (2009): 74–84.

50. R. Amit and P. J. H. Schoemaker, "Strategic assets and organizational rent," *Strategic Management Journal* 14, no. 1 (1993): 33–46.

51. Tom Wesson, *Foreign direct investment and competitive advantage* (Cheltenham, UK: Edward Elgar Publishing, 2004).

52. Shige Makino, Chung-Ming Lau, and Rhy-Song Yeh, "Asset exploitation versus asset seeking," *Journal of International Business Studies* 33, no. 3 (2002): 403–421.

53. J. J. Valenzuela, "Bitcoin remittances to Mexico see huge potential," *Cointelegraph*, 2015, https://cointelegraph.com/ news/bitcoin-remittances-to-mexico-see-huge-potential.

54. Valenzuela, "Bitcoin remittances."

also help respond to some of the major environmental, social, and governance (ESG) issues that companies typically face in dealing with the B4B. This issue is becoming more significant today as investors are increasingly using ESG compliance to screen potential investments. This is understandable; according to some estimates, of the 2.7 billion unbanked and underbanked people, about 1 billion are supply 5,000 global brands in some way.[55] Media coverage has shed light on highly unethical practices in mineral and metal extraction activities (especially cobalt), including child labor, human rights violations, and environmental damages.

Blockchain in combination with other technologies can help in ESG compliance (In Focus 12.1). The proponents of this technology have claimed that blockchain-based traceability initiatives are more effective than major other nonblockchain traceability programs. The traceability-as-a-service (TaaS) provider Circulor has developed solutions based on blockchain and other 4R technologies to reduce unethical behaviors in the supply chains of minerals such as cobalt and tantalum (In Focus 12.1).

Likewise, Anheuser-Busch InBev's legal and compliance teams worked with BanQu to help reduce contract workers' harassment or discrimination risk. In a pilot project in South Africa, a distributed ledger platform captured secure, traceable information on risks faced by mostly female brand promoters.[56]

Multinational firms' can also bring their resources such as technologies and skills in LMICs, which can result in a win-win situation. Some Western companies with expertise in 4R technologies have developed solutions to address economic, environmental, and societal challenges in LMICs. To take an example, in 2017, the government of Andhra Pradesh, India, collaborated with the Swedish startup ChromaWay to implement a blockchain-based land recording project in the capital city of Amaravati. ChromaWay had gained significant experience working with Sweden's land registry authority, Lantmäteriet.[57] Regarding the involvement of ChromaWay, it is worth noting that a complaint raised against the company is that it has merely proved the concept of blockchain-based land records again and again, failing to show traction and scalability.[58] ChromaWay's project in Andhra Pradesh attained some degree of scalability that had been missing.

Likewise, LMICs are suffering from a severe skill shortage. According

55. Paynter, "This digital ledger."

56. Yasmin Lambert, "In-house legal teams: pandemic tests speed and adaptability," *Financial Times*, December 10, 2020, https://www.ft.com/content/ac78ed05-e3b3-4c9a-84e7-106c7baea97d.

57. Asia Insights, "This Indian city."

58. Fintech, "Using blockchain."

to a study by the IFC, 230 million jobs in Africa need some level of digital skills by 2030.[59] This requirement also creates 650 million training opportunities and an estimated US$130 billion market.[60] Western firms are well positioned to lead efforts to help these countries develop required skills.

In Focus 12.1: Circulor's Blockchain-Based Cobalt Supply Chain Traceability Solution

Circulor operates a blockchain platform across the supply chain of China's Contemporary Amperex Technology Co.[61] The automotive company Volvo is a participant in CATL's supply chain, operated by Circulor.

Circulor utilizes the Oracle's blockchain platform (OBP), based on Linux Hyperledger Fabric. Oracle is a BaaS provider. OBP sets up, manages, and maintains the blockchain platform for enterprises.[62]

OBP is combined with AI algorithms to perform due diligence and identify data anomalies and actions that need further investigation. Data captured include the ore's origin, attributes (e.g., weight, size), chain of custody, and information to establish supply chain participants' compliance with globally recognized supply chain guidelines.[63]

When a registered mining company that has a concession applies

59. International Finance Corporation, *Digital skills in Sub-Saharan Africa spotlight on Ghana*, https://www.ifc.org/wps/wcm/connect/ed6362b3-aa34-42ac-ae9f-c739904951b1/Digital+Skills_Final_WEB_5-7-19.pdf.?MOD=AJPERES.

60. Alejandro Caballero and Sajitha Bashir, "Africa needs digital skills across the economy—not just the tech sector," *World Economic Forum*, October 22, 2020, https://www.weforum.org/agenda/2020/10/africa-needs-digital-skills-across-the-economy-not-just-tech-sector/.

61. Niclas Rolander, "Volvo cars goes for blockchain tech to avoid unethical cobalt," *Bloomberg*, November 5, 2019, https://www.bloomberg.com/professional/blog/volvo-cars-goes-for-blockchain-tech-to-avoid-unethical-cobalt/.

62. Vivek Acharya, "Oracle Blockchain Platform (OBP)—A driver in proliferating blockchain adoption," Government Blockchain Association, September 10, 2019, https://www.gbaglobal.org/oracle-blockchain-platform-obp-a-driver-in-proliferating-blockchain-adoption/.

63. Rachel Wolfson, "Understanding how IBM and others use blockchain technology to track global food supply chain," *Forbes*, July 11, 2018, https://www.forbes.com/sites/rachelwolfson/2018/07/11/understanding-how-ibm-and-others-use-blockchain-technology-to-track-global-food-supply-chain/#6f7249602d1e.

to use Circulor's mine-to-manufacturer traceability, the coordinates of the mine's operations and its historical production are entered in the system. Satellite data is used to verify that the mine is working[64] Circulor's plan is to use machine learning models and aerial imagery to ensure that child labor has not been used in the production process.[65]

The application's field test was carried out for tantalum mined in Rwanda and then for cobalt used in Volvo Cars' electric vehicle batteries. For the project in Rwanda, Circulor teamed up with the government of Rwanda and Power Resources Group (PRG), which has mining and refining operations in Rwanda and Macedonia.[66] In 2014, Rwanda accounted for 50% of the production of global tantalum concentrates.[67] As of November 2019, Volvo, CATL, and other supply chain participants were reported to record about 28 million material scans and other production events per month on the Oracle platform.[68]

The combination of AI and blockchain can be an effective way to address information and knowledge gaps, which represent a major challenge for supply chains.[69] There is the lack of reliable, authentic, and credible information about sustainability impacts at various phases of supply chains. Reliability and authenticity of data in the first mile of the supply chain, the most crucial step in assuring the quality of the ore,[70] are a key challenge. For instance, blockchain sys-

---

64. Matthew Burbidge, "Proving provenance," *Brainstorm Magazine*, May 15, 2019, http://www.brainstormmag.co.za/business/14571-proving-provenance.

65. Michael Kapilkov, "Volvo invests in blockchain startup to trace cobalt in its batteries," *Cointelegraph*, July 8, 2020, https://cointelegraph.com/news/volvo-invests-in-blockchain-startup-to-trace-cobalt-in-its-batteries.

66. Emmanuel Côme Mugisha, "Rwanda keen on accelerating development of blockchain," *New Times*, December 12, 2019, https://www.newtimes.co.rw/news/rwanda-keen-accelerating-development-blockchain.

67. Katharine Sanderson, "Concerns raised over tantalum mining," *Nature*, December 14, 2015, https://www.nature.com/news/concerns-raised-over-tantalum-mining-1.19023.

68. Wolfson, "Understanding."

69. M. Boström, A. M. Jönsson, S. Lockie, A. P. J. Mol, and P. Posterveer, "Sustainable and responsible supply chain governance: Challenges and opportunities," *Journal of Cleaner Production* 107 (2015): 1–7.

70. Fritz Brugger, "Blockchain is great, but it can't solve everything: Take conflict minerals," *African Arguments*, April 23, 2018, https://africanarguments.org/2019/04/23/blockchain-is-great-but-it-cant-solve-everything-take-conflict-minerals/.

tems can be corrupted if the government agents whose role is to tag bags collude with smugglers and enter incorrect data.[71] In Circulor's system, miners enter the data, and their identities are confirmed by facial recognition software.[72]

Future Research Implications

In this section, I suggest several potentially fruitful avenues for future research. Many big companies have clearly benefited from the use of 4R technologies. Zambian Breweries streamlined its buying process and optimized supply chain. Its revenue increased by 17% to reach US$114.8 million 2019 from US$98 million in 2018.[73] In Colombia, AB InBev's subsidiary Bavaria Breweries uses BanQu's technology to meet its sustainability goal: to use 100% returnable or recyclable packaging in its products by 2025.[74] However, regarding the question of whether technological solutions designed and initiated by these companies have also benefited marginalized groups, these companies tend to provide cherry-picked and sometimes misleading data. For instance, Zambian Breweries noted that smallholder farmers that supplied the company earned more than US$1.5 million in 2019.[75] Zambian Breweries has not reported earning per farmer, but its plan was to have 4,500 cassava farmers by the end of 2019,[76] which translated to per-farmer earnings of about US$333. Zambian Breweries also noted that a 53-year-old farmer and mother of 13 increased sales from 3,800 kilos of cassava to more

71. Joeri Cant, "Block firm helps Congo mine fight against blood diamonds," *Cointelegraph*, October 2, 2019, https://cointelegraph.com/news/blockchain-firm-helps-congo-mine-to-fight-against-blood-diamonds.

72. Aaron Ross and Barbara Lewis, "Congo mine deploys digital weapons in fight against conflict minerals," *Reuters*, October 1, 2019, https://www.reuters.com/article/us-congo-mining-insight/congo-mine-deploys-digital-weapons-in-fight-against-conflict-minerals-idUSKBN1WG2W1.

73. Natalie Marchant, "This start-up is using blockchain to help smallholder farmers prosper," *World Economic Forum*, May 28, 2021, https://www.weforum.org/agenda/2021/05/banqu-financial-inclusion-sustainability/.

74. BanQu, *Gaining visibility and ensuring ROI on returnables and recyclables*, 2021, https://banqu.co/use-cases/gaining-visibility-and-ensuring-roi-on-returnables-and-recyclables/.

75. BanQu, *Eliminating supply chain blindspots, boosting the bottom line*, 2021, https://banqu.co/use-cases/eliminating-supply-chain-blindspots-boosting-the-bottom-line-1/.

76. Paynter, "This digital ledger helps."

than 12,000 kilos in a year.[77] A few success stories such as this do not offer adequate information to assess the potential of 4R technologies in improving the welfare of the B4B. Without more detailed information such as the amount Zambian Breweries paid to the farmers before implementation of the system, earnings per farmer, and knowledge of whether farmers actually used their blockchain-based sales records to connect with NGOs, local cooperatives, MFIs, and banks to receive loans, it is not possible to assess the real impact of these solutions. Thus, what remains to be fully explored is the extent of true benefits of such solutions to farmers and other marginalized groups. This is among the issues to be considered in future research.

There has been arguably a lack of a reliable theory of growth to help countries navigate the transition from middle- to high-income status.[78]The proper utilization of 4R technologies may help middle-income countries maintain a high growth trajectory to help their transition to a high-income status. To provide a systematic understanding of the potential of the 4R in helping countries get out of the middle-income trap, future research is needed to further develop theory and enrich the understanding of middle-income countries' development of their capacity to benefit from the 4R.

One issue that was raised in this book is how 4R technologies are likely to be used differently by low-income and rich people. An intriguing avenue for future research is to more systematically examine how populations in different categories (e.g., poverty market, submerged market and exchange market or poor, vulnerable to falling into poverty, middle class and rich) differ in how they use different 4R technologies explored in this book such as blockchain, IoT, and remote sensing. The benefits that these different groups of people perceive from these technologies are also a worthwhile target of study.

Technology solutions for the B4B discussed in this book have been initiated and supported by various entities and actors. These include foreign technology companies, local startups, and international actors such as UN agencies. Future research also needs to evaluate the varying competences, capabilities, resources, and motivations of these different groups in designing solutions based on 4R technologies to serve the B4B.

The 4R has led to the emergence of new economic activities in LMICs. One such example is the data-labeling industry. Social and ethical issues in

77. BanQu, *Eliminating supply*.

78. Indermit S. Gill and Homi Kharas, "The middle-income trap turns ten," *Social Science Research Network*, August 26, 2015, https://papers.ssrn.com/sol3/papers.cfm?abstract_id=2651983.

this new industry has not been examined. For instance, what is not clear is whether data-labeling firms in developing countries are operating in more or less ethical ways compared to other foreign firms operating in these countries. Some critics have claimed that this industry has features that are akin to slavery. While data-labeling firms claim to engage in activities that have positive social impacts, it is not easy to assess the validity of such claims. Data labeling companies have their own definitions of ethical and fair practices. Moreover, the definitions vary widely across them. Thus, we may not be able to take self-reported information by data-labeling firms as proof that these firms are creating more positive social impacts in developing countries than other foreign firms. In many cases, the problem of assessing such claims is made more complex by the fact that they do not publish any information or fail to provide relevant information on such impacts. Future researchers might conduct systematic and in-depth analyses of these new industries vis-à-vis other modern industrial sectors such as the business process outsourcing industry.

Final Thoughts

Currently, major 4R technologies have various limitations, such as bias and lack of explainability in AI and of interoperability and standardization in blockchain solutions. Most LMICs also lack technological expertise and absorptive capacity to benefit from these technologies. A related point is that many technology startups in these economies perform poorly in indicators relating to level and quality of entrepreneurial activity. 4R technology projects also suffer from the lack of education and information and opposition from key actors. It is reasonable to expect that over time, many of these challenges are likely to diminish. 4R technologies thus hold the potential to fight the extreme poverty that has resulted from corruption, mismanagement, inefficiency in public infrastructure, and weak institutions in LMICs.

4R technologies can help firms in LMICs to utilize direct and observable quality signals of their products. Technologies such as blockchain provide mechanism to reward socially and environmentally responsible activities and make it possible for the B4B populations to participate in the global market. However, the adoption of these technologies among the B4B and economic productivity achieved are far from the full potential.

The analysis of this book highlights both positive and negative social impacts associated with 4R technologies. Negative impacts of 4R solutions have led to social problems as serious as a large number of suicides. Factors such as immature technologies and the lack of regulatory oversights have contributed to these adverse outcomes. For instance, it is likely that the algorithms currently used by lenders favor potential borrowers that are more susceptible to shaming and not necessarily those with ability to pay loans. The shaming strategy to collect debt is likely to be more effective if the borrowers have a large number of people in their networks. In addition to economic harm, the strategies used by fintech companies to collect debts entails many indirect costs, including emotional stress for borrowers. This means that in many cases, some marginalized groups are better off not borrowing money at all. Governments should work with technology companies to address these challenges. For instance, it is critical to develop AI algorithms and build models for evaluating financial literacy and entrepreneurial skills of borrowers. P2P lending regulations must be put in place to ensure that lenders refrain from offering loans to financially illiterate individuals or those that lack entrepreneurial skills.

Big companies' needs are often the key determinant when technological solutions are developed to connect the poor. Big companies have skills, resources, and financial capacity to benefit from these technologies. However, those benefits are conditional on poor people having skills to utilize the technologies and the existence of opportunities for them. Most LMICs also have lower awareness and underdeveloped ethical systems related to data privacy and ownership, which usually benefit big technology companies. The lack of a well-developed ecosystem around these technologies (e.g., payment systems) also hinders low-income people's ability to take full advantage of these technologies.

Developed world firms can easily access resources from LMICs, but not vice versa. For instance, as discussed in chapter 9, some industrialized countries' actions such as denial of visas for researchers and scholars to attend scientific conferences have inhibited the free flow of ideas that is essential for scientific progress and long-term development of the 4R technologies.

Various examples discussed in this book, such as Babyl in Rwanda, indicate that some firms in the Global North are exporting their solutions to LMICs despite the products' poor performance in their home countries. The functionality worsens because data are wrong or irrelevant in a different setting. Among the most important lesson is that LMICs must develop their own digital solutions and rely less on rich countries. This is because most solutions imported from industrialized countries tend to have a low degree of usefulness and high costs in the context of LMICs. This requires introducing 4R-friendly policies, intensifying R&D activities, and developing local digital manpower.

Glossary

3D printing
A computer-aided design (CAD) model to build a three-dimensional object, usually by adding material layer by layer.

Agricultural technology
The use of technology such as sensors, devices, machines, and AI in agriculture to improve yield, efficiency, and profitability.

Algorithm
A process or set of rules and instructions, typically used to solve specific problems or to perform a computation.

Anti-money laundering
The activities and processes that financial institutions perform to comply with laws and regulations intended to stop criminal financial activities.

Application programming interface
A set of programming codes that works as an intermediary to allow two applications to talk to each other.

Artificial intelligence
Simulation of human intelligence by machines to perform tasks that seemed to be possible only with human thinking and logic before.

Augmented reality
The use of sensors and algorithms to help a computer determine the position and orientation of a camera from whose viewpoint 3D graphics are

made available and computer-generated images are superimposed into a user's view of the real world.

Big data
Huge amount of structured and unstructured data that can be processed to enhance insight and improve decision-making.

Blockchain
A decentralized ledger that maintains digital records of a transaction simultaneously on multiple computers and mathematically connects block of records entered into the ledger to other blocks to form a chain of immutable records.

Business process outsourcing
Long-term contracting of a firm's noncore business processes to an external service provider.

Central bank digital currency
A country's fiat currency's digital form, similar to cryptocurrencies in many ways but not necessarily based on blockchain.

Chatbot
A software application that simulates human conversations via text messages on chat using artificial intelligence and natural language processing to understand what a user wants.

Cloud computing
Hosting applications on servers and delivering services via the internet (software, platforms, and infrastructure) for users to access computing power and resources and pay for services based on usage.

Country of origin
The country where a product is manufactured, produced, or designed.

Cryptocurrency
A currency that uses blockchain and is encrypted using cryptography and functions like money, which means that it defines value, serves as a value transfer and can be used for making and receiving payments.

Data labeling
Adding target attributes to training data and labeling them to make it possible for a machine learning model to learn the predictions it is expected to make and prepare data for supervised machine learning.

Data divide
The gap among countries of various levels of economic development in the availability of data needed for scientific research and decision-making.

Data privacy
The use and governance of personal data, which is concerned with individuals' rights to control their personal information and how such information is used.

Digital colonialism
The use of digital technologies to dominate another nation or territory politically, economically, and socially.

Digital twin
Providing a virtual representation of and describing a physical entity accurately both at micro and macro levels.

Environmental, social, and governance criteria
A set of standards often used by investors to assess a company's operations in terms of environmental criteria (e.g., pro-environmental initiatives undertaken in managing natural resources and the natural environment), social criteria (relationships with employees, suppliers, customers, and communities) and governance (e.g., leadership, executive pay, shareholder rights).

Ethereum network
A public blockchain-based open software platform, in which each node can be discovered by and known to other nodes in the network.

Explainable artificial intelligence
Processes and methods that make it possible for human users to comprehend the results and output that machine learning algorithms create.

Financial technology
Computer programs and other technologies such as mobile money or contactless payments, nontraditional credit-scoring models based on information from consumers' social media behaviors, cryptocurrencies, and AI-enabled chatbots used by financial institutions to help customers that aim to improve and support the delivery and use of banking and financial services.

Fourth Industrial Revolution
Merging of advanced and emerging technologies such as artificial intelligence, autonomous vehicles, and the internet of things with humans' physical lives.

Gateway
A network node that connects two networks that have different transmission protocols.

General Data Protection Regulation
Regulation in the European Union to governs personal data in the member countries.

General-purpose technology
A technology that possesses potential to generally transform a wide range of household as well as business activities.

Genome editing
Making changes to the DNA of a cell or an organism.

Global North
An alternative term to refer to developed countries.

Global South
A value-free alternative to "Third World" used to refer to economies in Africa, Latin America, and developing Asia including the Middle East.

Group of Twenty (G20)
An intergovernmental forum that consists of 19 countries and the European Union.

High-income economies
According to the World Bank, economies that had a per capital gross national income of $12,696 or more in 2020.

Hyperledger fabric
An open-source blockchain platform from the Linux Foundation, provided by IBM as "Blockchain as a Serv ice" and is targeted at businesses.

Industrial internet of things
The extension internet of things applications such as smart sensors and actuators in industrial sectors and applications.

Information and communications technology
A technological tool and resource that can be used to transmit, store, create, share, or exchange data and information.

Internet of things
The network of physical objects or "things" (e.g., machines, devices, appliances, animals, people) embedded with electronics, software, and sensors, which are provided with unique identifiers and possess the ability to transfer data across the web with minimal human intervention.

Know Your Customer
The process of verifying identity and risks of customers to ensure compliance with money-laundering laws.

Leapfrogging
Occurs when a developing country skips the stages of the path taken by industrial nations and hence catches up sooner in terms of technology adoption or economic growth.

Least developed countries
Low-income countries that are vulnerable to economic and environmental shocks and have low levels of human assets and thus face severe structural barriers to sustainable development.

Light-emitting diode
A light source made of semiconductor materials that emits light when current flows through it.

Long-term Evolution
A wireless broadband communication standard for mobile devices and data terminals, which is based on the GSM/EDGE and UMTS/HSPA standards

Low-income economies
Economies that, according to the World Bank, had a per capita gross national income of $1,045 or less in 2020.

Low-Power, Wide Area Networks
A low power wide area network protocol promoted by the LoRa Alliance, which is built on top of Semtech Corporation's long-range wireless radio frequency.

Mainnet
A blockchain protocol that is fully developed and deployed for actual cryptocurrency transactions with value, which means that transactions are broadcasted, verified, and recorded on a blockchain

Median voter theory
A theory arguing that competition between political parties would lead to an outcome that tends to favor the median voter.

Microfinance institution
A provider of loans that are smaller than those traditional banks normally grant.

Middle-income economies
Economies that, according to the World Bank, had a per capita gross national income between $1,046 and $12,695 in 2020 (specifically, between $1,046 and $4,095 as lower middle-income economies and between $4,096 and $12,695 as upper-middle-income economies).

Middle-income trap
A situation in which countries that have got out of the poverty trap and grew to middle-income levels subsequently fail to grow further to achieve the levels advanced countries.

Multinational corporation
A corporation that is registered and operates in two or more countries.

Natural language processing
A subfield of artificial intelligence that deals with the ability of computer programs to understand process and analyze spoken and written human languages, referred to as natural language.

Parametric insurance
An insurance model in which payouts are made based on the occurrence of clearly defined events.

Peer-to-peer lender
Helps connect borrowers to entities that are willing to fund their loans.

Permissionless blockchain.
An open blockchain platform functioning like a shared database, which

means that anyone can join, everyone can read everything, and a user cannot control who can write.

Poverty trap
Mechanisms that serve to strengthen themselves, which cause poverty to persist.

Proof of history
A mechanism to let users create a historical record, which provides a way to cryptographically verify a passage of time between two events.

Proof of stake
A consensus model in blockchains in which only a small group of nodes can validate transactions.

Proof of work
A consensus model in blockchains in which all users can compete to verify transactions.

Quick response code
A type of two-dimensional barcode, that is, a machine-readable optical label which contains information about the item to which it is affixed.

Remote sensing
Use of cameras installed in satellites or aircrafts to detecting and monitor an area's physical characteristics with radiation that is reflected and emitted from the area.

Robotics
Design, construction, and use of robots.

Sensor
A device that may contain some or all elements of a computer (e.g., processor, memory, storage, inputs and outputs, software) and responds to some type of stimulus input (e.g., location, atmospheric pressure, altitude, velocity, light, heat, temperature, pressure, illumination, motion, moisture, power, humidity, blood sugar, air quality, soil moisture, vehicular movement) from the physical environment by emitting a signal.

Small and medium-sized enterprises
Businesses whose number of employees fall below certain limits, typically fewer than 250.

Smallholder farmers
Farmers that operate fewer than two hectares of cropland.

Smart contract
A contract that executes automatically when certain conditions are met.

Stablecoins
Cryptocurrencies whose values are pegged to a traditional fiat currency such as the US dollar.

Stockholder centric bias
The tendency to take actions that favor stockholders more compared to other stakeholder groups

South-South cooperation
The exchange of resources, technology, and knowledge among economies in the Global South.

Supervised machine learning
Training or teaching computers regarding what to look for.

Third Industrial Revolution
The shift from mechanical and analogue technology to digital electronics such as digital computers and digital record keeping that began in the second half of the twentieth century.

Total factor productivity
A measure of productive efficiency that focuses on the amount of output that can be produced from a given amount of inputs

Traceability-as-a-service
A blockchain-based traceability solution, which is often combined with the internet of things to enable the traceability of products from source to consumer by connecting all relevant points in a supply chain.

Validator
A network node in a blockchain that helps process and validate transactions so that they can be added to the permanent ledger.

Virtual reality

The use of computer technology to place the user inside an experience with the simulation of senses such as vision, hearing, touch and smell by creating a simulated environment or experience, which could be similar to or different from the real world.

Wearables

Electronic devices powered by microprocessors with the ability to send and receive data via the internet, which can be worn as accessories, embedded in clothing or implanted in a human body.

Wide area network

A large network of information that is used to facilitate communication and sharing of information between devices that are not in a single location.

About the Author

Nir Kshetri is a professor at University of North Carolina-Greensboro. He has authored ten books and about 200 academic articles, which have been translated into Arabic, Chinese, German, Spanish, French, Japanese, Portuguese, and other languages. He has been ranked among the world's top 0.03% researchers according to Elsevier BV and Stanford University's "Updated science-wide author databases of standardized citation indicators" published in August 2021 (https://elsevier.digitalcommonsdata.com/datasets/btchxktzyw/3) . Nir's work has been featured by hundreds of media outlets, such as *BBC, Barron's, Economist, Al Jazeera, Wall Street Journal, Foreign Policy, Public Radio International, Scientific American*, and *Bloomberg* TV. He has provided consulting services to Asian Development Bank, the Commonwealth Secretariat, various UN agencies, and a number of private companies. His editorial roles include computing economics editor of *Computer*, IT economics editor of *IT Professional*, and associate editor of *Electronic Commerce Research*. He is a two-time TEDx speaker about the roles of emerging technologies such as artificial intelligence (https://www.youtube.com/watch?v=W6da0kEfBsY&t) and blockchain in fighting poverty (https://www.youtube.com/watch?v=WDo_Jlov9R4).

Index

ACRE, 193–94, 217–19, 221, 239–40, 247, 315–16
Ada Health, 165–66, 232
Afreximbank, 17–21
Africa Centers for Disease Control and Prevention, 175
African Information Communication Technology Alliance, 323
African Union, 18, 284, 306
Agryo, 183; Brazil's, 183
AgTech, 201–2, 241
AI, xiii, 3, 9, 11–12, 14, 18–19, 22–26, 31, 37–41, 45–71, 90, 99, 109–11, 115, 122–24, 126, 128, 142, 152, 157, 165–66, 168, 170, 175, 177, 180, 183–84, 188, 190, 193–95, 201, 206, 208–12, 222, 227, 235–36, 238, 241–46, 248–53, 255, 258, 262–64, 269–70, 272–73, 277–86, 290, 296–99, 301–2, 308–11, 313–16, 320, 322–25, 327–28, 330–31, 334–35, 337–39; auditability of, 251; benefits of, 157, 299; effects of, 55, 62, 64; explainability in, 322, 334; explainability of, 251; use of, 52, 55, 66, 280. 284–85, 313
Alice, 45, 201, 241; Solinftec's AI assistant, 45–46, 201
AMRU Rice, 266–67
Anemia, 158
Angola, 257–60, 278, 283, 301; oil-rich, 257
Anheuser-Busch, 188–89, 254, 329
API, 93, 130, 183, 212, 216, 219; Agryo, 183; Bext360's, 111; Flutterwave's, 241; Migo's, 211
Apollo Agriculture, 192–93
Apollo Hospitals, 171, 321
Aravind Eye Hospital, 38–39, 168, 316
Arbol,217, 219–21, 268, 316
artificial intelligence, 209, 290, 337–38; explainable, 339
authoritarian, 253, 298, 301, 303, 315, 319
Azuri Technologies, 142, 152

Babylon, 166–68, 320
Baidu, 274–75, 279–80; Chinese search engine, 279
Baidu Busca, 279
balloons, 299, 300, 321; Google's, 299; loon, 299, 321
Bank Verification Number, 322
Bankymoon, 72, 81
BanQu, 15–17, 94, 188–90, 198–99, 238, 254, 329, 332
Bboxx, 131. 133–34 ; UK-based pay-as-you-go solar provider, 133
Belt and Road Initiative, 296
Benin, 20, 41–43, 161, 179
Bext360, 111
Bextmachine, 111
bioethanol, 147–49

bitcoin, 27, 72, 86, 91, 217, 295
BitFury, 86
Bitland, 81, 86
BitPesa, 295
blockchain, xiii, 3, 9, 15–16, 19, 25–27. 31, 61, 72–77, 79–86, 88–102, 110–11, 164, 169–72, 177, 183, 188, 193–94, 197–99, 204–5, 212, 214, 216–27, 232–33, 238–40, 244–45, 249, 252, 254, 262, 265, 268, 288–90, 295, 316, 318, 327–31, 333–34, 338–40, 342–44; applications of, 9; implementations of, 85; use of, 27, 76, 84, 95, 238, 265, 289; WeChat's, 84
BlocRice, 266–67
Botnar, 166
Brazil, 45, 48, 51–52, 55–56, 62, 91, 113, 120, 165, 183, 199, 201–2, 241, 257, 279, 298, 303, 317

Cambridge Analytica, 68–69
carbon dioxide, 48, 161
Carcafe, 196
Care Quality Commission, 167
Cashew, 41–43
CATL, 330–31
CBDC, 93
CCP, 271, 273–75, 279
CDMA, 134
Central bank digital currency, 93, 338
Central Bank of Nigeria, 322
Century Microfinance Bank, 182
Chained Finance, 81, 88–89
Chainlink Community Grant, 218
Chainlink, 183, 217, 219, 221; decentralized oracle network company, 217
chatbot, 24, 26, 50–51, 70, 166, 210, 314, 338–39; Ada Health's, 165; Ant Group's AI-based customer service, 208; banking, 210; conversational, 26; customer-service, 24, 50, 52; Diamond Bank's, 314; multilingual, 290; United Bank's, 323
Chekkit, 170
Child Labor Monitoring System, 199
child labor, 75, 198–99, 329, 331
Circulor, 3303–2; TaaS provider, 329
Clean-Burning Stove, 136, 160–62
CloudWalk, 297; Chinese company, 296
COLCO, 185–87, 200
Colombia, 27, 112, 185, 260–61, 288–89, 332; bitcoin in, 27; EcoProMIS project in, 292, 316; monetary poverty in, 260; palm growers in, 185; Procuraduría General de, 289; rural cocoa-growing regions of, 261; Ruta-N in, 186
ConsenSys, 100
cost of enforcement, 79, 140, 143
cost of measuring, 22, 79, 140
cost of networking, 73
cost of verification, 73
costs of measurement, 22, 140
counterfeit drug, 172, 177
COVID-19, 11, 13, 31, 32, 39, 40, 87, 103, 108–9, 112–34 116–21, 125, 151, 158, 164, 174–77, 222, 224, 251, 316, 325
cupping, 195–96

data divide, 64, 105, 245, 339
data labeling, 57–59, 70, 263–64, 334, 338; AI, 258; global market for, 57
data protection and privacy legislation, 284
data-sharing, 305–6
De Beers, 75–76
deep learning, 44, 52–53, 62, 109–10, 113, 116, 118, 126, 157, 200, 209, 236, 246, 321
DeFi, 217
Demetria, 193, 195–96, 316
Democratic Republic of Congo, 8, 78, 137, 179
diabetic retinopathy, 39, 45, 158, 316
digital colonialism, 255, 295, 339
digital neocolonization, 298
Digital Society of Zimbabwe, 297
DNA, 163–64, 340
dosifier, 159–60
drone, 41–42, 110, 113, 122, 168, 172–74, 176–77, 184–85, 236, 287, 299, 313, 317, 327; commercial, 173, 287; crop-spraying, 287; delivery, 287
due diligence, 75, 330; customer, 21, 309; high costs of, 209

EcoProMIS, 185–86, 292, 316
ED&F Man, 196
EHR, 167
E-Livestock Global, 197
Embu, 145–46, 152
entrepreneur, 77–78, 100, 162, 204, 235, 262, 271; energy, 162
entrepreneurial activities, 85, 98, 143, 203, 313, 319; high levels of, 233; IoT's impacts on, 130; productive, 283
Environmental Protection Agency, 294
ESG, 329
espionage, 300, 311
Ethanol, 45, 146–48; KOKO's IoT solution for, 147
Etherisc, 193, 218–21, 240; smart contract developer, 217–18
ethical, 33, 63, 65, 74–75, 284–86, 304–5, 310–11, 333–35
ethics dumping, 67, 311
ethics shirking, 67, 69, 311
ethics, 65, 276, 286, 304

facial recognition, 66–67, 211, 252–53, 272, 278, 284–85, 296–97, 299, 301–2, 327, 332
fair-trade goods, 304
Fantom, 170–71
FAO 47, 77, 180
farmer, 9, 17, 20, 23, 41, 43, 46–47, 95–96, 110–11, 122–23, 131–35, 151–52, 180–94, 197–202, 213–19, 221, 227, 232–33, 239–40, 245, 247, 254, 261–63, 265–68, 292–94, 313, 316, 319, 327, 332–33, 344; African, 134; app for, 196; barley, 189–90; Cambodian, 233, 265; cassava, 189, 332; coffee, 193, 196; commercial, 194, 197; earnings per, 333; female, 17; illiterate, 267; Indian, 318; indigenous Andean, 104; informal sector, 187; large commercial, 189; organic , 266; paddy-field, 221; Rwandan, 17; rice, 266–67; small-scale, 181, 189, 246; smallholder, 30, 41, 46, 96, 122, 131–33, 152, 179–82, 184, 187–88, 190–91, 193–95, 202, 213, 216–19, 221, 227, 239–40, 327, 332, 344; tea, 95; U.S., 293–94
farming, 45, 52, 74, 96, 122, 132, 180, 183, 187, 201–2, 214, 217, 247, 263, 265, 267, 292; cattle, 317; contract, 193; high-risk, 183; livestock, 292; precision, 122, 187, 292; smart, 150; sustainable, 95–96, 110, 183, 197
Federación Nacional de Cafeteros, 196
financial literacy, 264, 328, 335
5G, xiii, 3, 157, 249, 258, 280–81, 325
Flutterwave, 231, 241–42
flying geese paradigm, 32, 256, 282
Fortification, 9. 159–60; remote, 9; small-Scale, 159–60
Fourth Industrial Revolution, 3, 340; Rwanda's Centre for the, 286

Gaofen-1 satellites, 122
GCC economies, 313
GDPR, 272, 294
GEE, 107–8
genome editing, xiii, 3, 163, 340
genome profile, 175
Genome Sequence Database, 246
genome, 163–64, 246; human, 163, 305
GISC, 305
Global Alliance for Clean Cookstoves, 162
Global Burden of Disease, 160–61
global economic pyramid, 7, 31
Global Fishing Watch, 125
Global North, 67, 73, 323, 325, 335, 340
Global South, 67. 89, 94, 240, 298, 320–21, 323–24, 340, 344
GPT, 9, 25. 31, 37
Gro Intelligence, 201, 231–32
GSM, 29, 134, 138, 149–50, 161, 231, 341
GSMA, 129
G20, 73, 309, 340

Haraka, 211, 316
Harambee, 54–55
Healthcare, 4, 8, 13, 31–32, 49, 52, 123, 126, 150, 157–59, 164–66, 168, 169, 171, 174, 176–78, 225, 231–32, 245–46, 292, 314, 316, 320; access to, 177;

Healthcare (*continued*)
affordable, 168; barrier to,13; cost-effective, 32; costs of, 168; digital, 157, 166; 5G-enabled, 157; overburdened, 165, 172, 177; quality of, 172
hearX, 174
Hefei Bitland Information Technology, 281
hemoglobin, 158
hepatitis C, 164
high-income countries, 7, 157, 179, 280, 284, 292, 315, 325
HIV, 164
HOT, 113, 117–18
ID4D, 94
IDs, 318; blockchain-based, 100; tamper-proof personal, 327
Illumina, 163
iMerit, 57–58, 263, 305
Industry 4.0, 3
informal personal exchange, 22
insurable event, 92
insurance broker, 220
insurance industry, 169, 205, 205, 213
insurance payout, 205, 216–17, 227, 239–40
insurance policy, 92–93, 193, 240
insurance premium, 110
insurance product, 92–93, 181, 217, 227, 316
insurance sector, 169, 193, 205, 217
insurance, 5, 32; 51–53, 57, 78, 81, 92, 101, 122, 166, 181, 193, 203–4, 208, 213–20, 227, 239, 247, 262, 294, 316, 322, 342; agricultural, 213–14, 262, 316; agriculture, 8; blockchain-based, 92; crop, xiii, 180, 193–94, 199, 205, 213–14, 216–21, 232, 239, 246–47, 268, 312, 316, 319; decentralized, 92, 218–19; farm, 213 268; health, 78, 101, 283; life, 93; index-based, 218, 246; medical, 169; parametric, 216–17, 342; vehicle, 53
International Organization for Migration, 73–74
IoT, 9, 10, 22, 28–29, 31, 45,92, 100, 111, 122, 127–43, 145–53, 159, 161, 171, 176, 184–87, 195, 198, 201, 216, 227, 236, 238, 262, 272, 294, 314, 316, 322–23
I-STEM, 175

JD.com, 201, 262, 276, 308
JTI, 198–99
Jubilee Insurance, 323

Kadena, 164
Kenya, 12, 16, 29, 40, 46–47, 53, 58–59, 68, 78, 98, 112, 127, 132–34, 136–39, 141–42, 145–49, 151, 153, 164, 170, 172, 174–76, 181–82, 192–93, 201–2, 204, 218–19, 221, 223–24, 231, 234, 241, 247–48, 252, 257–60, 264, 278, 283, 286, 288, 295–96, 300–301, 308, 313–16, 319, 322
kinship ties, 22
Kiva field partners, 93
Kiva, 93, 95, 100
KOKO Cooker, 147–48
KOKO Networks, 145–47, 152
KOKOpoints, 146–49
KopaGas, 136–37, 231, 319
KYC, 16, 19, 100, 212

land registry, 81, 85, 233, 329; blockchain-based, 86
Latin America, 8, 51, 56, 91, 116, 128–29, 177, 246, 251, 286, 310, 340
LC, 87, 97
LDCs, 64–5, 107, 151–52, 284
leapfrogging, 32, 152, 256, 268, 270, 312–13, 318, 325, 341
liquefied petroleum gas, 136, 161
liquid crystal monitor, 281
LMIC, xiii, 4, 7, 9, 15, 26, 29, 32–33, 37–40, 45, 49–50, 53–55, 57–58, 62–64, 67, 69, 72, 73, 75, 78, 80–81, 94, 98–99, 101, 129, 135, 140, 149, 151–52, 157–60, 165–66, 168–70, 172, 174, 177–81, 184, 191, 200–206, 208, 212, 217, 221–22, 227, 231–35, 237, 240, 242, 244–45, 247–50, 252–53, 255, 257, 164, 278, 280, 283–85, 287–88, 292–96, 300, 307, 309–15, 319, 321–25, 327–29, 333–35

LocalBitcoins, 27
Lockheed Martin, 110; global security and aerospace company, 109
LoRa Alliance, 341
LoRa, 139, 150
low-income countries, 158, 282, 341
LPG cylinder 136–38; IoT Solution for, 137
LPG distribution, 136–37
LPG metering, 138
LPG, xiii, 136–38, 160–61, 231, 241, 319; bottled, 231; cleaner-burning, 160, 312
LPWAN, 150
LTE, 134, 150–51, 259
LTE-M, 139, 150

machine learning, 13, 38, 41–44, 116, 184, 192, 211, 245–46, 331, 338–39; supervised, 43, 344
Malaria, 10–14, 162, 169, 175; intensive 11; preventing 11; fighting 11
Malawi, 95–96, 139, 176, 199, 210, 316; Lujeri Tea Estate in, 96
MANSA, 19–21
marginalized communities, 284, 327
marginalized demographic groups. 55
marginalized groups, 16, 23, 70, 78, 93–94, 239, 254, 311, 315, 332–33, 335
marginalized individuals, 16
marginalized population, 122
Marine Fishing Industry, 74, 124
market creation, 144
Mezzanine, 176
MFI, 93, 188–89, 333
M-Gas, 136–37, 231
microinsurance, 81, 93, 217, 220
middle-income countries, 4–5, 64–65, 255, 271, 274, 278, 280, 283, 324, 333; low- and, 4, 9; upper-, 5, 8, 40
middle-income trap, 32, 256, 270–71, 273–74, 277, 283, 324, 333, 342
M-KOPA, 29, 141–42, 145, 152–53, 314
ML, 38–39, 41–44, 46–47, 51, 54, 56–57, 59, 61, 63, 70, 109, 115–16, 118, 121, 126, 141–43, 152, 179, 188, 193, 195, 206, 209, 242–44, 247, 250–51, 253, 255, 296, 314
MonetaGo, 81, 83
Mongolia, 164
Monsanto, 292
M-PESA, 29, 145, 147–48, 159, 162, 219, 223, 231
MSME, 98
multinational corporation, 62, 342
MyBucks, 210–11, 316

nanotechnology, 3
NASA, 46. 107, 141. 121–22, 299
National Bank of Dubai, 26
NB IoT, 150
Network for Genomic Surveillance in South Africa, 175
NeurIPS, 245, 277
Nexleaf Analytics, 162
NGO, 7. 10, 96, 112, 117, 119, 124, 185, 189, 292, 325, 333
NigComSat-1R, 268
Nigeria, 20, 40, 55, 105, 135–36, 140, 175–76, 184, 204, 210–11, 231, 241–42, 268, 291, 295, 314, 322–23
NigeriaSat-2, 268
Niti Aayog, 171
NLP 20, 250–52
No. 1 Central Document, 100
nonfungible token, 217
Nupe Energy, 231
Nuru, 46–47; AI assistant, 46; PlantVillage, 46, 122, 238
Nvidia, 277
Nzoma, 258

OBP, 330
Off.Grid:Electric, 145
OKO, 214–15, 316; insurance startup, 214
open banking, 212–13, 322
Oracle, 110, 171, 198, 210–11, 330–31
oracle, 217, 227; off-chain, 227
Oracle's blockchain platform, 330
Organisation for Economic Co-operation and Development, 75

pandemic 103, 106–7, 112–15, 118–21, 125–26, 176, 224, 325; COVID-19, 31–32, 39, 87, 103, 108–9, 112, 158, 175–76, 224; predicting and preventing, recovery from, 113, 118, stopping the spread of, 113, 115
pandemic preparedness, 4, 32, 157, 316
PAPSS, 19–20
PayGo, 136–38, 231, 241
People's Bank of China, 94
Pezesha, 98
Philippine's Bonifacio Global City, 278, 302
Plasmodium falciparum, 11
Poladrone, 190
poverty, 4–5, 15, 39, 61, 64, 69, 76–77, 116, 128, 225, 245, 259, 261–63, 265, 271, 283, 291, 306, 319, 322–23, 327, 333, 343; basin of, 282; dimension of, 8; eliminating, 93; extreme, 18, 103, 113, 116, 129, 222, 262, 326, 334; falling into, 6, 261, 283, 333; family in, 103; fighting, 86, 312; households in, 222; monetary, 260; move out of, 30; people living in, 126, 257, 288; people lifted from, 305; predictor of, 86; pull poor people out of, 32; rural, 179, 271; push poor people into further, 264; vulnerable to, 5, 103, 262; vulnerable to falling into, 265
poverty alleviation, 116, 271
poverty data, 244
poverty level, 40
poverty line, 8, 61; the World Bank's, 8
poverty market, 6–7, 10, 15. 22, 30, 94, 222, 312, 333
poverty trap, 32, 77, 256, 259–60, 262, 270, 282–83, 343; escape the, 283; fall deeper into the, 283; get out of the, 282, 342; low equilibrium, 191; pulling people out of the, 261; pushing poor people into the, 264
Power Resources Group, 331
Project Healthy Children, 159–60, 238, 316
property rights, 77, 86, 181; enforcement of, 77; protection of, 84
P2P, 27, 98, 264, 288, 308, 335
P2P lender, 88, 288, 308, 322

QR code, 76, 170

Recommendations for Data and Biospecimen Governance in Africa, 306
refugee camp, 27; Jordan's, 225; Kenya's Dadaab, 16; Malawi's Dzaleka, 210
refugee, 16. 226–27; Financial Services to, 210; food vouchers to, 225; Syrian, 226
Relative Wealth Index, 39–40, 117
remittance, 27–28, 90, 294; cross-border, 90; Crypto, 27; international, 27, 72, 89; cost of, 28, 89
remote sensing, xiii, 3, 49. 59–60, 103–5, 108–9, 111, 126, 184, 193, 238, 316, 333, 343
RFID, 135, 197
RippleNet, 90
Rwanda, 54, 112, 139, 166–68, 173, 176, 178, 200, 218, 235–56, 241, 247, 286–68, 290–91. 313, 315, 331, 335
Rwanda's Centre for the Fourth Industrial Revolution; 286
RxAll, 170, 231
Rymedi, 164

SAB, 141, 143, 152–53
Sanku, 159–60, 238, 316
Sarafu, 223–25, 316
satellite data, 41, 105–7, 110, 112, 114, 121, 123–24, 185, 200, 205, 214, 218, 237–39, 247, 262, 317, 324, 325, 331
satellite image, 42, 44, 107, 109–13, 116–20, 234, 247, 317
satellite imagery, 3, 12, 14, 31, 39, 42, 103–4, 107–14, 116, 118–20, 122, 125, 190, 222, 236, 316, 325–26
satellite navigation, 13, 113, 115
satellite network, 214
satellite technology, 106, 113, 115, 122–23, 125, 236, 268, 324
Sigfox, 149–51, 294
SIM, 29, 133, 176, 187, 212; Orange, 215; Vodafone's IoT, 159
Simpa, 141, 318

smallholders, 183, 192–94, 267
smart contract, 92, 101, 169–67, 183, 193, 197–98, 204, 217–20, 227, 240, 265–66, 344; blockchain-based, 110, 193–94, 218, 227, 239; enforce, 217; ethereum, 220; green, 198
smart vision spectacles, 39, 169
SME, 15, 17. 21, 78, 89, 94, 102, 204, 206, 209, 274
SMIC, 272, 276
SMS, 47, 146, 172, 182, 189, 215
Solinftec, 45, 52, 55, 201, 241
South African Breweries, 314
South African Civil Aviation Authority, 287
South Asia, 7–8, 128, 151, 174, 250
South-South Cooperation 32, 240, 344
SSA, 8–9, 12, 18, 29, 129, 131, 136, 140, 184, 249, 257, 292, 295
Stockholder centric bias, 254, 344
submerged market, 7, 22, 312, 333
Sub-Saharan Africa, 8, 106, 128–29, 152, 174, 181, 213, 257
SunCulture, 131–34, 201–2; solar irrigation company, 132
supply chain, 16, 19, 50, 73, 75–76, 81, 88, 95–96, 110–11, 121, 137, 147, 164, 170–71, 176, 186, 188–89, 191, 196, 204, 238, 254, 266, 330–32, 344; blockchain-based cobalt, 330; cattle, 197; drug, 171; farm, 292; pharmaceutical, 171–72, 316, 329; temperature-controlled, 176; tracking of coffee, 111
supply chain financing, 88
supply chain transparency, 27, 189
surveillance, 64, 66, 175–76, 253, 273, 280, 291, 298–302; AI, 301–3; AI-based, 65, 301–2, 311; COVID-19, 175; facial recognition, 302; mass, 64, 278, 297, 302; the Chinese government's extensive, 301; workplace, 306
sustainability, 95–96, 112, 184, 197–98, 237, 312, 331–32; environmental, 255

Tanzania, 17. 78, 127, 135–37, 139, 145, 148, 159, 165–66, 176, 199, 218, 232, 295, 319
TechnoServe, 41–43
Tether, 296
Third Industrial Revolution, 3, 344
third-party provider, 216
threshold cost, 191, 194
Togo, 40, 103, 116, 131, 134, 222; COVID-19 lockdown in, 103; fine-grained map of, 116; the government of, 104, 116, 146, 322
traceability, 26, 76, 185, 187, 189, 196–97, 254, 316, 329, 331, 344; blockchain solutions for, 172; blockchain-based, 329, 344; blockchain-based cobalt supply chain, 330
traceability-as-a-service, 329, 344
Trade Information Porta, 18–19
tribal violence, 68

Ubenwa, 175
Uganda, 114, 137, 145, 148, 176, 182, 189, 215, 235, 241, 254, 278. 286, 295, 301, 316
unbanked, 16, 188, 203, 212, 329
underbanked, 328–29
unethical act, 255
unethical behavior, 255, 329
unethical practice, 67, 284, 329
UNICEF, 227, 309
Uthabiti, 172, 316
Uyghurs, 64–67, 297, 301

vaccines, 173–74, 176–77, 169, 171–74, 288, 313; COVID-19, 40, 118, 164, 176–77; hepatitis C, 164; time sensitive, 178
value chain, 32, 56, 146, 181, 187, 271, 277; agricultural, 32, 184. 197; cassava crop, 189
virtual reality, 3, 345
Vodafone, 159, 176
Volvo, 331

Water Productivity through Open Access of Remotely Sensed Derived Data, 46, 122
WFP, 10, 225–27, 238
WHO, 11, 38, 160, 164, 169–71, 174–75, 227

WII, 239–40
WiMAX, 150
World Food Program, 10
World Health Organization, 38, 161, 170
Worldcoin, 296

Zambian Breweries, 189, 332–33, Anheuser-Busch's local business, 189
Zenvus, 292
Zika, 162
Zimbabwe, 194, 197, 296–97
Zipline, 173–74, 177, 287, 313; American medical product delivery company, 173
Zzapp, 11–14
ZzappMalaria, 11, 13

Printed and bound by CPI Group (UK) Ltd, Croydon, CR0 4YY

20/08/2025

14722032-0004